244
Topics in Current Chemistry

Editorial Board:
A. de Meijere · K. N. Houk · H. Kessler · J.-M. Lehn · S. V. Ley
S. L. Schreiber · J. Thiem · B. M. Trost · F. Vögtle · H. Yamamoto

Topics in Current Chemistry

Recently Published and Forthcoming Volumes

Anion Sensing
Volume Editor: Stibor, I.
Vol. 255, 2005

Organic Solid State Reactions
Volume Editor: Toda, F.
Vol. 254, 2005

DNA Binders and Related Subjects
Volume Editors: Waring, M.J., Chaires, J.B.
Vol. 253, 2005

Contrast Agents III
Volume Editor: Krause, W.
Vol. 252, 2005

Chalcogenocarboxylic Acid Derivatives
Volume Editor: Kato, S.
Vol. 251, 2005

New Aspects in Phosphorus Chemistry V
Volume Editor: Majoral, J.-P.
Vol. 250, 2005

Templates in Chemistry II
Volume Editors: Schalley, C.A., Vögtle, F., Dötz, K.H.
Vol. 249, 2005

Templates in Chemistry I
Volume Editors: Schalley, C.A., Vögtle, F., Dötz, K.H.
Vol. 248, 2005

Collagen
Volume Editors: Brinkmann, J. Notbohm, H., Müller, P. K.
Vol. 247, 2005

New Techniques in Solid-State NMR
Volume Editor: Klinowski, J.
Vol. 246, 2005

Functional Molecular Nanostructures
Volume Editor: Schlüter, A.D.
Vol. 245, 2005

Natural Product Synthesis II
Volume Editor: Mulzer, J.
Vol. 244, 2005

Natural Product Synthesis I
Volume Editor: Mulzer, J.
Vol. 243, 2005

Immobilized Catalysts
Volume Editor: Kirschning, A.
Vol. 242, 2004

Transition Metal and Rare Earth Compounds III
Volume Editor: Yersin, H.
Vol. 241, 2004

The Chemistry of Pheromones and Other Semiochemicals II
Volume Editor: Schulz, S.
Vol. 240, 2005

The Chemistry of Pheromones and Other Semiochemicals I
Volume Editor: Schulz, S.
Vol. 239, 2004

Orotidine Monophosphate Decarboxylase
Volume Editors: Lee, J.K., Tantillo, D.J.
Vol. 238, 2004

Long-Range Charge Transfer in DNA II
Volume Editor: Schuster, G.B.
Vol. 237, 2004

Long-Range Charge Transfer in DNA I
Volume Editor: Schuster, G.B.
Vol. 236, 2004

Spin Crossover in Transition Metal Compounds III
Volume Editors: Gütlich, P., Goodwin, H.A.
Vol. 235, 2004

Spin Crossover in Transition Metal Compounds II
Volume Editors: Gütlich, P., Goodwin, H.A.
Vol. 234, 2004

Spin Crossover in Transition Metal Compounds I
Volume Editors: Gütlich, P., Goodwin, H.A.
Vol. 233, 2004

Natural Products Synthesis II
Targets, Methods, Concecpts

Volume Editor: Johann Mulzer

With contributions by
U. Beifuss · T. J. Heckrodt · M. Kalesse · H.-J. Knölker · P. Metz ·
J. Mulzer · U. Nubbemeyer · M. Tietze

Springer

The series *Topics in Current Chemistry* presents critical reviews of the present and future trends in modern chemical research. The scope of coverage includes all areas of chemical science including the interfaces with related disciplines such as biology, medicine and materials science. The goal of each thematic volume is to give the nonspecialist reader, whether at the university or in industry, a comprehensive overview of an area where new insights are emerging that are of interest to a larger scientific audience.

As a rule, contributions are specially commissioned. The editors and publishers will, however, always be pleased to receive suggestions and supplementary information. Papers are accepted for *Topics in Current Chemistry* in English.

In references *Topics in Current Chemistry* is abbreviated Top Curr Chem and is cited as a journal.

Visit the TCC content at springerlink.com

Library of Congress Control Number: 2004112161

ISSN 0340-1022
ISBN 3-540-21124-1 **Springer Berlin Heidelberg New York**
DOI 10.1007/b94542

Springer is a part of Springer Science+Business Media
springeronline.com

Printed in The Netherlands

Cover design: KünkelLopka, Heidelberg/design & production GmbH, Heidelberg
Typesetting: Fotosatz-Service Köhler GmbH, Würzburg

Printed on acid-free paper 02/3141 xv – 5 4 3 2 1 0

Volume Editor

Professor Dr. Johann Mulzer
Institut für Organische Chemie
Universität Wien
Währingerstr. 38
1090 Wien, Austria
johann.mulzer@univie.ac.at

Editorial Board

Prof. Dr. Armin de Meijere
Institut für Organische Chemie
der Georg-August-Universität
Tammannstraße 2
37077 Göttingen, Germany
ameijer1@uni-goettingen.de

Prof. Dr. Horst Kessler
Institut für Organische Chemie
TU München
Lichtenbergstraße 4
85747 Garching, Germany
kessler@ch.tum.de

Prof. Steven V. Ley
University Chemical Laboratory
Lensfield Road
Cambridge CB2 1EW, Great Britain
svl1000@cus.cam.ac.uk

Prof. Dr. Joachim Thiem
Institut für Organische Chemie
Universität Hamburg
Martin-Luther-King-Platz 6
20146 Hamburg, Germany
thiem@chemie.uni-hamburg.de

Prof. Dr. Fritz Vögtle
Kekulé-Institut für Organische Chemie
und Biochemie der Universität Bonn
Gerhard-Domagk-Straße 1
53121 Bonn, Germany
voegtle@uni-bonn.de

Prof. Kendall N. Houk
Department of Chemistry and Biochemistry
University of California
405 Hilgard Avenue
Los Angeles, CA 90024-1589, USA
houk@chem.ucla.edu

Prof. Jean-Marie Lehn
Institut de Chimie
Université de Strasbourg
1 rue Blaise Pascal, B.P.Z 296/R8
67008 Strasbourg Cedex, France
lehn@chimie.u-strasbg.fr

Prof. Stuart L. Schreiber
Chemical Laboratories
Harvard University
12 Oxford Street
Cambridge, MA 02138-2902, USA
sls@slsiris.harvard.edu

Prof. Barry M. Trost
Department of Chemistry
Stanford University
Stanford, CA 94305-5080, USA
bmtrost@leland.stanford.edu

Prof. Hisashi Yamamoto
Arthur Holly Compton Distinguished
Professor
Department of Chemistry
The University of Chicago
5735 South Ellis Avenue
Chicago, IL 60637
773-702-5059, USA
yamamoto@uchicago.edu

Topics in Current Chemistry also Available Electronically

For all customers who have a standing order to Topics in Current Chemistry, we offer the electronic version via SpringerLink free of charge. Please contact your librarian who can receive a password for free access to the full articles by registration at:

springerlink.com

If you do not have a subscription, you can still view the tables of contents of the volumes and the abstract of each article by going to the SpringerLink Homepage, clicking on "Browse by Online Libraries", then "Chemical Sciences", and finally choose Topics in Current Chemistry.

You will find information about the

- Editorial Board
- Aims and Scope
- Instructions for Authors
- Sample Contribution

at springeronline.com using the search function.

Preface

From its early days, the total synthesis of complex molecules, especially those that are natural products, has been the king's discipline in organic chemistry. The reasons for this are manifold: the challenge lying in a novel and intricate molecular architecture or the difficulty encountered when isolating the substance from its natural sources, or the possibility of finding a wide test ground for established methodology or the incentive to invent new methodology when the old one has failed, or simply the art and elegance which is so typical of a truly efficient synthetic sequence. In any case, everybody will agree that total synthesis is the best way to train young chemists and to prepare them for any kind of later employment.

In these two volumes, the contributions of a number of organic synthetic chemists from the German speaking area have been collected and it can easily be seen that their expertise is on a par with any other synthetic community. It is thus the hope of the authors and the editor that these articles, which highlight all the various aspects of organic synthesis, will provide not only an insight into the basic strategy and tactics but also the purpose of organic syntheses.

Vienna, September 2004 Johann Mulzer

Contents

Contents of Volume 243
Natural Product Synthesis I
Targets, Methods, Concepts

Volume Editor: Johann Mulzer
ISBN: 3-540-21125-X

Top Curr Chem (2005) 244: 1–41
DOI 10.1007/b96886

Marine Natural Products from *Pseudopterogorgia elisabethae*: Structures, Biosynthesis, Pharmacology, and Total Synthesis

Thilo J. Heckrodt · Johann Mulzer (✉)

Department of Organic Chemistry, University of Vienna, Währingerstrasse 38, 1090 Wien, Austria
johann.mulzer@univie.ac.at

1	**Introduction**	2
2	**Natural Products from *Pseudopterogorgia Elisabethae***	3
2.1	Occurrence and Isolation	3
2.2	Structures	6
3	**Biosynthesis and Pharmacology**	13
3.1	Biosynthetic Pathways	13
3.2	Pharmacological Properties	19
3.2.1	Anti-inflammatory Activity	19
3.2.2	Antituberculosis Activity	20
3.2.3	Anticancer Activity	20
3.2.4	Antiplasmodial Activity	21
4	**Recent Partial and Total Syntheses: New Challenges for the Diels–Alder Reaction**	21
4.1	Pseudopteroxazole (Corey, 2001)	21
4.2	Elisabethin C (Yamada, 2002)	23
4.3	Elisabethin A (Mulzer, 2003)	25
4.4	Elisapterosin B (Rawal, 2003)	29
4.5	Elisapterosin B/Colombiasin A (Rychnovsky, 2003)	31
4.6	Colombiasin A (Nicolaou, 2001)	34
4.7	Synthetic Studies Toward Colombiasin A (Harrowven, 2001, and Flynn, 2003)	36
5	**Conclusions**	39
References		39

Abstract In the late 1990s, the Caribbean octocoral *Pseudopterogorgia elisabethae* became the target of extensive chemical investigations leading to the isolation and characterization of a remarkable number of diterpenoid secondary metabolites. Most of these newly discovered compounds are based on so far unprecedented carbon skeletons and often feature unusual structural characteristics. Besides their exciting structures, many of these marine natural products display potent pharmacological activity against various diseases, for instance tuberculosis or cancer. Although at first glance it is not always evident, all structures are consistent with a biosynthesis pathway starting from geranylgeranyl phosphate to deliver via serrulatane intermediates an enormous variety of diterpenoid natural products. Thus, the organism *Pseudopterogorgia elisabethae* is capable of performing a kind of diversity-oriented synthesis creating stereochemical and structural complexity from simple precursors. The intricate molecular architecture of these natural products also drew the attention of synthetic chemists. Over the last few years considerable synthetic efforts have been made resulting in several total syntheses. Therefore, this class of diterpenoids is now also synthetically accessible, with cycloaddition reactions proving to be the ultimate tool for the construction of the carbon skeletons.

Keywords Pseudopterogorgia elisabethae · Marine natural products · Diterpene biosynthesis · Total synthesis · Diels–Alder reaction

Abbreviations

AIBN	2,2′-Azobisisobutyronitrile
9-BBN	9-Borabicyclo[3.3.1]nonane
Boc	*tert*-Butoxycarbonyl
CAN	Ceric ammonium nitrate
CNS	Central nervous system
DBU	1,8-Diazabicyclo[5.4.0]undec-7-ene
DCC	Dicyclohexylcarbodiimide
DIBAL-H	Diisobutyl aluminum hydride
DMAP	4-Dimethylaminopyridine
DMP	Dess–Martin periodinane
GGPP	Geranylgeranyl pyrophosphate
HMPA	Hexamethylphosphoramide
HOBt	1-Hydroxybenzotriazole
IBX	*o*-Iodoxybenzoic acid
IMDA	Intramolecular Diels–Alder reaction
LAH	Lithium aluminum hydride
LDA	Lithium diisopropylamide
MCPBA	*m*-Chloroperbenzoic acid
MOM	Methoxymethyl
NaHMDS	Sodium hexamethyldisilazide
NBS	*N*-bromosuccinimide
PCC	Pyridinium chlorochromate
PDC	Pyridinium dichromate
TBAF	Tetrabutylammonium fluoride
TBS	*tert*-Butyldimethylsilyl
TEA	Triethylamine
TES	Triethylsilyl
TFA	Trifluoroacetic acid
Tr	Trityl (triphenylmethyl)

1 Introduction

Natural products play a dominant role in the discovery of leads for the development of drugs. Thus, it is always of importance to identify organisms which could be a source of novel natural products. In this regard, the octocoral fauna of the West Indies is unique as it provides a multitude of chemically rich gorgonian corals (also known as sea whips, sea fans, or sea plumes) which could be a potential supplier of biologically active compounds. Faunistically, extensions of the West Indian region reach into the Gulf of Mexico, all the Antilles, the Bahamas, the Florida Keys, the Bermudas, the islands of the Caribbean, and south along the northeast coast of South America to the reefs of Brazil. All over this region the gorgonian octocorals flourish as they do nowhere else in the world [1]. In the West Indies the gorgonians belong to the most abundant octocorals and they are conspicuous members of most tropical and subtropical marine habitats. They represent an estimated one third of the known fauna [2]; so far over 100 species have been documented from this major family. The first studies of the natural product chemistry of this interesting group of marine invertebrates were carried out in the late 1950s. Numerous studies of the chemistry of gorgonian corals have been published and summarized since these early investigations [1]. Gorgonian metabolites possess novel structures that are largely unknown from terrestrial sources. In recent years the species *Pseudopterogorgia elisabethae* has attracted particular attention since it was shown to contain a huge variety of novel bioactive secondary metabolites possessing a promising pharmacological profile.

2
Natural Products from *Pseudopterogorgia Elisabethae*

Since the pioneering work of the Fenical group in the 1980s, natural products from the Caribbean gorgonian *Pseudopterogorgia elisabethae* (Octocorallia) have continued to capture the attention of natural products chemists. Even more effort was put in when the Rodríguez group from Puerto Rico started in the late 1990s to report the isolation, structural elucidation, and biological properties of an increasing number of novel marine metabolites from *Pseudopterogorgia elisabethae*. The structural variety found among the many terpenoid natural products isolated from this gorgonian, as well as the large spectrum of biological activities exhibited by many of these compounds, is indeed quite remarkable.

2.1
Occurrence and Isolation

Pseudopterogorgia species are among the most common of the Caribbean species with over 15 documented. Chemical studies of *Pseudopterogorgia* species began in 1968 with investigations of the sesquiterpene hydrocarbons from the most common representative of this genus, *Pseudopterogorgia americana*. As mentioned, since 1998 extensive attention has been given to the chemically rich species *Pseudopterogorgia elisabethae*. Its classification within the systematic taxonomy of the animal kingdom is shown in Fig. 1.

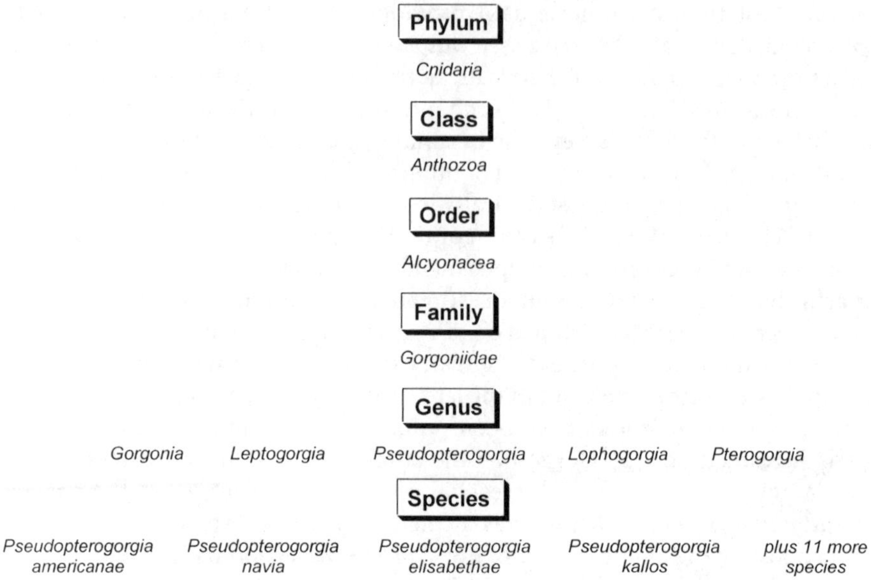

Fig. 1 *Pseudopterogorgia elisabethae* within the animal kingdom

Fig. 2 Caribbean Sea and collection sites

The species *Pseudopterogorgia elisabethae* (Bayer) is found in the tropical western Atlantic, including Florida, Bahamas, Cuba, Jamaica, Honduras, Belize, and Mexico. Its distribution is moderately widespread and more commonly on deeper reef communities as well as hardbottom areas, including forereef zones and intermediate reefs; it occurs in a depth range typically from 40 to 70 m but may be found as shallow as 20 m. Collection sites of *Pseudopterogorgia elisabethae* used for chemical investigations are located for instance in the central Bahamas and in the Florida Keys. Samples of the animal which were used by the Rodríguez group had been collected during an underwater expedition to the eastern Caribbean Sea in the deep waters near San Andrés Island, Colombia. The collection sites throughout the Caribbean Sea are indicated with blue crosses in Fig. 2.

Fig. 3 Caribbean octocorals of the genus *Pseudopterogorgia*

Gorgonians of the genus *Pseudopterogorgia* are best characterized as "sea plumes" based upon their large, highly finely branched (plumose), and physically soft forms (Fig. 3). The Caribbean sea whip *Pseudopterogorgia elisabethae* was collected by hand using scuba at depths of 25–30 m. The gorgonian was sun-dried and kept frozen prior to its extraction. The dry animal was blended with $MeOH/CHCl_3$ (1:1), and after filtration, the crude extract was evaporated under reduced pressure. The extract was then partitioned between hexane and H_2O, and the resulting organic extract was concentrated in vacuo to yield an oil. Extensive chromatography (e.g., silica gel columns, HPLC, Bio-Beads SX-3 columns) afforded a large number of compounds (see next section) in varying but small quantities. The yields based on the dry weight of the prepurified hexane fractions were between 0.001 and 0.01%.

2.2
Structures

The following gives an overview of the natural products isolated to date from *Pseudopterogorgia elisabethae*. The structures of these compounds were proposed on the basis of comprehensive spectral analyses, chemical transformations, and X-ray crystallographic analyses.

Compounds from *Pseudopterogorgia elisabethae* can be classified according to their carbon skeletons. So far 15 carbon skeletons have been identified which are based on the serrulatane skeleton. Various cyclizations (see also Sect. 3.1, Scheme 1) of this precursor lead to new polycyclic structures (e.g., amphilectanes, elisabethanes, etc.) that are starting points for new degradation products (*seco*-, *nor*-, *bisnor*-, etc. compounds).

Figure 4 shows compounds possessing the serrulatane carbon skeleton. They stem mainly from samples of *Pseudopterogorgia elisabethae* collected in the central Bahamas Islands. Important members of this group are the *seco*-pseudopterosins A–D (**3**, aglycone) [3]. These compounds are arabinose glycosides possessing aglycones of the serrulatane class of diterpenoids encountered in composite plants of the genus *Eremophila*. *Seco*-pseudopterosins E–G possess the same aglycone (**3**), but were found to bear fructose residues [4]. The structures of the *seco*-pseudopterosins were suggested on the basis of comprehensive spectral analyses and upon chemical transformations. Closely related, but less hydroxylated, are erogorgiaene (**1**) and hydroxyerogorgiaene (**2**) [5]. Elisabethadione (**4**) [4] also exists in two hydroxylated forms, either *tert*-hydroxyelisabethadione (**5**) [6] or *sec*-hydroxyelisabethadione (**6**) [7]. Elisabethamine (**7**) is an example of a diterpene alkaloid [8], while *seco*-pseudopteroxazole (**8**) contains an uncommon benzoxazole moiety [9]. Dimerization of two hydroxyerogorgiaene (**2**) molecules leads to the bisditerpene **9** [5].

A C9–C14 cyclization forms the ring C and leads to the amphilectane skeleton which is found for instance in pseudopterosins A–F (**10**, aglycone) [10] isolated from *Pseudopterogorgia elisabethae* extracts stemming from a Bahamian collection site (Fig. 5). This class of natural products can be characterized as

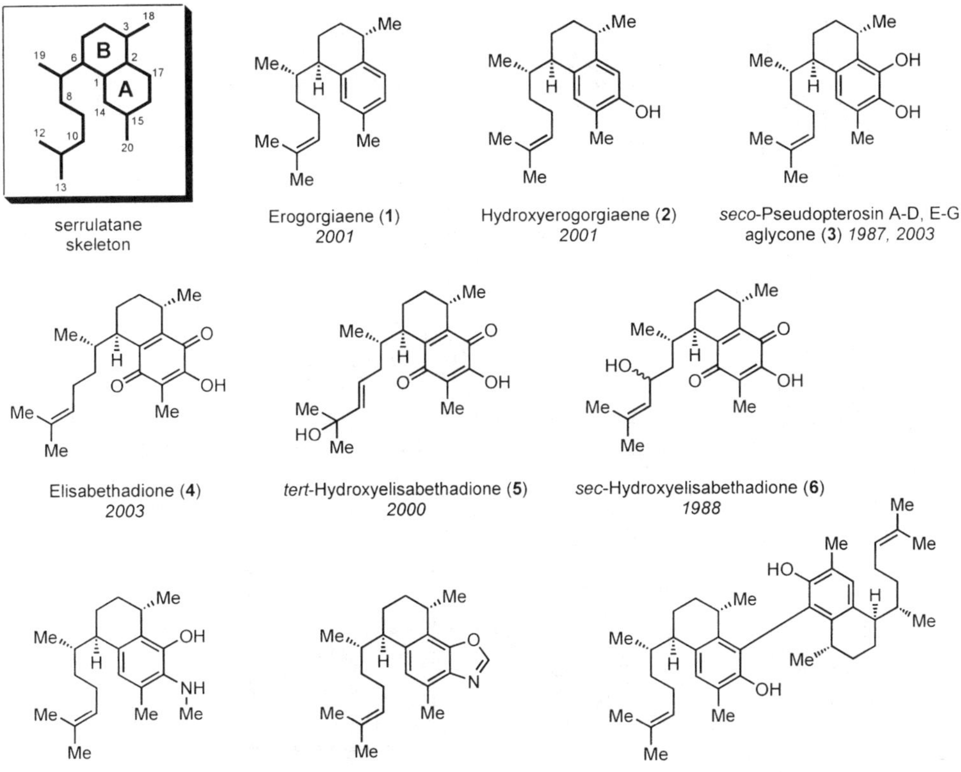

serrulatane skeleton

Erogorgiaene (**1**) *2001*

Hydroxyerogorgiaene (**2**) *2001*

seco-Pseudopterosin A-D, E-G aglycone (**3**) *1987, 2003*

Elisabethadione (**4**) *2003*

tert-Hydroxyelisabethadione (**5**) *2000*

sec-Hydroxyelisabethadione (**6**) *1988*

Elisabethamine (**7**) *2000*

seco-Pseudopteroxazole (**8**) *1999*

Hydroxyerogorgiaene bisditerpene (**9**) *2001*

Fig. 4 Compounds possessing the serrulatane skleleton

diterpene pentose glycosides. An X-ray crystallographic study, together with degradation to obtain D-xylose and relation by chemical interconversion, gave the structure of pseudopterosins A–F (**10**, aglycone). A Bermudian collection of *Pseudopterogorgia elisabethae* contained pseudopterosins G–J (**11**, aglycone) [11], while pseudopterosins M–O (**11**, aglycone) [4] were found in samples from the Florida Keys. On the other hand, pseudopterosins K and L (**12**, aglycone) [11] were obtained from a Bahamas collection. Interestingly, the stereochemistry of the pseudopterosins seems to vary with the collection site. This phenomenon appears to be a general fact, as quantitative and qualitative variations in metabolite composition are often associated with the geographic location of the source organism. The 3,6-stereocenters of pseudopterosins G–J (**11**, aglycone) were originally assigned to be inverted in comparison with the shown structures; total syntheses revealed the correct structures [12]. This error in configurational assignment caused further misassignments in newly discovered structurally related natural products (e.g., pseudopteroxazoles, helioporins). The structure of the pseudopterosin aglycone **11** closely resembles the helioporins,

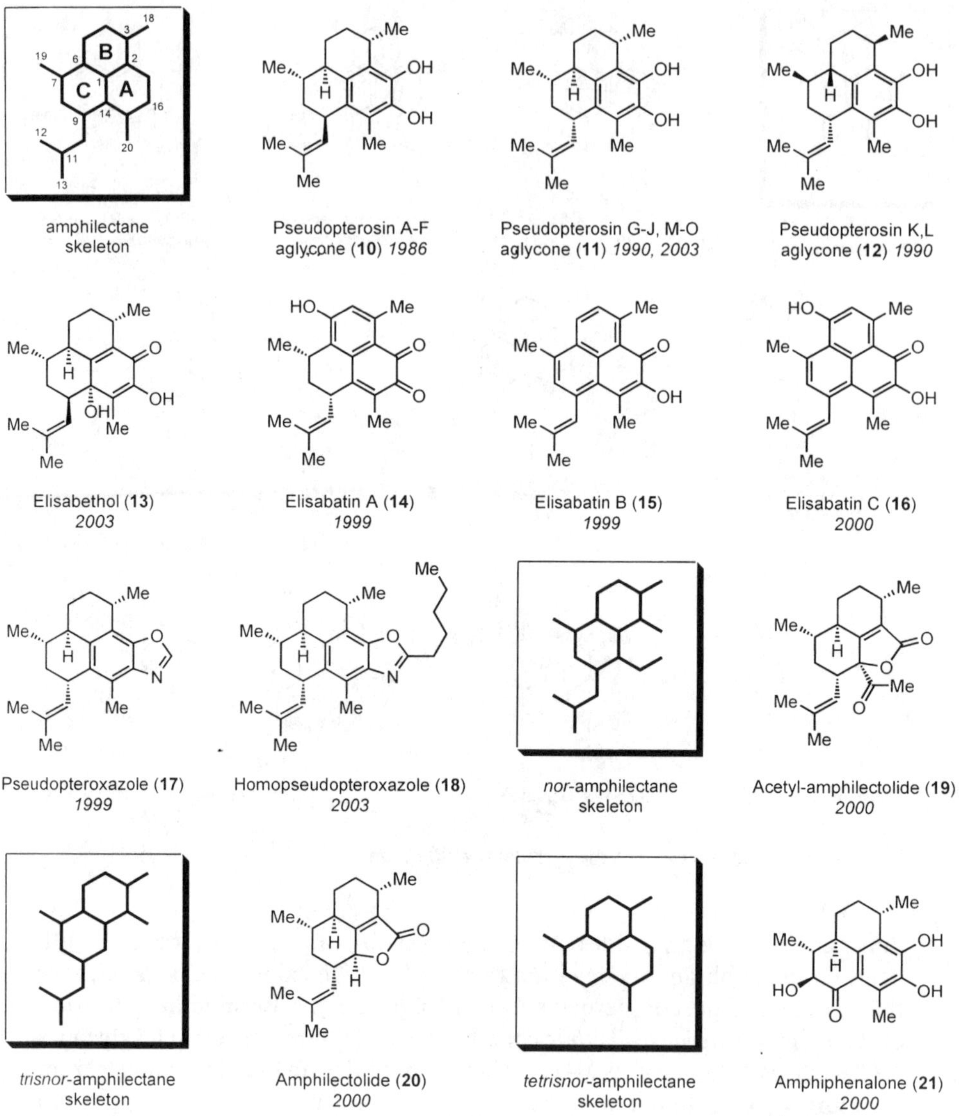

Fig. 5 Compounds of the amphilectane class

featuring the same relative stereochemistry as is found in this compound class [13].

Elisabethol (**13**) [4] shows the same stereochemistry as the pseudopterosin aglycone **10**. Elisabatins A–C (**14, 15, 16**) [14] have the same amphilectane skeleton found in the aglycone portion of the pseudopterosins. The elisabatins are unique among the amphilectane-type diterpenoids because they possess an unusually high unsaturation number that leads to extended aromatic conjugation.

Pseudopteroxazole (**17**) [9] and homopseudopteroxazole (**18**) [15] contain an uncommon benzoxazole moiety. Acetyl amphilectolide (**19**) [6] belongs to the *nor*-amphilectane class having a rearranged norditerpene skeleton. This compound has a logical structure from a biosynthetic viewpoint. Thus, acetyl amphilectolide (**19**) could be envisioned as a precursor of amphilectolide (**20**) [16] via deacetylation with concomitant loss of two carbons. Amphiphenalone (**21**) [6], a novel tetrisnorditerpene, was isolated as colorless plates. Interestingly, although it appears that compound **21** contains an unprecedented carbon skeleton, the overall NMR evidence indicated that **21** possesses some structural features reminiscent of the amphilectane skeleton found in the aglycone portion of the pseudopterosins. Notwithstandingly, comparison of their molecular formulae showed that amphiphenalone (**21**) lacked the four carbons typically ascribed to the isobutenyl side chain at C9. In its place, a ketone carbonyl function now appears in **21**, suggesting loss of the C4 alkenyl side chain by oxidative cleavage of an amphilectane-based precursor (Fig. 5).

Since 1998 the Rodríguez group in Puerto Rico has focused considerable attention on *Pseudopterogorgia elisabethae* found in the southeast Caribbean Sea. Their collections from San Andrés Island, Colombia, resulted in the isolation of terpenoid metabolites which, unlike the pseudopterosins, lacked a sugar moiety. Moreover, the analytical data revealed that the majority of the compounds found in this specimen possessed distinctively different carbon frameworks quite unlike those already described for the aglycone portion of the pseudopterosins (see above). This gives further support to the assumption that metabolism and production of terpenoid secondary metabolites in *Pseudopterogorgia elisabethae* varies from site to site. The novel tricyclic carbon skeleton of elisabethin A (**22**) [17], which was named elisabethane, is formed by connecting C9 and C1 of a serrulatane-based precursor; it represents a new class of diterpenes (Fig. 6). Elisabethin A (**22**) was crystallized from $CHCl_3$ after successive normal-phase and reversed-phase column chromatography. While the spectral data of compound **22** were in full accord with the proposed structure for the molecule, an unambiguous proof of the structure was highly desirable. The structure of **22** was, therefore, confirmed by a single-crystal X-ray diffraction experiment (see Fig. 6) which also yielded its relative stereochemistry. The absolute configuration of the molecule was not determined in the X-ray experiment; this task was accomplished by the total synthesis of **22** (see Sect. 4.3). Hydroxylation at C2 leads to elisabethin D (**23**) [18], while elisabetholide (**24**) [16] represents a compound possessing a *seco*-elisabethane skeleton. Loss of either one or two carbon atoms results in the *nor*-elisabethane (elisabethin B, **25**) [17] or the *bisnor*-elisabethane skeleton (elisabethin C, **26**) [17]. The absolute configuration of **26** was determined later by total synthesis (see Sect. 4.2).

The tetracyclic elisapterane skeleton (Fig. 7) is undescribed so far and constitutes a new class of C_{20} rearranged diterpenes. C10–C15 cyclization of an elisabethane-based precursor accounts for the formation of the six-membered ring D. Starting from elisapterosin B (**27**) [18] a series of oxidation and cy-

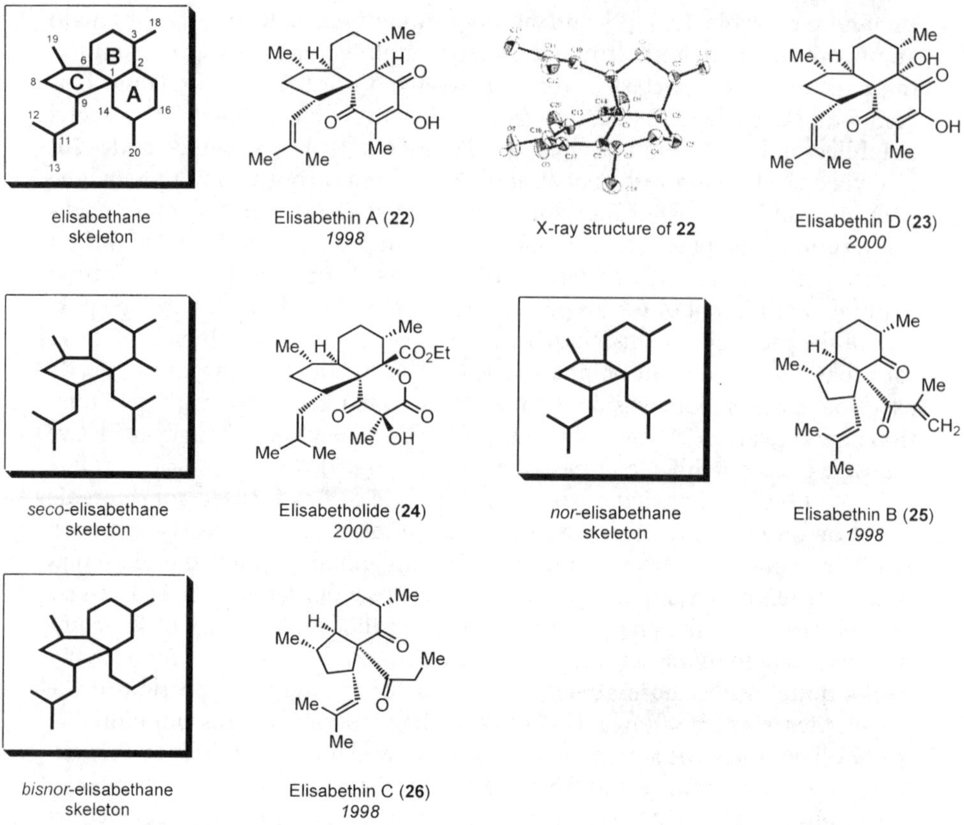

Fig. 6 The elisabethane-based compound class

clization reactions produces quite a large number of different secondary metabolites (**28–31**) possessing the elisapterane skeleton. The structures of these compounds, which until now have remained as the only known examples of this rare family of marine natural products, were established by spectroscopic and chemical methods. Subsequently, the structure of elisapterosin B (**27**) was confirmed by a single-crystal X-ray diffraction experiment (see Fig. 7). In addition to their complex carbocyclic array, some compounds of this class possess several unusual structural features. For instance, in elisapterosin D (**29**) [19] there exists a fully substituted 1,2,4-cyclohexatrione ring moiety; moreover, in all elisapterosins the carbonyl functionality about the cyclopentanone unit is flanked by quaternary carbons. Elisapterosin A (**30**) [18] and elisapterosin E (**31**) [19] contain an *oxo*-heterocyclic ring formed by intramolecular cyclization. Loss of one carbon leads to the *nor*-elisapterane (elisabane) skeleton represented by elisabanolide (**32**) [17]. The structure of **32** was confirmed by X-ray crystallography (see Fig. 7), but since the X-ray experiment did not define the absolute configuration the enantiomer shown is an arbitrary choice.

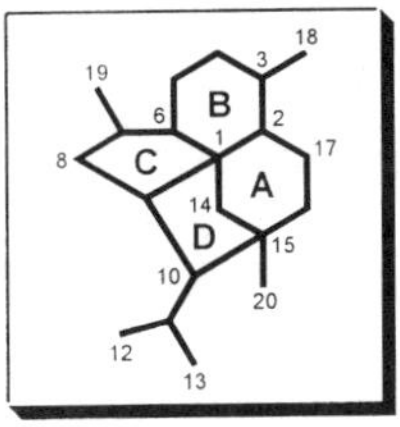

elisapterane skeleton

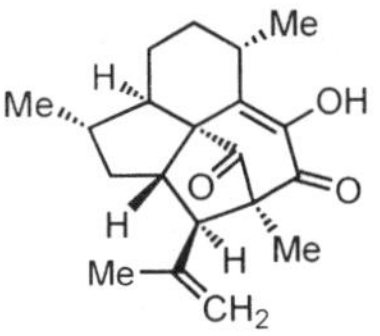

Elisapterosin B (**27**)
2000

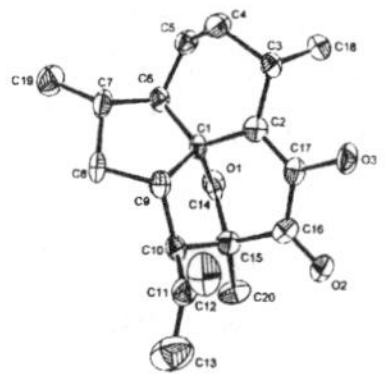

X-ray structure of **27**

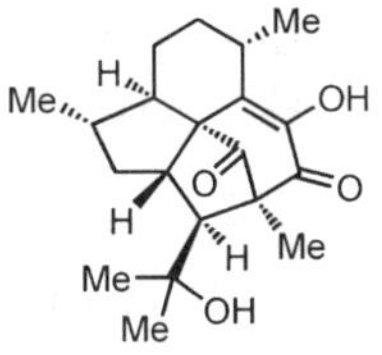

Elisapterosin C (**28**)
2000

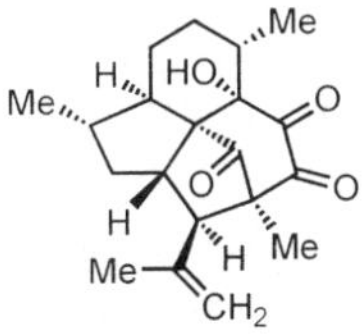

Elisapterosin D (**29**)
2003

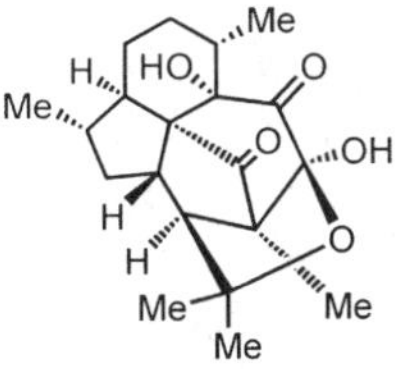

Elisapterosin A (**30**)
2000

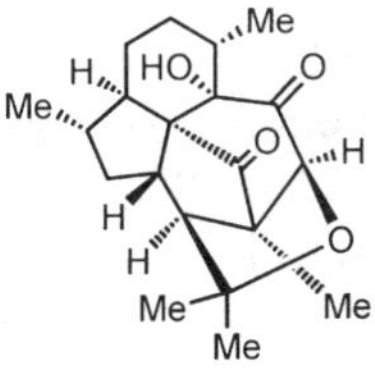

Elisapterosin E (**31**)
2003

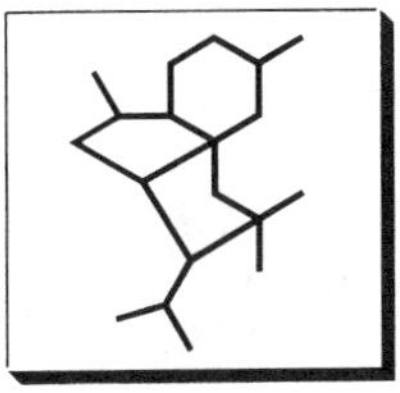

nor-elisapterane (elisabane) skeleton

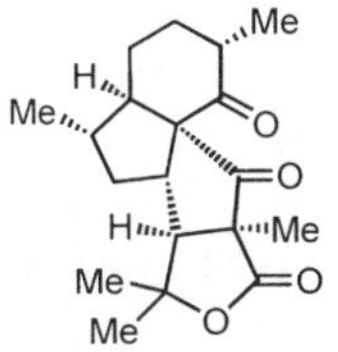

Elisabanolide (**32**)

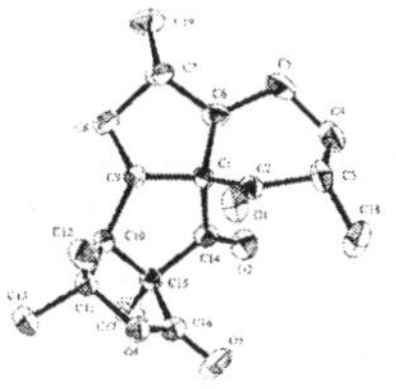

X-ray structure of **32**

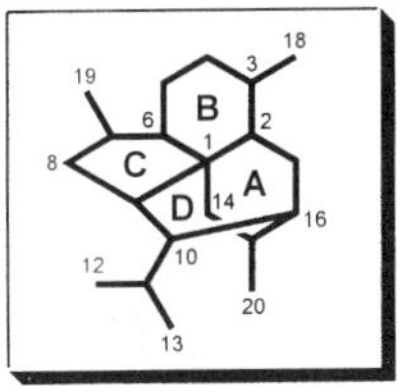

cumbiane skeleton

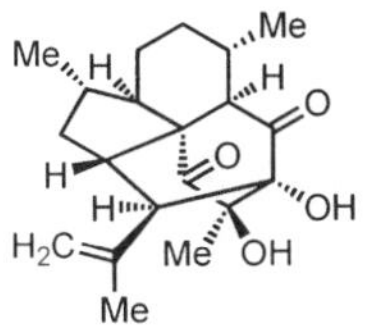

Cumbiasin A (**33**)
2000

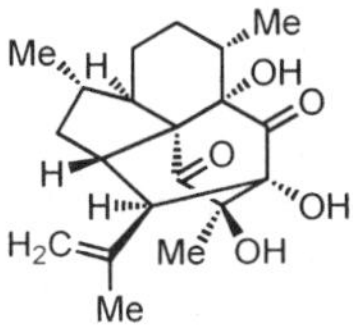

Cumbiasin B (**34**)
2000

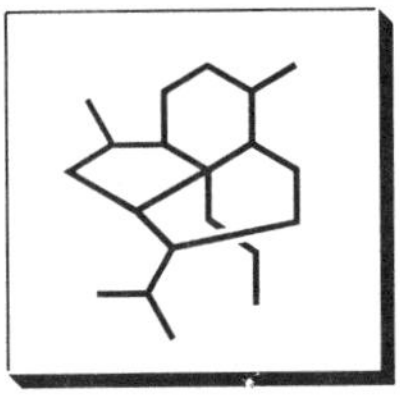

seco-cumbiane skeleton

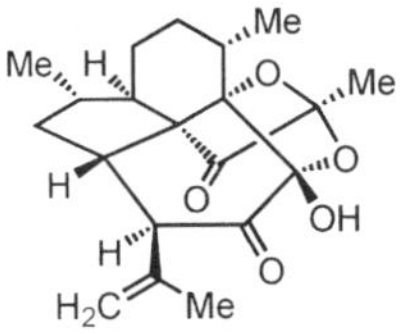

Cumbiasin C (**35**)
2000

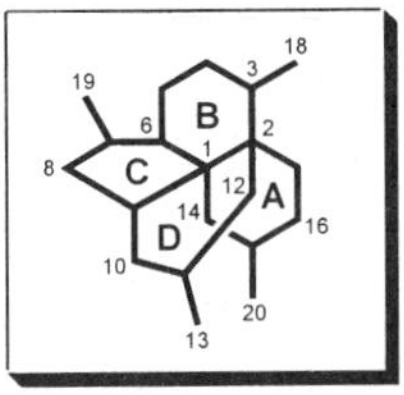

colombiane skeleton

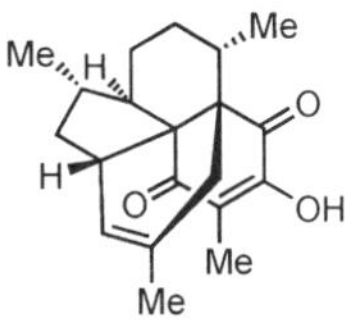

Colombiasin A (**36**)
2000

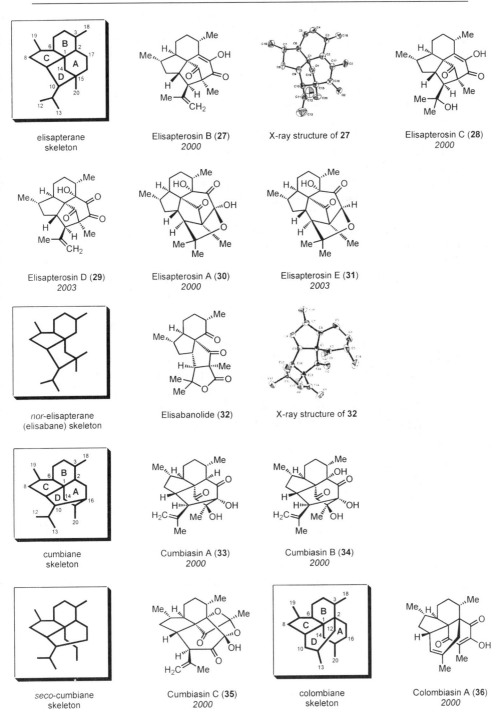

Fig. 7 Compounds based on the elisapterane, colombiane, and cumbiane skeleton

Another tetracyclic carbon skeleton, named cumbiane, has been isolated from *Pseudopterogorgia elisabethae*. Its representatives are the diterpenoids cumbiasin A (**33**) and B (**34**) [20]; their structures and relative configurations were elucidated by interpretation of a combination of spectral data. The six-membered ring D was formed by connecting C10 and C16 of an elisabethane carbon skeleton. The carbocyclic skeleton of the cumbiasins is unprecedented and represents a new class of C_{20} rearranged diterpenes. The tricyclic *seco*-cumbiane skeleton is derived from the cumbiasins by cleavage of the C15–C16 bond. Due to intramolecular cyclizations two additional *oxo*-heterocycles are present in cumbiasin C (**35**) [20] (Fig. 7).

A further tetracyclic metabolite from *Pseudopterogorgia elisabethae* possessing the unprecedented cage-like colombiane carbon skeleton is represented by colombiasin A (**36**) [21]. The ring D of the colombiane skeleton arises from a C12–C2 cyclization of an elisabethane-based precursor. Experimental data were not obtained to define the absolute stereochemistry of **36**; this task was later accomplished by total synthesis (see Sect. 4.6).

The ring C of the tricyclic ileabethane skeleton was formed by C8–C14 cyclization of a serrulatane-based precursor (Fig. 8). Besides a new carbon skeleton, ileabethin (**37**) [22] possesses several unusual structural features. The substitution pattern of the fully substituted aromatic ring resembles that of the pseudopterosins class diterpene glycosides, while the isobutenyl side chain found typically in many *Pseudopterogorgia elisabethae* metabolites appears in **37** masked as a spiro *gem*-dimethyl dihydrofuran moiety. The pseudopterane skeleton found in acetoxypseudopterolide (**38**) [23] differs significantly from the previously described carbon skeletons. This is the first example of a pseudopterane-type diterpene isolated from *Pseudopterogorgia elisabethae*; so far

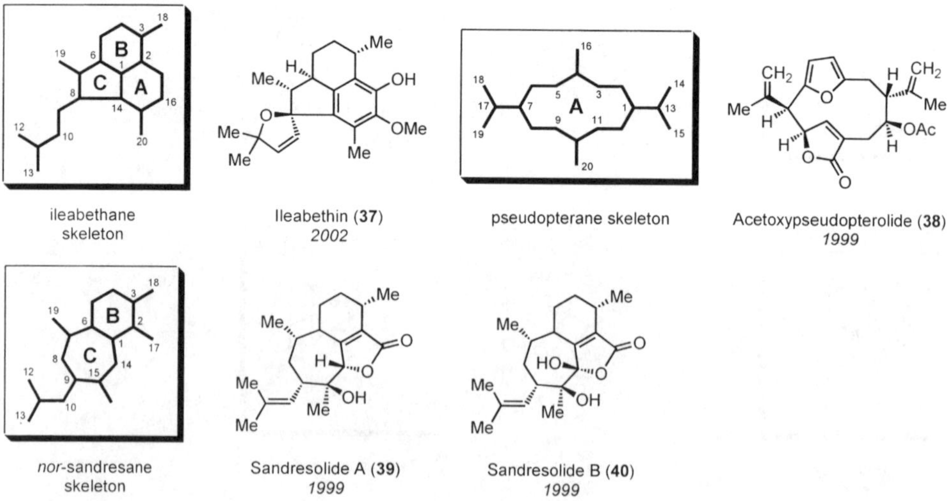

Fig. 8 Compounds possessing the ileabethane, pseudopterane or *nor*-sandresane skeleton

there is no obvious structural or biosynthetic interrelation between **38** and the previously described compounds. The two novel *nor*-diterpenes sandresolide A (**39**) and B (**40**) [24] feature the *nor*-sandresane carbon skeleton, which is also quite different from the other frameworks. The ring B is here fused to a seven-membered ring that could stem from the adventitious rearrangement of an amphilectane-based diterpene involving loss of a carbon atom and concomitant 1,2-alkyl shift with ring expansion (Fig. 8).

3 Biosynthesis and Pharmacology

3.1 Biosynthetic Pathways

As shown in the previous section, the structural diversity found among the plethora of natural products isolated from *Pseudopterogorgia elisabethae* is indeed remarkable. Scheme 1 gives an overview of the biosynthetic pathways to the different carbon frameworks produced by *Pseudopterogorgia elisabethae*.

Starting from geranylgerane, a series of cyclizations leads to the serrulatane skeleton which is the central intermediate in the biosynthetic pathways to all other carbon frameworks. The biogenetic origins of the amphilectane, ileabethane, and elisabethane ring systems could be traced back to this intermediate. Further cyclizations, bond cleavages or losses of carbons result in the wide variety of carbon frameworks shown. So far, 15 serrulatane-derived carbon skeletons have been isolated from *Pseudopterogorgia elisabethae* in addition to other skeletal types, for instance the pseudopterane [23] and aristolene skeletons.

Recently it was shown by radiolabeling studies that the formation of the serrulatane skeleton is catalyzed by the pseudopterosin diterpene cyclase, which can be considered as a key enzyme in terpene biosynthetic pathways (Scheme 2). The elisabethatriene cyclase is a monomer with a molecular mass of 47 kDa [25].

A potential mechanism accounting for the formation of elisabethatriene (**42**) from geranylgeranyl pyrophosphate (GGPP, **41**) is described in Scheme 3. In this proposed biosynthesis, the initial loss of the pyrophosphate group from **41** initiates a ring closure, generating a tertiary carbocation following isomerization of the C2/C3 double bond from *E* to *Z*. Hydride shifts at C1 and C6, respectively, introduce the stereochemistry at C6 and C7 and subsequently form an allylic carbocation. After a second ring closure, further rearrangement followed by abstraction of the hydrogen at C20 accounts for the formation of the bicyclic triene **42** [25].

Pseudopterosins coexist with the *seco*-pseudopterosins, suggesting that these two classes of diterpenes are produced from a single cyclase product. Elisabethatriene (**42**) undergoes aromatization to erogorgiaene (**1**); presumably a series

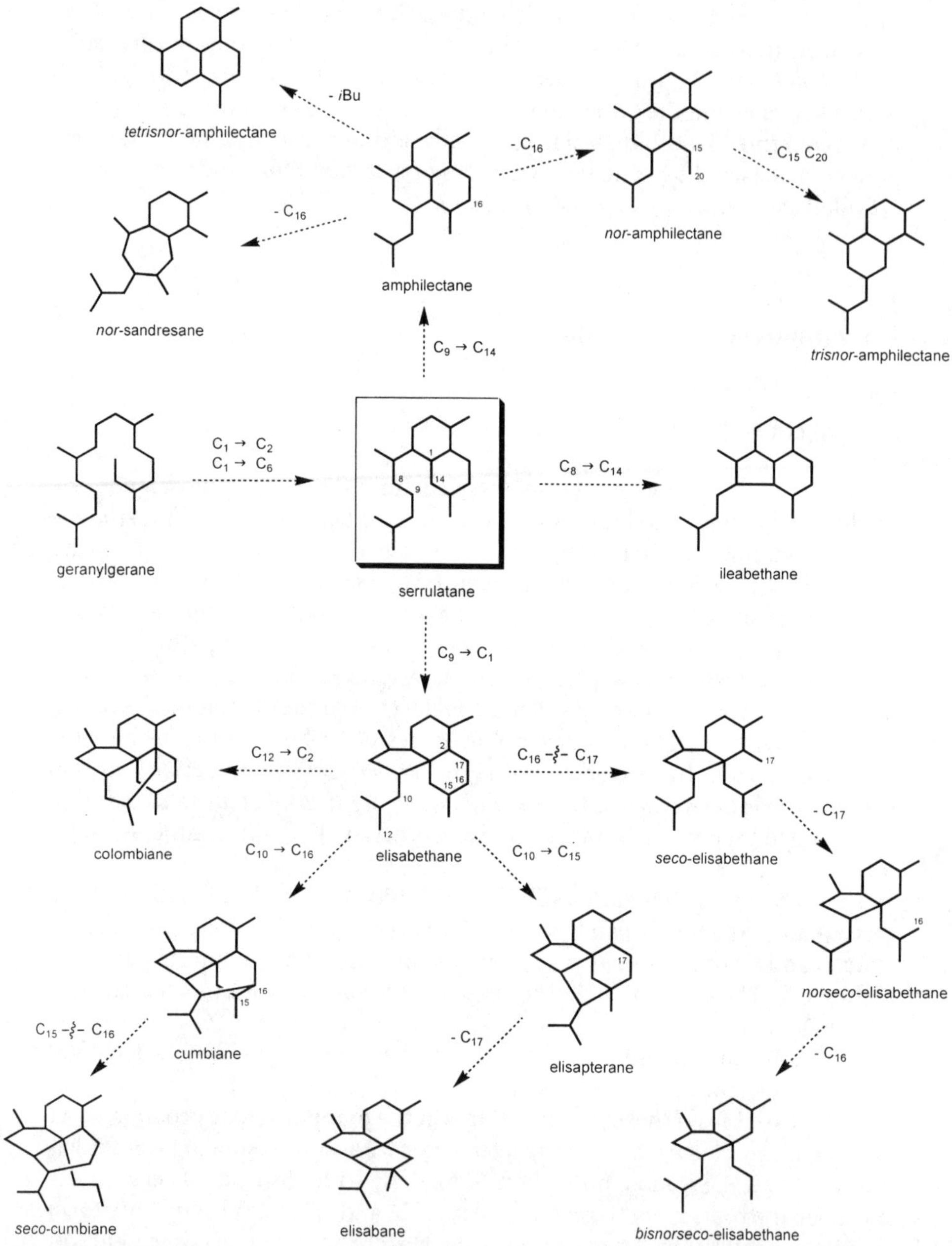

Scheme 1 Biosynthetic pathways to carbon skeletons produced by *Pseudopterogorgia elisabethae*

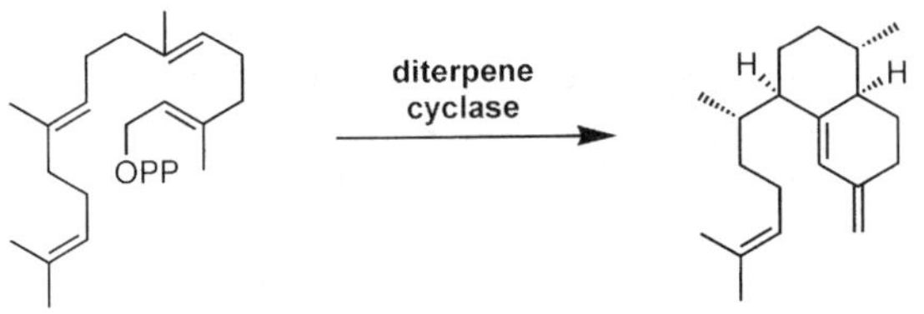

GGPP (**41**) Elisabethatriene (**42**)

Scheme 2 Diterpene cyclase-catalyzed cyclization to serrulatane skeleton

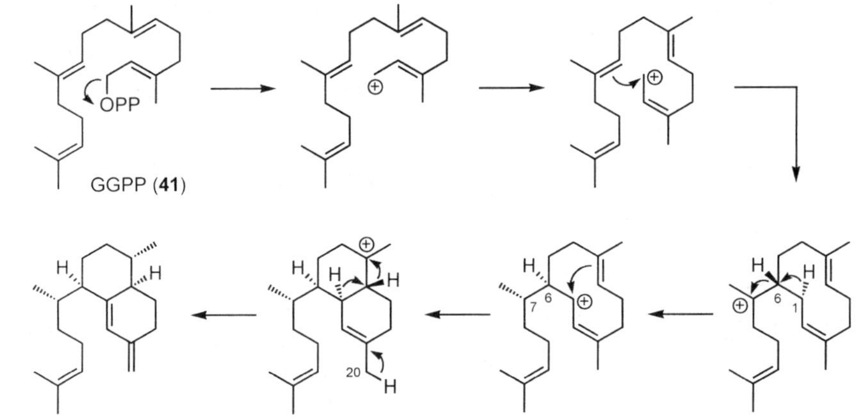

Elisabethatriene (**42**) Hydride shift Hydride shift

Scheme 3 Potential mechanism accounting for the formation of elisabethatriene (**42**)

of oxidations leads then via the known hydroxyerogorgiaene (**2**) or its regioisomer **43** to the *seco*-pseudopterosin aglycone **3** (Scheme 4). Glycosylation at either position would provide one of the *seco*-pseudopterosins. The biosynthetic pathway to the pseudopterosins appears to diverge at intermediate **43** and it is postulated that this compound undergoes oxidation to a hydroquinone which would then be oxidized to yield elisabethadione (**4**). Subsequent ring closure forms elisabethol (**13**) followed by aromatization to afford pseudopterosin aglycone **10**. Finally, glycosylation accounts for the ultimate production of the pseudopterosins [25].

The biosynthesis of the elisabethins possessing the elisabethane carbon skeleton also seems to proceed via elisabethatriene (**42**) and elisabethadione (**4**). Allylic oxidation of **4** results in known *sec*-hydroxyelisabethadione **6** followed by subsequent phosphorylation of the allylic alcohol (Scheme 5). Upon reduction of the quinone, the highly active species **44** is generated containing an *o,p*-activated position. This very strong nucleophile is ready to undergo immediate intramolecular C1–C9 cyclization to intermediate **45**. Further tautomerization finally leads to formation of elisabethin A (**22**).

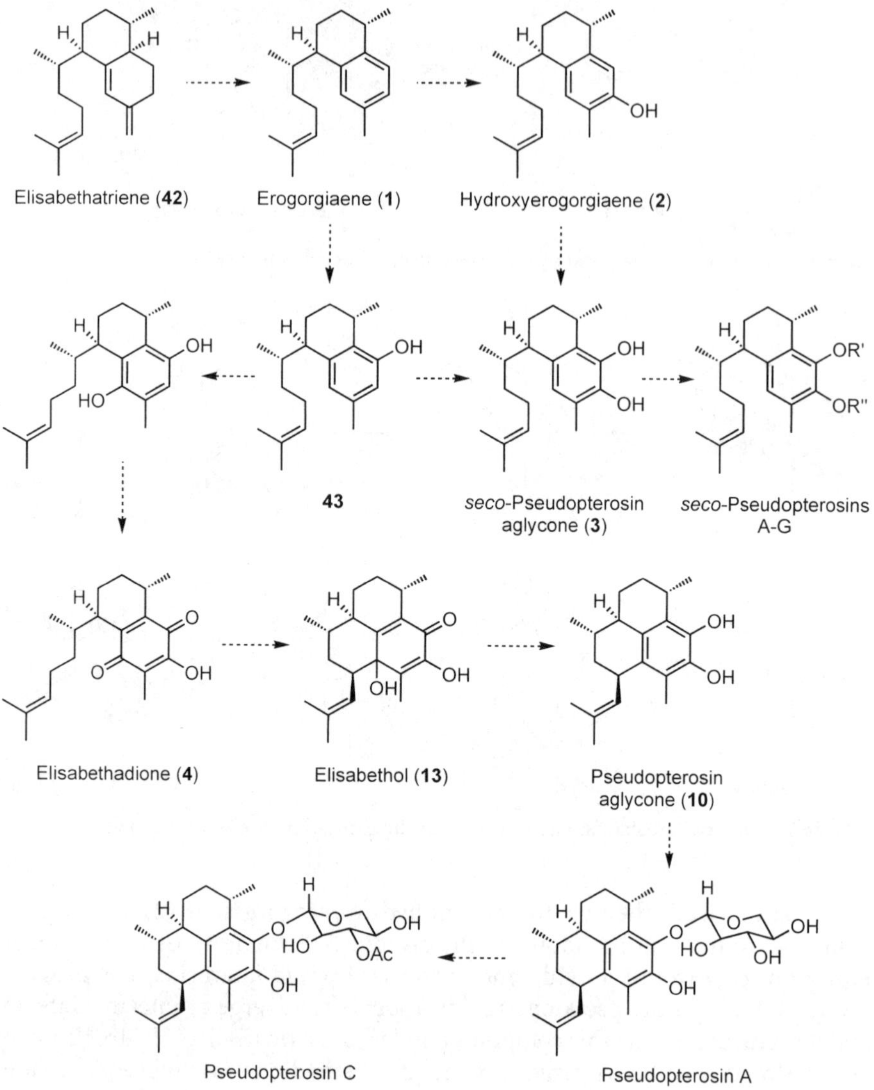

Scheme 4 Plausible biosynthetic pathway of the *seco*-pseudopterosins and pseudopterosins

Interestingly, the apparent mechanism by which diterpene **22** is produced is consistent in metabolites **32**, **25**, and **26** (Scheme 6), even though each has a completely different carbon skeleton [17]. Elisabethin B (**25**) and elisabethin C (**26**), for instance, appear to be synthesized from **22** by the following reactions. The cyclohexanone ring of **32**, **25**, and **26** is formed by oxidative cleavage of the bond between C2 and C17 in **22** after tautomerization. Loss of CO_2 followed by oxidation at C16 produces a key bicyclic β-ketocarboxylic acid which, after reduction and dehydration across C15 and C16, gives **25** or alternatively, upon

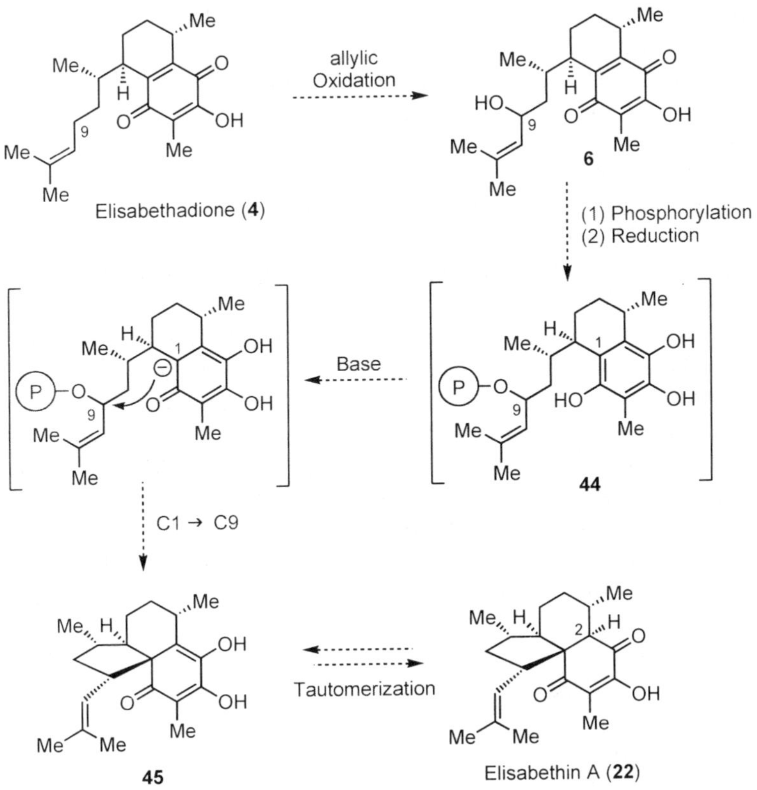

Scheme 5 Proposed biosynthesis of elisabethin A

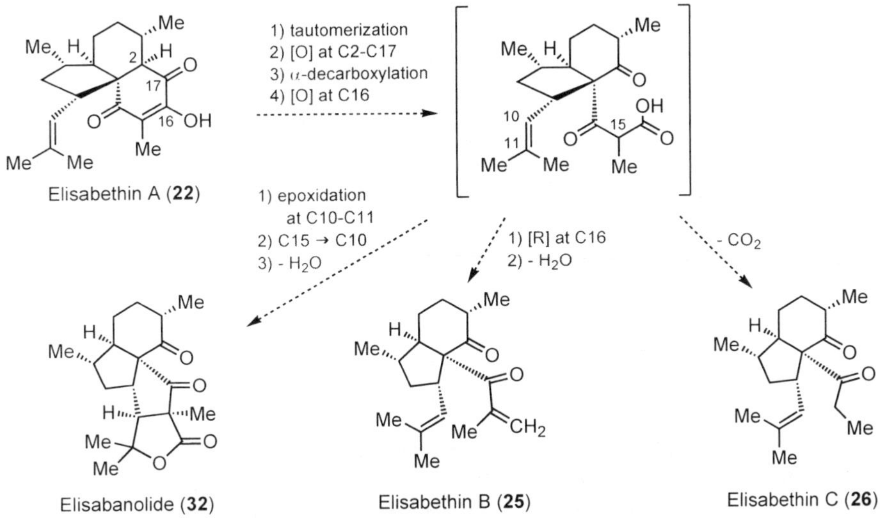

Scheme 6 Suggested biosynthesis of the elisabethins

decarboxylation, yields **26**. In turn, elisabanolide (**32**) appears derived from subsequent condensation and annelation of the proposed bicyclic *β*-ketocarboxylic acid precursor. Thus, the cyclopentanone ring of **32** is formed by attack of the anion of the active methine at C15 (adjacent to the C14 and C16 carbonyls) on the C10 position of an epoxide group between C10 and C11. Intramolecular esterification of the carboxylic acid functionality with a tertiary carbinol at C11 gives the five-membered lactone moiety in **32** [17] (Scheme 6).

Although the biosynthesis of colombiasin A (**36**) and the elisapteranes is not known, the fact that these diterpenes occur together in *Pseudopterogorgia elisabethae* suggest a common biosynthetic pathway. This concomitance strongly implies that elisabethin A (**22**) is a biogenetic precursor of colombiasin A (**36**) and the elisapterosins. In such a case, **22** must first undergo allylic oxidation at C12 followed by phosphorylation or protonation of the newly introduced oxygen. Base-catalyzed removal of the proton on C2 and intramolecular alkylation of the resulting enolate would result in **36** [21] (Scheme 7). In contrast, deprotonation of the hydroxy group in **22** produces the *β*-diketoenolate, which reacts intramolecularly in an S_N2' fashion to give elisapterosin B (**27**). Hydration of the external double bond would lead to elisapterosin C (**28**) and subsequent hemiketal formation accounts for the presence of elisapterosin A (**31**).

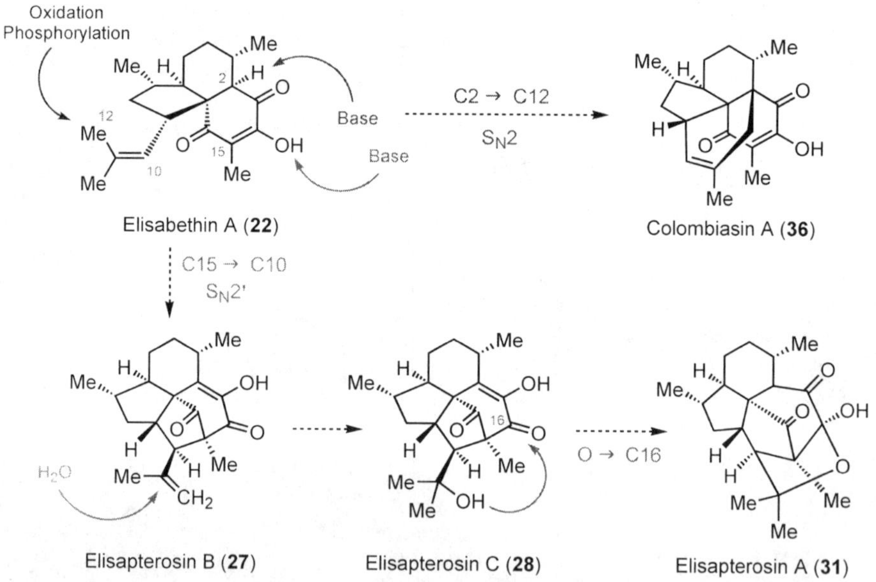

Scheme 7 Putative biosynthetic pathways to tetracyclic derivatives

An alternative pathway (Scheme 8) could start from the two regioisomers **5** and **6**, both isolated from *Pseudopterogorgia elisabethae*. Thus, after phosphorylation or protonation followed by elimination to dienylquinone **46**, intramolecular cyclization would generate zwitterionic intermediate **47**. Hydride

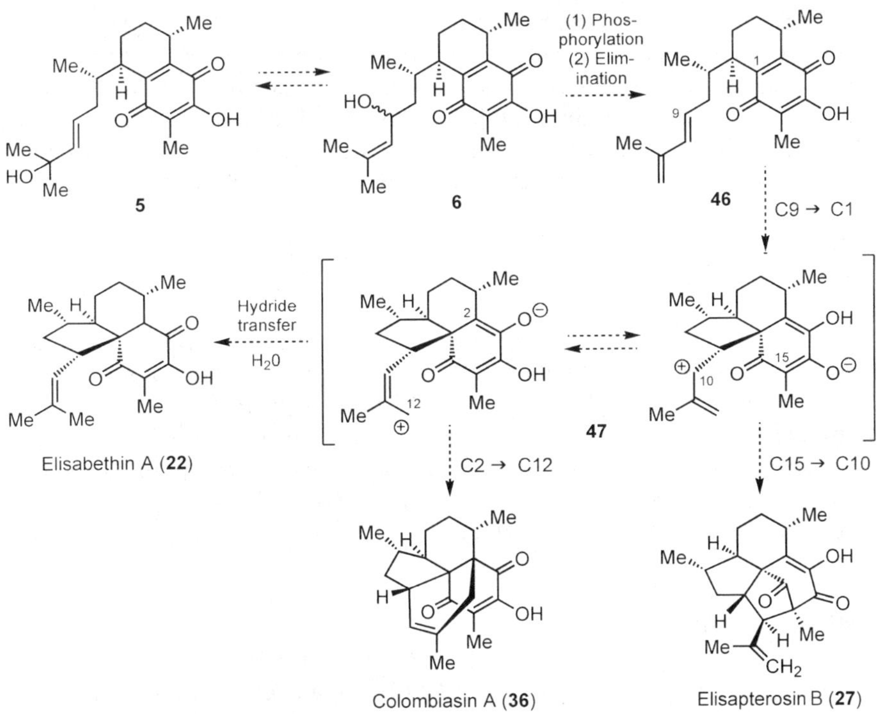

Scheme 8 Alternative pathway to the tricyclic elisabethane skeleton via a zwitterionic species and further cyclizations

transfer to **47** would result in the formation of elisabethin A (**22**), whereas 2,12-cyclization would give colombiasin A (**36**) and 10,15-cyclization would lead to the assembly of elisapterosin B (**27**). In fact, these cyclizations have been observed in vitro on treating **46** with Lewis acid (see Sect. 4.5). Therefore an equally evident interpretation for the formation of **36** and **27** would involve intramolecular cycloaddition reactions. Thus, commencing with **46**, a [4+2] cyclization would result in the colombiane skeleton, while a [5+2] cycloaddition would deliver the elisapterane scaffold (see also Sects. 4.5 and 4.6). In contrast, the cumbiasins might be generated from **47** via a SET process forming a C10 radical which then attacks the C16 position (Scheme 8).

3.2 Pharmacological Properties

3.2.1 Anti-inflammatory Activity

The pseudopterosins and *seco*-pseudopterosin isolated from Bahamian specimens of *Pseudopterogorgia elisabethae* display strong anti-inflammatory and analgesic activity. These diterpene glycosides had potencies superior to those of existing drugs such as indomethacin in mouse ear models [26]. These compounds appear to act by a novel mechanism of action; they are not active against PLA_2, cyclooxygenase, and cytokine release, or as regulators of adhesion molecules. Evidence suggests the pseudopterosins block eicosanoid release rather than biosynthesis in murine macrophages. Purified pseudopterosins (pseudopterosin E) have been commercialized in the skin cream Resilience [27].

3.2.2 Antituberculosis Activity

Many compounds of the serrulatane, amphilectane, elisabethane, and elisapterane class as well as further derivatives exhibit antituberculosis activity. Typical values for growth inhibition of *Mycobacterium tuberculosis* range between 20 and 60%. However, some of these marine diterpenoids show a very high growth inhibition rate. Among the most effective compounds are elisapterosin B (**27**) [18], homopseudopteroxazole (**18**) [15], erogorgiaene (**1**), [5] and pseudopteroxazole (**17**) [9] (Fig. 9).

3.2.3 Anticancer Activity

In biological screenings, elisapterosin A (**30**) [18] and elisabatin A (**14**) [14] indicated moderate (nonselective) in vitro cancer cell cytotoxicity in the NCI's 60 cell-line tumor panel. Elisabethin B (**15**) [17] was found to have significant in vitro cancer cell cytotoxicity with concentrations of 10^{-5} M eliciting significant differential responses at the GI_{50} level from all the renal, CNS, and leukemia cancer cell lines. Elisabethamine (**7**) [8] exhibited activity against lung cancer (LNCap) and prostate cancer (Calu) cell lines (IC_{50} values of 10.35 and 20 µg/ml, MTT assay). Mild anticancer activity against a prostate cancer cell line was also shown by acetoxypseudopterolide (**38**) [23] (IC_{50} value of 47.9 µg/ml, MTT assay).

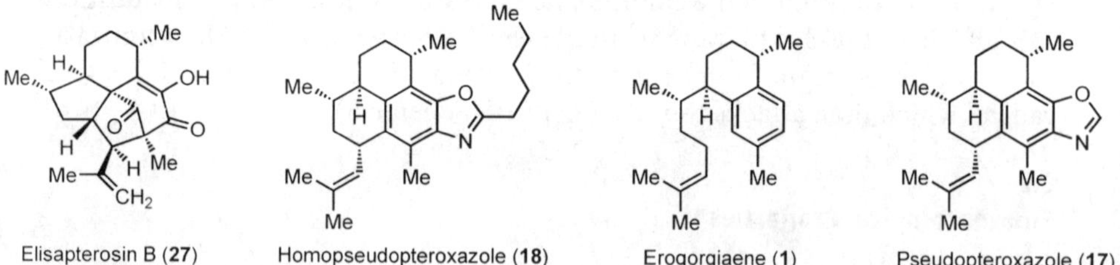

Elisapterosin B (**27**) 79% | Homopseudopteroxazole (**18**) 80% | Erogorgiaene (**1**) 96% | Pseudopteroxazole (**17**) 97%

Fig. 9 Compounds displaying the highest growth inhibition rates against *Mycobacterium tuberculosis* H37Rv (12.5 µg/ml)

3.2.4 Antiplasmodial Activity

Very recently elisapterosin A (**30**) [18, 19] was found to exhibit strong antiplasmodial activity (IC_{50} 10 μg/ml) against *Plasmodium falciparum*, the parasite responsible for the most severe forms of malaria.

4 Recent Partial and Total Syntheses: New Challenges for the Diels–Alder Reaction

The promising pharmacological profiles and fascinating structures of the natural products from *Pseudopterogorgia elisabethae* have inspired considerable synthetic efforts to gain access to this important compound class. As most of the natural products were isolated and structurally elucidated over the past few years, the number of partial and total syntheses is still limited. The following gives an overview of synthetic strategies to access structures isolated from *Pseudopterogorgia elisabethae* published to date.

Except for Yamada's synthesis, all total syntheses feature the venerable Diels–Alder reaction as a key transformation, in either an intermolecular or intramolecular fashion or in a combination of both.

4.1 Pseudopteroxazole (Corey, 2001)

Corey's retrosynthetic concept (Scheme 9) is based on two key transformations: a cationic cyclization and an intramolecular Diels–Alder (IMDA) reaction. Thus, cationic cyclization of diene **50** would give a precursor **49** for *epi*-pseudopteroxazole (**48**), which could be converted into **49** via nitration and oxazole formation. Compound **50** would be obtained by deamination of compound **51** and subsequent Wittig chain elongation. A stereocontrolled IMDA reaction of quinone imide **52** would deliver the decaline core of **51**. IMDA precursor **52** should be accessible by amide coupling of diene acid **54** and aminophenol **53** followed by oxidative generation of the quinone imide **52** [28].

The following work presented by the Corey group describes the enantiospecific synthesis of the originally proposed structure of pseudopteroxazole (**17**). As the synthesized compound **48** was not identical with **17**, the results of this study led to a structural revision yielding the correct stereochemistry of **17** [28]. The Corey group has also published a total synthesis of the pseudopterosin aglycones **11** [12] and **10** [29].

The actual synthesis (Scheme 10) commenced with the coupling of diene acid **54** and aminophenol **53** to provide diene amide **55**. In situ generation of quinone monoimide **52** under oxidative conditions and subsequent intramolecular Diels–Alder (IMDA) reaction furnished an 8:1 mixture of *endo*/*exo*

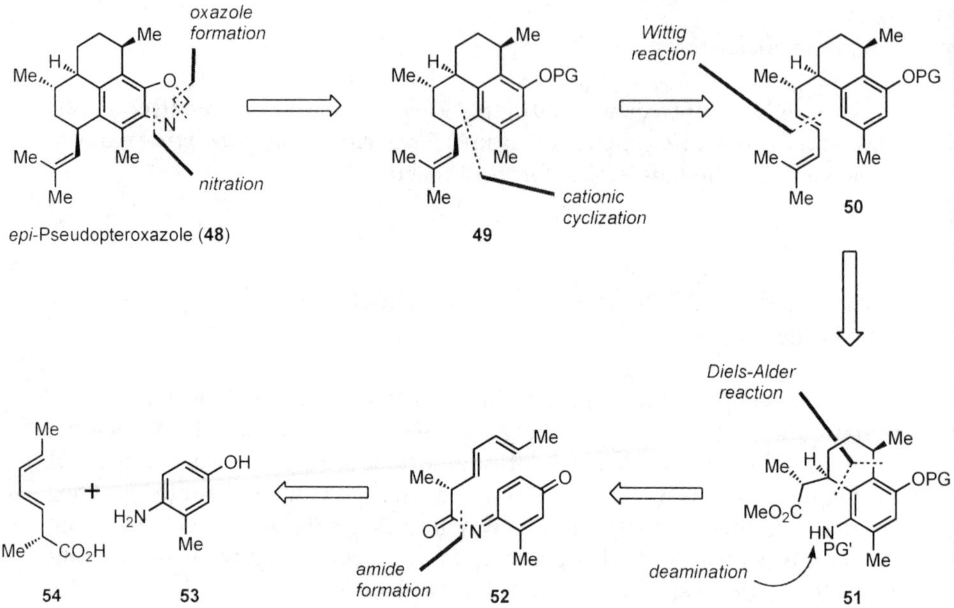

Scheme 9 Corey's retrosynthetic analysis of *epi*-pseudopteroxazole (**48**)

adducts (**56**). This IMDA cyclization appeared to be the first example of this type of internal cycloaddition involving an in situ generated quinone imide. Hydrogenation and benzylation produced benzyl ether **57**. Boc protection and methanolysis of the δ-lactam subunit afforded methyl ester **58**. Amino alcohol **59** was obtained by reduction of the ester function and subsequent cleavage of the boc group. Deamination and Dess–Martin oxidation delivered aldehyde **60**, which was subjected to a Wittig coupling to give (*E*)-conjugated diene **61**. Cationic cyclization to **62** was effected quantitatively using methanesulfonic acid as catalyst, but gave a mixture of diastereomers in a ratio of about 1:2, which were separated chromatographically. The stereochemical outcome of this reaction can be rationalized considering the steric interference of the (*E*)-methyl part of the isopropylidene group and the methyl group attached to C3. Benzyl ether cleavage followed by nitration furnished nitrophenol **63**, which was then treated with Zn dust to effect conversion to the corresponding aminophenol. Upon reaction with methyl orthoformate the desired tetracyclic oxazole **48** was formed. In summary, the synthesis of 3,6-*epi*-pseudopteroxazole (**48**) comprises 16 linear steps and was achieved in an overall yield of 3.2% (Scheme 10).

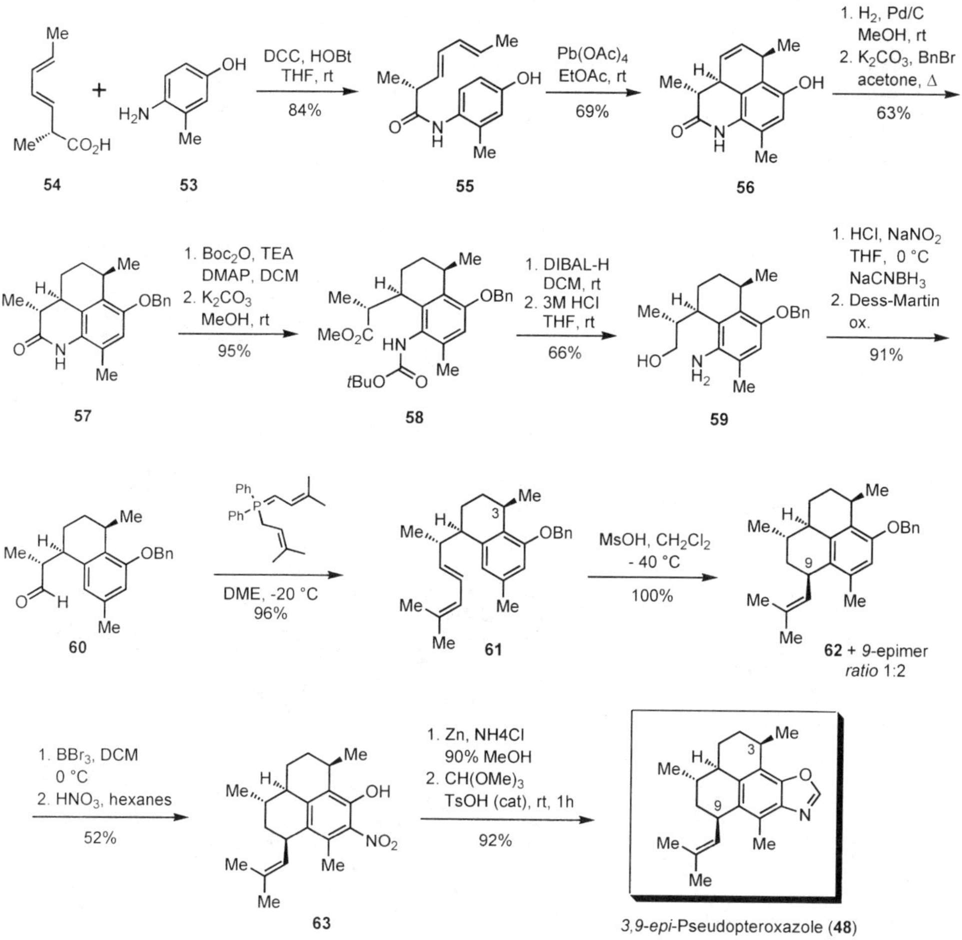

Scheme 10 Corey's total synthesis of 3,9-*epi*-pseudopteroxazole (**48**)

4.2 Elisabethin C (Yamada, 2002)

In Yamada's retrosynthetic analysis (Scheme 11), elisabethin C (**26**) is traced back to lactone **64** which would be converted into **26** by deoxygenation and chain elongations. Key intermediate **64** could be obtained by a stereoselective Dieckmann cyclization. The required ester lactone precursor **65** would be accessible from **66** by a series of oxidation reactions. Further disconnection would lead to commercially available (+)-carvone (**67**). Stereoselective successive alkylation of **67** and reduction of the enone should deliver **66** [30].

The linear sequence starts from (+)-carvone (**67**) (Scheme 12). Establishing the quaternary center at C1 was achieved via a successive enolate alkylation/hydroxyalkylation to furnish hydroxyketone **68**. Reduction of the enone

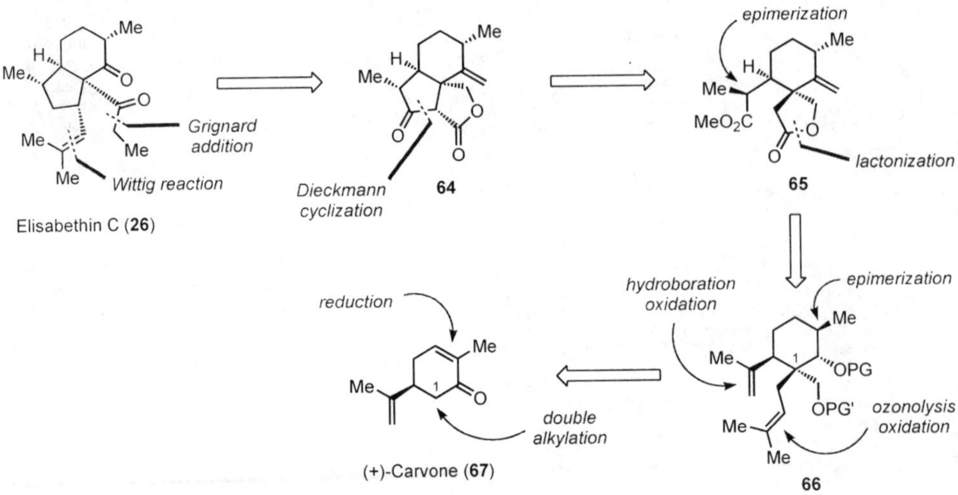

Scheme 11 Yamada's retrosynthetic analysis of elisabethin C (26)

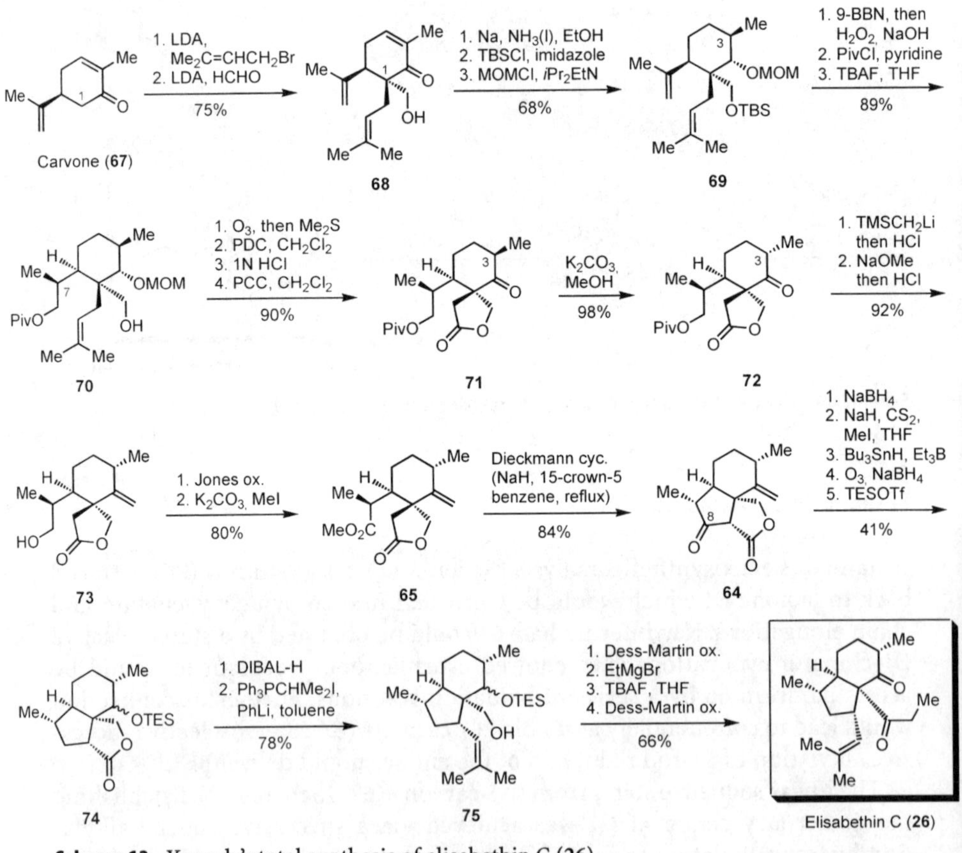

Scheme 12 Yamada's total synthesis of elisabethin C (26)

(resulting in the wrong C3 configuration), TBS protection of the primary and MOM protection of the secondary alcohol led to **69**, which was then converted into **70** by stereocontrolled hydroboration/oxidation, pivaloyl protection, and desilylation. The stereogenic center at C7 was generated in the undesired configuration, but this was rectified later by epimerization under the Dieckmann cyclization conditions. Compound **70** was transformed into ketolactone **71** by a succession of routine operations. The undesired configuration at C3 of **71** was corrected to **72** by base-catalyzed epimerization, and then the keto function was protected as the *exo* methylene group in **73**. Jones oxidation and methylation gave the desired key intermediate **65**, and after Dieckmann cyclization with concomitant epimerization, the ketolactone **64** was obtained. The superfluous 8-keto functionality was removed reductively, and the *exo*-methylene group was cleaved via ozonolysis and reduction to the alcohol. TES protection furnished silyl ether **74** as an epimeric mixture. The isopropylidene appendage was introduced via reduction of the lactone to the lactol and subsequent Wittig olefination. The resulting intermediate olefin **75** was then transformed into elisabethin C (**26**) via four additional steps. All told the sequence has about 29 linear steps and an overall yield of 5.2% (Scheme 12).

4.3
Elisabethin A (Mulzer, 2003)

In 2001, the Nicolaou group reported some model experiments directed toward the synthesis of elisabethin A (**22**) (Scheme 13) [31]. They found that oxidation of anilide derivatives with Dess–Martin periodinane (DMP) resulted in the formation of *o*-azaquinones, which they decided to apply as substrates for IMDA reactions. Among a number of examples, starting compound **76** was converted in a three-step sequence into amide **77** and further processed to diene **78**. DMP oxidation gave **79** as a transient intermediate, whose further reactions followed two divergent pathways. In the first, *o*-azaquinone **79** spontaneously underwent an IMDA reaction to form the polycyclic compound (±)-**80** and after hydration (±)-**81**. Interestingly ketohydroxyamide (±)-**81** closely resembles the pseudopterosin natural products. Alternatively, **79** was further oxidized to furnish *p*-benzoquinone **82**, whose IMDA reaction resulted in adduct (±)-**83**. Facile hydrolysis of (±)-**83** was accomplished upon exposure to K_2CO_3 to render semiquinone (±)-**84**, which embodies the tricyclic framework of elisabethin A (**22**) in racemic form. Apart from the low overall efficiency of the sequence, (±)-**84** is generated with the wrong relative configuration at C1/C6, as established via single-crystal diffraction analysis. This result was interpreted in terms of an *endo* transition state of the IMDA cyclization (Scheme 13).

A related IMDA key step, however with *exo* geometry, was applied in Mulzer's synthesis of elisabethin A (**22**) [32]. According to the retrosynthetic plan (Scheme 14) the elisabethane carbon skeleton would be assembled via an IMDA cyclization of quinone **86** which would be generated by oxidation of the corresponding hydroquinoid precursor. Selective hydrogenation, base-catalyzed

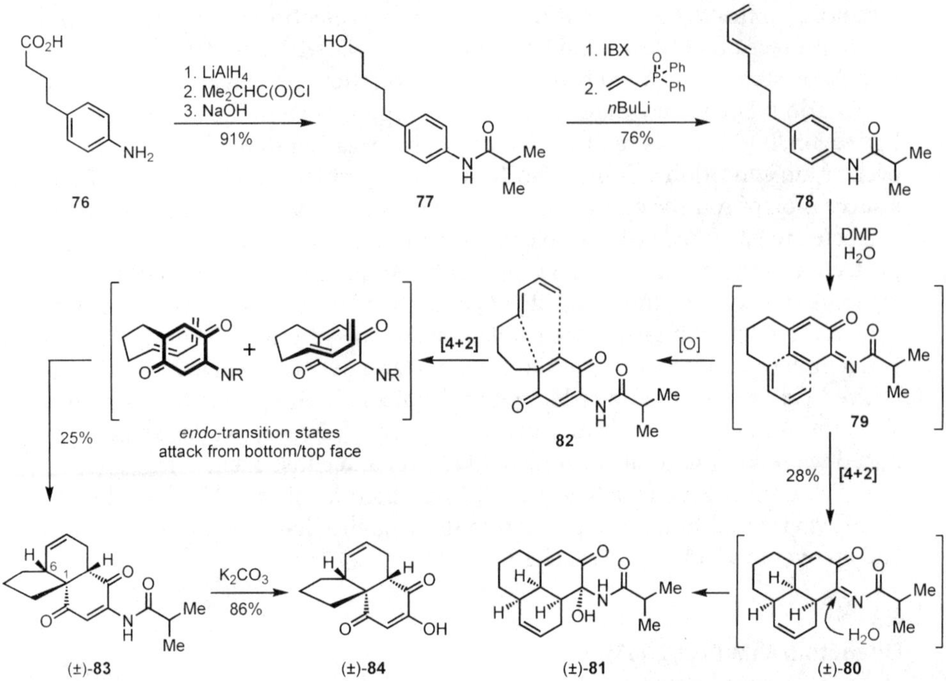

Scheme 13 DMP-mediated generation of Diels–Alder precursors and subsequent cycloaddition reactions

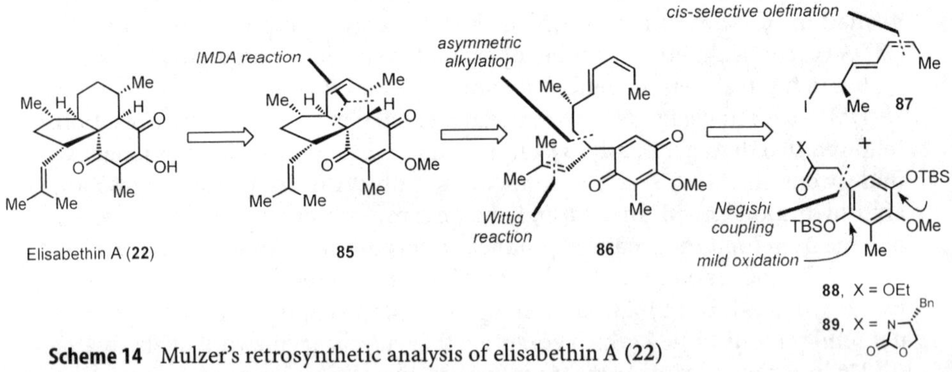

Scheme 14 Mulzer's retrosynthetic analysis of elisabethin A (22)

epimerization, and demethylation should then lead to **22**. The precursor for the key step would be available by stereoselective α-alkylation of aryl acetic acid derivatives **88** (ester) or **89** (Evans' auxiliary) and Wittig olefination. Chiral alkylation agent **87** would be obtained from commercially available (+)-Roche ester by a sequence of selective olefination reactions.

The diene fragment in the form of iodide **87** was synthesized from commercially available (+)-Roche ester **90** by tritylation, LAH reduction, and Swern oxidation (Scheme 15). Olefination of aldehyde **91** via the HWE reaction furnished the desired (*E*)-Weinreb enamide **92**, which was reduced to the *α*,*β*-unsaturated aldehyde **93**. The *Z* geometry of the second double bond was installed by a "salt-free" Wittig reaction to give isomerically pure diene **94**, which was deprotected and converted into iodide **87** by an Appel reaction.

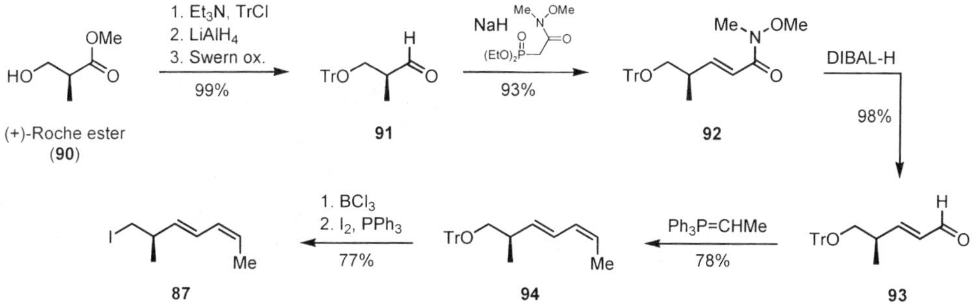

Scheme 15 Synthesis of the diene fragment

The synthesis of the benzenoid fragment (Scheme 16) started from commercially available aldehyde **95**, which was converted to phenol **96** by a Bayer–Villiger oxidation and subsequent hydrolysis. Selective 4-*O*-demethylation was achieved via an oxidation/reduction sequence to give hydroquinone **97** [33]. *O*-Silylation and regioselective bromination of the *o*,*p*-activated position with NBS led to aryl bromide **98**, which was then reacted with stannane **99** using a palladium-catalyzed Negishi–Reformatsky coupling [34] to give aryl acetic ester **88**. In order to obtain the Evans' oxazolidinone **89**, ester **88** was transformed into acid **100** in a three-step reduction/oxidation sequence, which was necessary as ester **88** remained unchanged even after refluxing in concentrated EtOH/NaOH. Acid **100** was converted to the desired imide **89** via the mixed anhydride. Diastereoselective alkylation of **89** was achieved by sodium enolate formation and subsequent addition of iodide **87** to furnish **101** with a *de* of 86%. Removal of the auxiliary, Swern oxidation to **102** and Wittig reaction delivered triene **103**. Upon treatment of **103** with TBAF followed by the addition of aqueous $FeCl_3$ solution, quinone **86** emerged (observed by TLC and NMR) and underwent the IMDA reaction in situ at room temperature to render adduct **85**. The less substituted double bond in **85** was then hydrogenated followed by base-catalyzed epimerization at C2 and demethylation to provide elisabethin A (**22**). The convergent synthesis was accomplished in 19 steps along the longest linear sequence with 7% overall yield (Scheme 16).

In contrast to Nicolaou's model system, the IMDA reaction of quinone **86** to adduct **85** proceeded through an *exo* transition state geometry. This difference is due to the two additional substitutes attached to the diene side chain of quinone **86**. A more detailed analysis (molecular modeling) of the four theoretically possible transition states (Fig. 10) reveals that an enophilic *endo*/*exo*

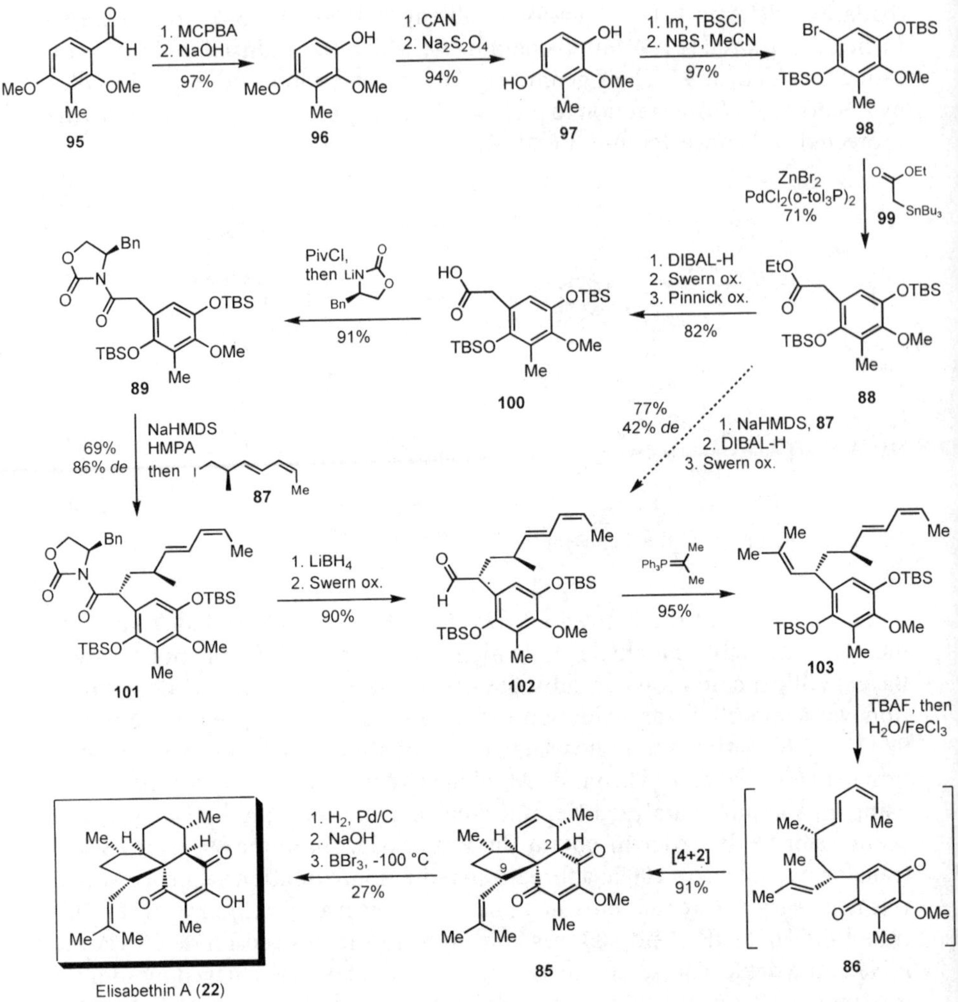

Scheme 16 Mulzer's total synthesis of elisabethin A (22)

attack from the top face (**104, 105**) shows strong steric interactions between the isobutenyl rest and the (*E*)-double bond of the diene unit. On the other hand, in the case of an *endo* bottom face attack (**106**), steric repulsion between the 19-methyl group and the (*E*)-double bond comes into play shifting the equilibrium to *exo* arrangement (**107**). These results make clear that unfunctionalized model systems are often not appropriate to predict stereochemical issues associated with IMDA cyclizations.

The significance of this IMDA reaction lies firstly in the use of a tethered (*E*,*Z*)-diene – the first case that has been successfully developed – secondly, in the unusually mild and virtually biomimetic conditions (aqueous medium, ambient temperature), and thirdly in the high yield and stereoselectivity.

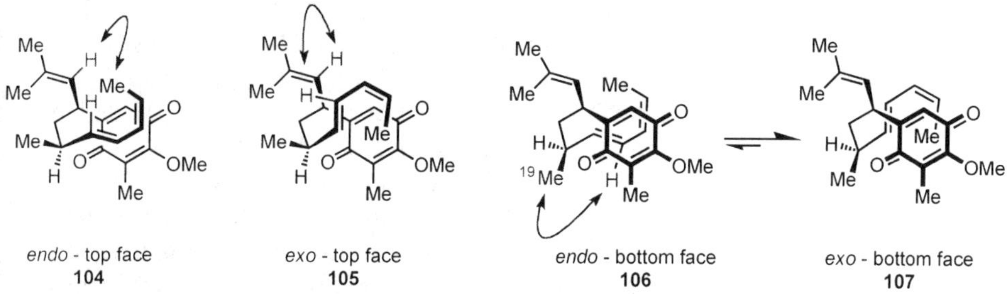

Fig. 10 Facial selectivity and *endo* vs. *exo* transition state

4.4 Elisapterosin B (Rawal, 2003)

In Rawal's synthetic plan (Scheme 17) to elisapterosin B (**27**) the tetracyclic elisapterane skeleton would be formed in an oxidative cyclization of an elisabethane precursor (**108**). Further disconnection via a retro-IMDA reaction would lead to intermediate **109** bearing a (*Z,E*)-diene unit and a quinone portion. The quinone would be generated by aromatic oxidation of a phenol while Pd-catalyzed cross-coupling should deliver stereoselectively the required diene moiety. Wittig olefination would introduce the isopropenyl unit in **109**. Aryl acetic ester **110** would be constructed by a pinacol-type ketal rearrangement of **111**, which should be obtainable from **112** by lactone reduction and acetylenation. Negishi coupling of acid chloride **114** and the Grignard reagent of aryl bromide **113** followed by alkylation would furnish aryl lactone **112**. The required stereochemical information for the asymmetric realization of the synthesis

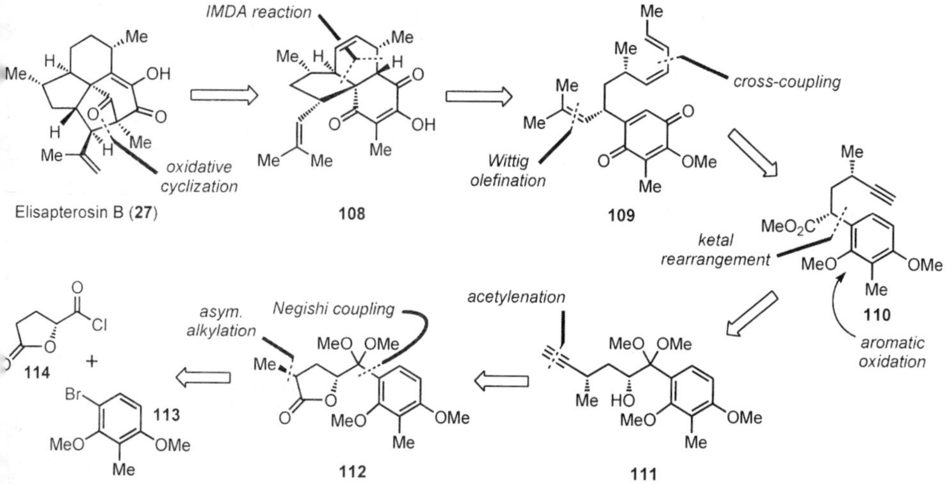

Scheme 17 Rawal's retrosynthetic analysis of elisapterosin B (**27**)

stems on the one hand from the chiral starting material **114**, and on the other hand from the geometrically defined (*Z*,*E*)-diene unit (Scheme 17) [35].

The total synthesis (Scheme 18) commenced with a Negishi coupling of commercially available acid chloride **115** (due to cost reasons it was decided to elect the opposite enantiomer resulting in an *ent*-**27** synthesis) to the Grignard reagent of aryl bromide **113**. Under ketal-forming conditions the expected methyl ketal **116** was accompanied by the lactone methanolysis product, making treatment of the crude material with *t*BuOK necessary to obtain a good overall yield. Stereoselective methylation (78% *de*) afforded the desired *trans* product **117**, which was reduced to the lactol and acetylenated using the Seyferth reagent [36]. Upon heating the mesylate of **118** in the presence of an acid scavenger ($CaCO_3$) the pivotal pinacol-type rearrangement via **119** took place to render methyl ester **120**. In preparation for the envisaged IMDA reaction the acetylene was converted to the (*Z*)-bromoalkene (**121**) and cross-coupled with (*E*)-bromopropene. The ester functionality in resulting compound **122** was reduced to the aldehyde and transformed into the 2-methylpropenyl side chain via Wittig olefination. Regioselective demethylation [37] of the more hindered methyl ether provided phenol **123** (Scheme 18).

The required quinone **124** for the IMDA cyclization was obtained by salcomine-catalyzed oxidation [38] (Scheme 19). Heating in toluene initiated a clean cycloaddition to furnish the *endo* adduct **126** as a single diastereomer. Of

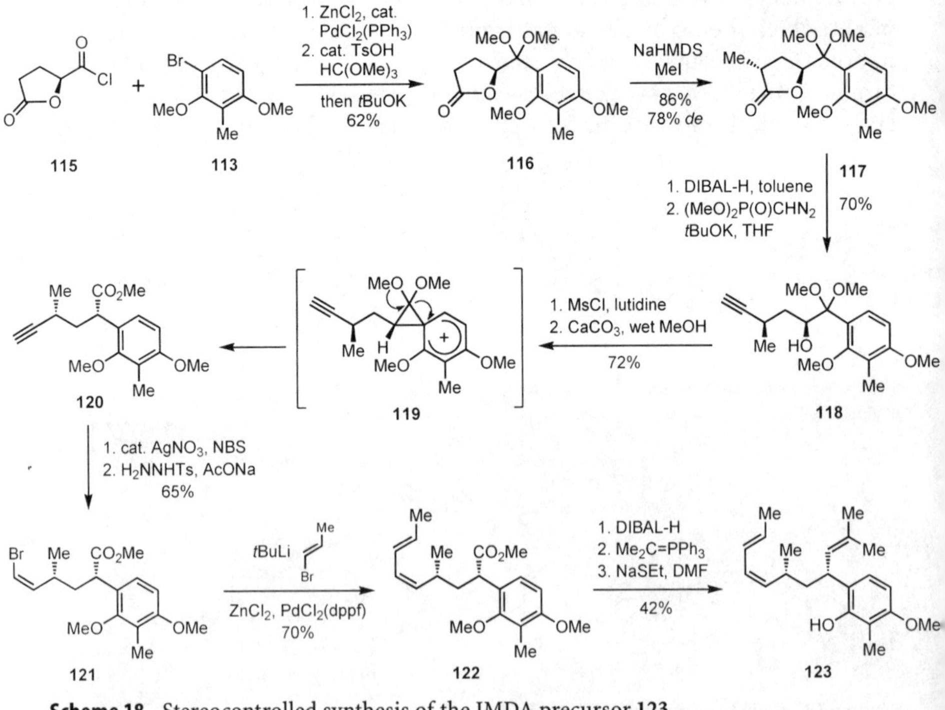

Scheme 18 Stereocontrolled synthesis of the IMDA precursor **123**

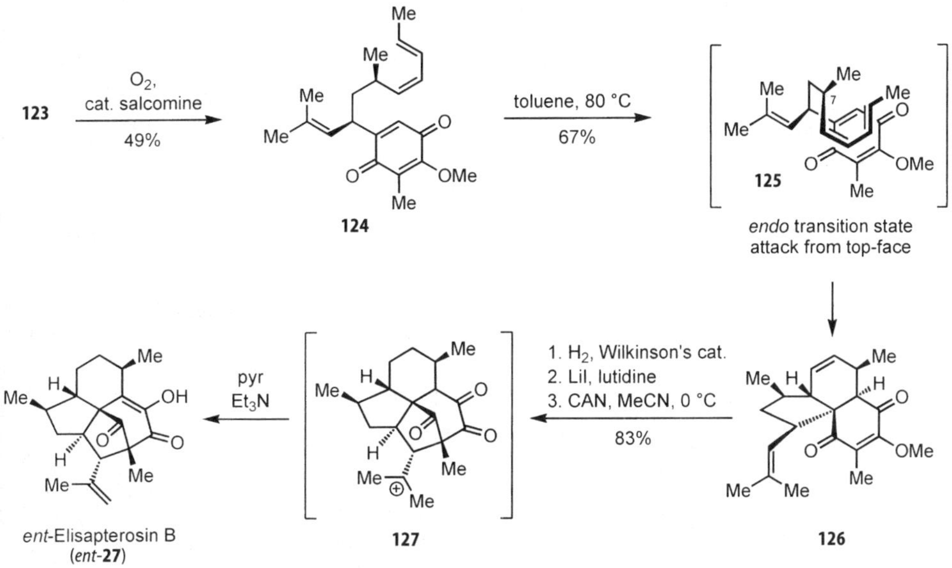

Scheme 19 Completion of Rawal's total synthesis of *ent*-elisapterosin B (*ent*-**27**)

the two possible *endo* transition states the one shown (**125**) avoids potentially severe allylic strain between the C7 Me group and the propenyl unit on the (*Z*)-double bond. The Diels–Alder adduct **126** was then selectively hydrogenated using Wilkinson's catalyst and subsequently demethylated under anionic conditions. Upon treatment with CAN, diketone intermediate **127** was formed in an oxidative cyclization and subsequently enolized with base to the desired target compound *ent*-**27**. In summary, the synthesis required 18 steps and gave 14% overall yield.

4.5 Elisapterosin B/Colombiasin A (Rychnovsky, 2003)

Another total synthesis of elisapterosin B (**27**), as well as colombiasin A (**36**) was reported by the Rychnovsky group [39]. The underlying concept of this approach was the proposed biosynthetic pathway shown in Scheme 8. Thus, the authors decided to prepare the putative metabolite **46** in *O*-methylated form **128** and subject it to Lewis acid conditions in the hope that cyclization might occur to either **27** or **36**, or both. The required precursor **128** would stem from an intermolecular Diels–Alder reaction between diene **129** and quinone **130** (Scheme 20).

The actual synthesis (Scheme 21) started with the stereoselective alkylation of Myers' hydroxy-amide **131** [40] followed by reductive removal of the auxiliary to give **132** in high yield and enantioselectivity. Wittig olefination furnished enoate **133**, which was then elaborated into the (*E*)-1-acetoxy-diene **129** using

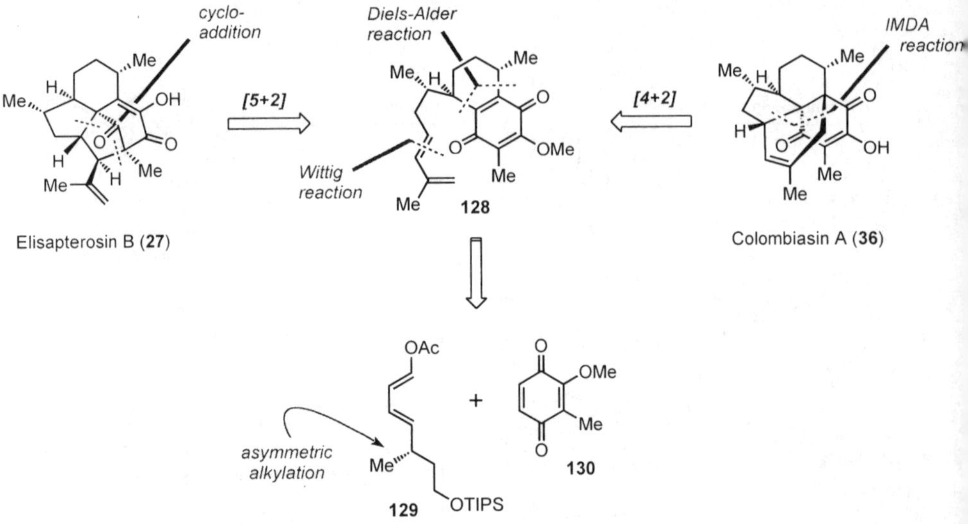

Scheme 20 Rychnovsky's retrosynthetic analysis of elisapterosin B (**27**)/colombiasin A (**36**)

Kowalski's one-carbon homologation protocol [41]. IntermolecularDiels–Alder reaction with quinone **130** required particular conditions (lithium perchlorate in ether) to produce an inseparable 1.7:1 mixture of the diastereomeric adducts **134a** and **134b** which was carried through the rest of the synthesis (separation in the final step). The moderate selectivity of this addition was interpreted in terms of the Felkin–Ahn transition state (see Scheme 21). Reduction of the carbonyl group at C17 in **134a/b** was necessary to prevent aromatization in the subsequent steps. The C3 methyl group was introduced via S_N2 displacement of the *O*-acetate to give **135a/b**. Hydrogenation and Dess–Martin oxidation returned enedione **136a/b**, which was converted to the quinone using DBU/O_2 followed by immediate reduction and acetate protection. Treatment with HF and oxidation with Dess–Martin periodinane delivered aldehyde **137a/b**. Wittig reaction (*E/Z*=3:1) and oxidation under mild conditions produced quinone **128a/b**. The [5+2] cycloaddition was induced with boron trifluoride and did indeed furnish **27a/b** as expected, though in moderate yield along with some **36a/b** methyl ether. Pure **27** was obtained upon chromatographic separation after the final step in the sequence. On the other hand, **36** was accessible via thermal [4+2] IMDA cyclization, demethylation, and diastereomer separation. This IMDA reaction was performed in analogy to Nicolaou's precedence discussed in the following section. Colombiasin A (**36**) was prepared in 17 steps and 3.9% yield while elisapterosin B (**27**) was synthesized in 16 steps and 2.6% overall yield (Scheme 21).

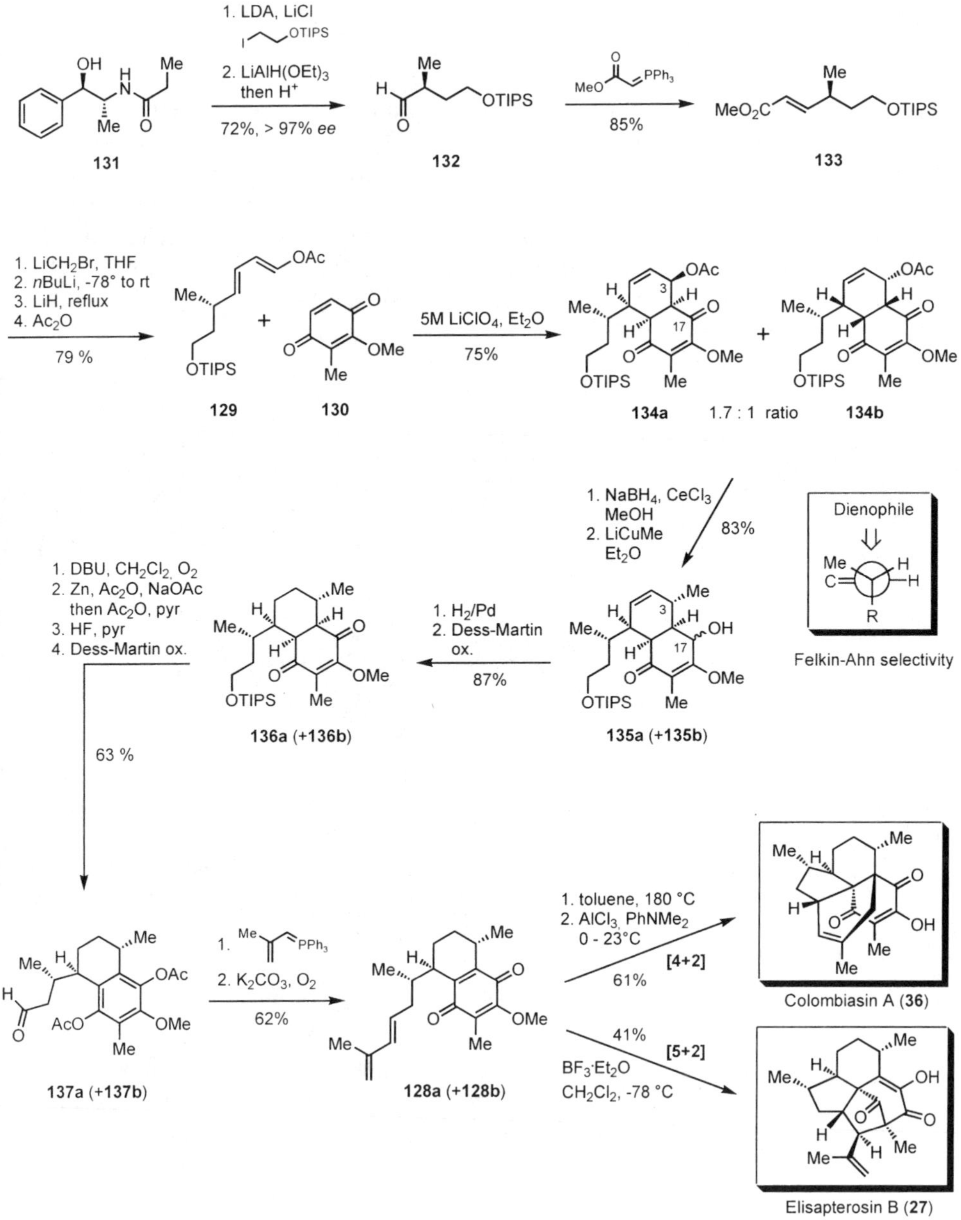

Scheme 21 Rychnovsky's total synthesis of elisapterosin B (27)/colombiasin A (36)

4.6 Colombiasin A (Nicolaou, 2001)

The retrosynthetic concept of the Nicolaou group is shown in Scheme 22. The target molecule **36** is disconnected via an IMDA cyclization of the diene quinone precursor **138**, which would be generated from the tetraline derivative **139** using Wittig chemistry followed by aromatic oxidation. A Claisen-type rearrangement would provide access to **139** whereby the side chain required for the rearrangement of **140** would be introduced by *O*-acylation. The core of **141** would be formed via an intermolecular Diels–Alder reaction between diene **142** and *p*-benzoquinone **130** [42].

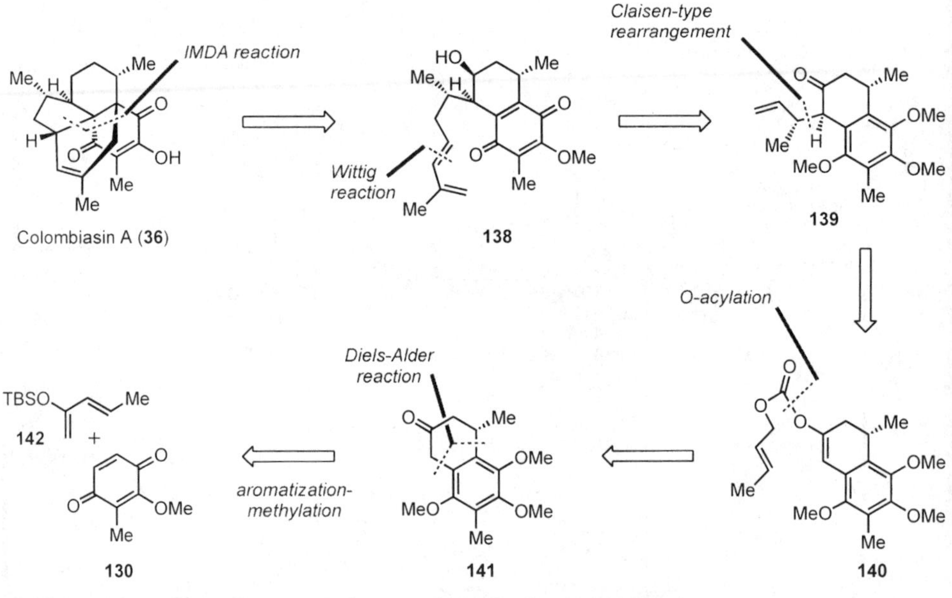

Scheme 22 Nicolaou's retrosynthetic analysis of colombiasin A (**36**)

In the synthetic direction (Scheme 23) **142** and **130** form under chiral catalysis [43] the labile *endo* adduct **143** regio- and stereoselectively. Compound **143** was aromatized with base and methyl iodide to furnish ketone **141** after desilylation. The appendage at C6 was introduced via a palladium-catalyzed Claisen-type rearrangement [44] of enol ester **140**, which was prepared by *O*-acylation of **141** with chloroformate **144**. Rearrangement product **139** was obtained with the correct configuration at C6 and wrong configuration at C7. To invert the stereogenic center at C7, a set of six reactions was applied. Specifically, the C5 ketone was protected as the silylated alcohol **145**. Then the 8,9-olefin was oxidized to aldehyde **146**, which was used for base-catalyzed epimerization at C7. A 2:1 mixture of the epimeric aldehydes was obtained and separated. The (7*R*)-aldehyde was converted to olefin **147**. The (7*S*)-epimer

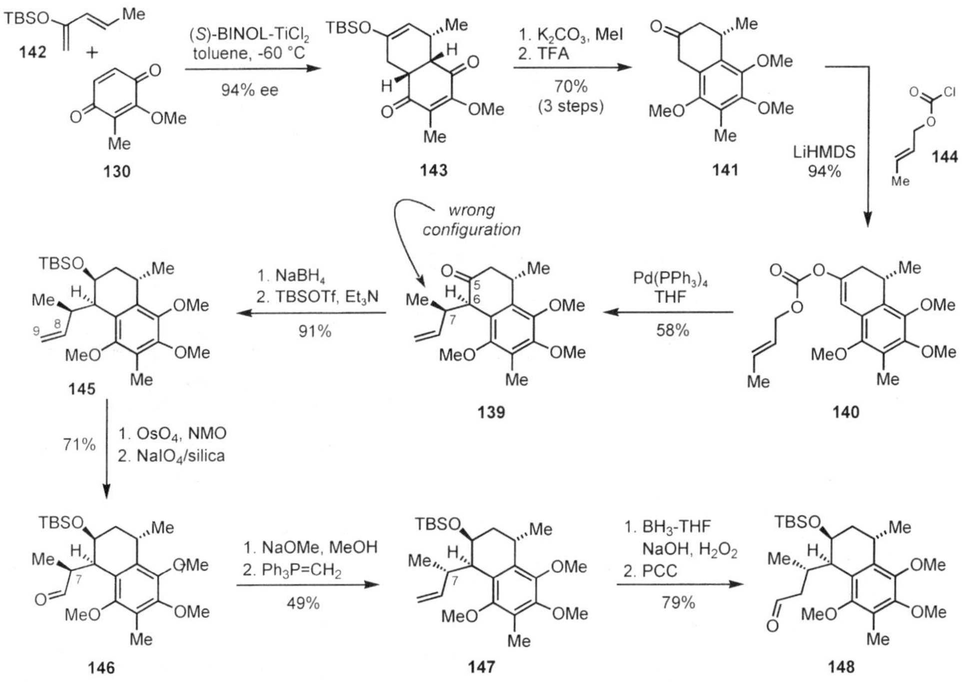

Scheme 23 Synthesis of the decaline core, attachment of side chain, and epimerization

was recycled, so that overall 49% of aldehyde (7*S*)-**146** was transformed into (7*R*)-**147**. Hydroboration and oxidation of the primary alcohol delivered aldehyde **148** (Scheme 23).

Wittig olefination of **148** furnished diene **149** as a 3.7:1 *E*/*Z* mixture (Scheme 24), which was to be processed toward the envisaged key Diels–Alder cyclization. However, the benzenoid ring could not be oxidized to the quinone without decomposition of the diene system. Hence, the diene was protected as the 1,4-sulfur dioxide adduct **150**, which was then successfully oxidized to the quinone. Under thermal conditions the diene system was regenerated via cheletropic elimination of SO_2 to give **138** and underwent in situ the IMDA reaction to form the colombiasin skeleton **152**. As expected by the authors, the Diels–Alder reaction did proceed via the *endo* transition state **151** and, thus, the correct configuration was established at C9. The observed facial selectivity can be rationalized considering the configuration of the C6 stereocenter which allows only an attack from the top face. The superfluous 6-OH functioniality was removed via Barton–McCombie deoxygenation [45] to furnish *O*-methyl colombiasin A **153**. The final *O*-demethylation of **153** with boron tribromide was, however, plagued by concomitant migration of the 10,11-double bond into the 11,12-position (use of *cis*-cyclooctene as a competitive olefin to suppress acid-induced migration). In summary, the synthesis of colombiasin A (**36**) required 20 steps and resulted in 0.02% overall yield (Scheme 24).

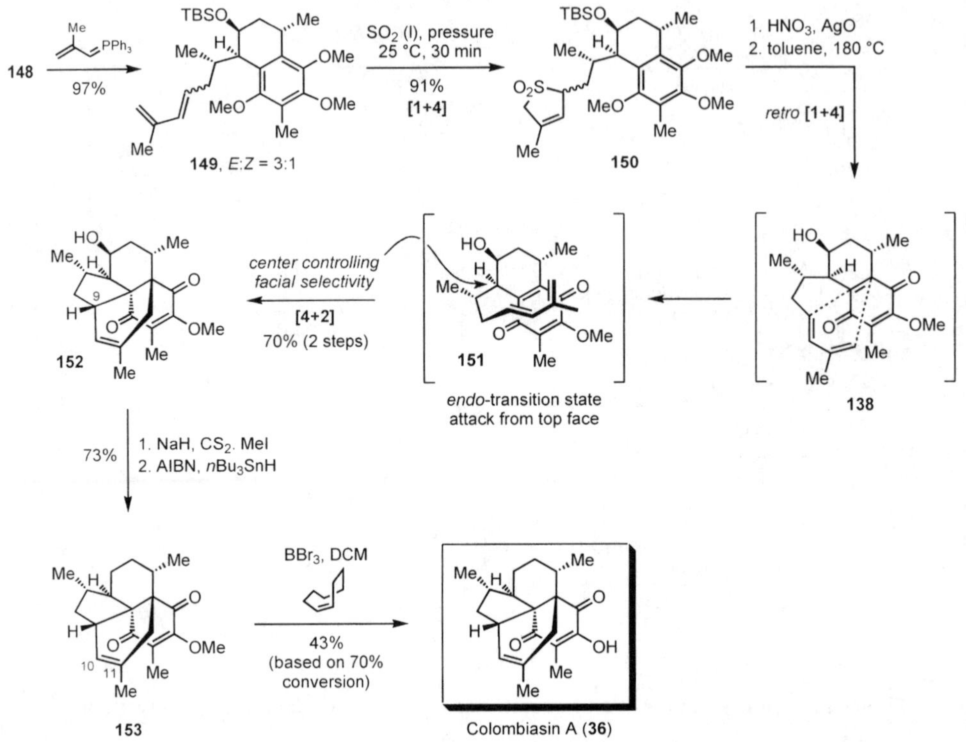

Scheme 24 Final stages and completion of the total synthesis of colombiasin A (**36**)

4.7 Synthetic Studies Toward Colombiasin A (Harrowven, 2001 and Flynn, 2003)

In an uncompleted synthetic attempt toward racemic colombiasin A (**36**) (Scheme 25) Harrowven et al. [46] started from racemic aldehyde **155** (prepared from dimethoxytoluene **154**, eight steps and 32% overall yield), which was connected with phosphonate **156** in a HWE olefination leading to a separable 3:2 *E*/*Z* mixture of **157**. Cationic cyclization furnished diastereoselectively *spiro*-lactone **159** (as a 5:1 epimeric mixture) along with the undesired condensation product **158**. On hydrogenation in methanol, lactone **159** was converted into ester **160**. Unfortunately, under these conditions only the C7 stereocenter was installed with the correct relative configuration while the C6 center of the six-membered ring kept the undesired stereochemistry (*cis* to C3-Me). Ester **160** was converted via an oxidation/reduction sequence followed by Wittig reaction to diene **161** as an *E*/*Z* mixture. As noted by Nicolaou, the benzene ring could not be oxidized to the quinone without decomposition of the material. Thus, the diene moiety was protected as its sulfur dioxide adduct **162** which could be converted into quinone **163**, albeit in low yield. The thermal IMDA cyclization furnished the *endo* adduct **165**, but in contrast to Nicolaou's synthesis, the

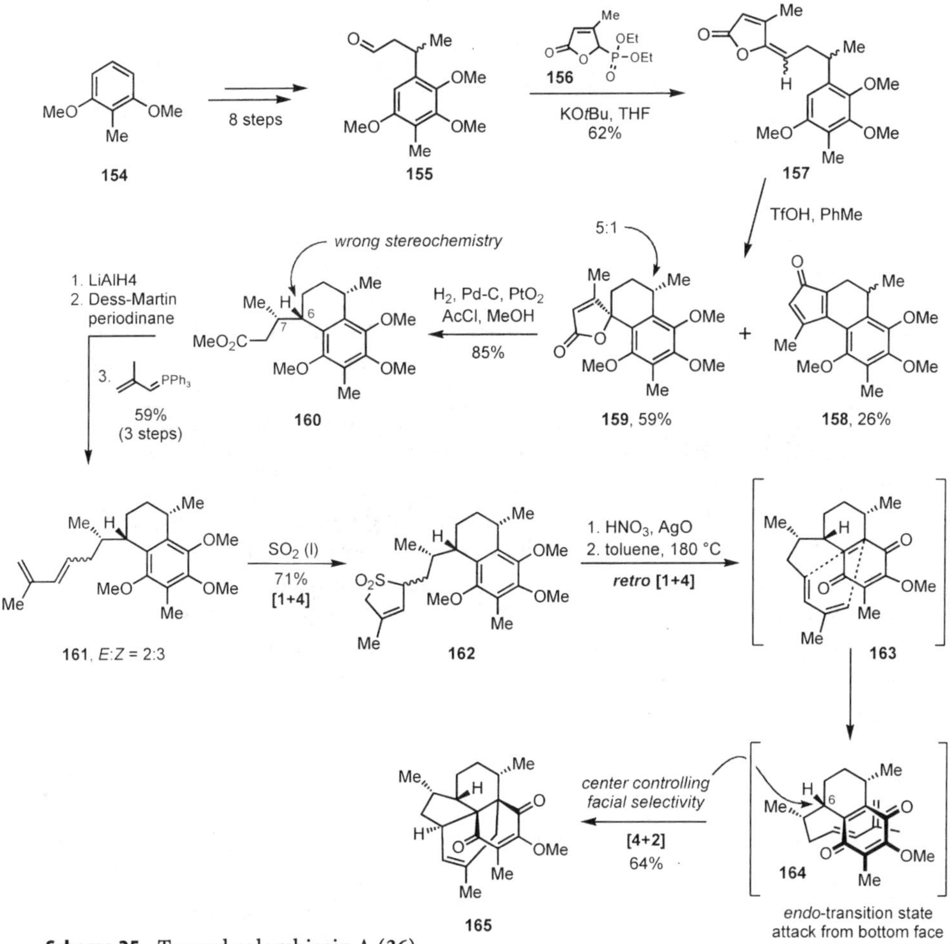

Scheme 25 Toward colombiasin A (**36**)

attack of the diene occurred from the bottom face (see **164**), which is due to the wrong configuration at C6 (Scheme 25).

In a rather elegant approach towards colombiasin A (**36**) Flynn et al. [47] would access the tetracyclic carbon skeleton through an enantioselective intermolecular Diels–Alder sulfoxide elimination–intramolecular Diels–Alder (DA-E–IMDA) sequence between double-diene **166** and quinone **167** (Scheme 26). A key element of the proposed approach would be the chiral sulfoxy group in **167** which controls both the regio and facial selectivity of the intermolecular Diels–Alder reaction and eliminates generation of the dienophile for the IMDA reaction.

The above-mentioned strategy was validated by the synthesis of tetracycle **176** (Scheme 27). The required double diene **170** was prepared from 2,5-dimethoxytetrahydrofuran (**168**) by treating it with acid and the HWE reagent.

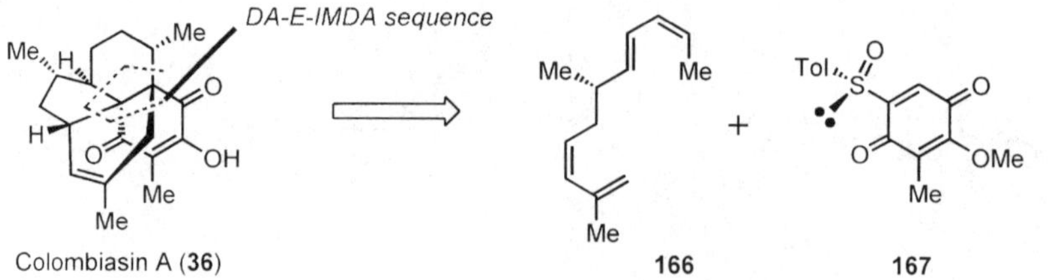

Scheme 26 Flynn's retrosynthetic analysis of colombiasin A (36)

Subsequent monosilylation and Wittig reaction furnished unsymmetrical double diene **170**. The synthesis of the other Diels–Alder partner started from bromophenol **173** (prepared in three steps from dimethoxytoluene), which was doubly metalated and reacted with (*S*,*S*)-menthyl *p*-toluenesulfinate **173**. CAN oxidation delivered quinone **171**, which underwent a Diels–Alder reaction with double diene **170** to give compound **175** possessing the correct regio- and stereochemistry. Upon heating in toluene the desired elimination occurred followed by IMDA reaction to adduct **176**, which was obtained in an excellent yield and enantioselectivity. Both Diels–Alder reactions proceeded through an *endo* transition state; the enantioselectivity of the first cyclization is due to the chiral auxiliary, which favors an *endo* approach of **170** to the sterically less congested face (top face) (Scheme 27).

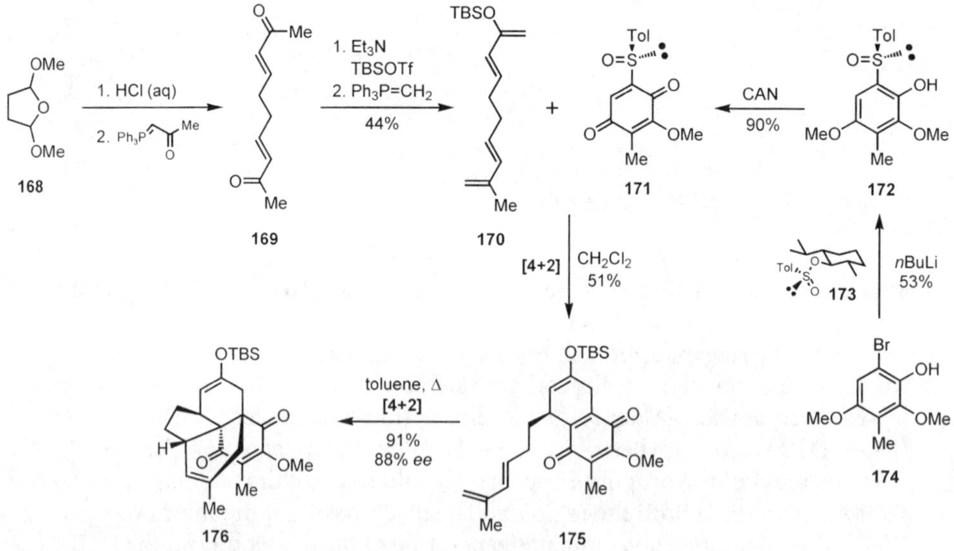

Scheme 27 Synthesis of the colombiane skeleton via double Diels–Alder approach

5 Conclusions

In summary, the total syntheses of the natural products from *Pseudopterogorgia elisabethae* demonstrate the power of modern annulation reactions. The classic intermolecular Diels–Alder reaction continues to be a popular tool – for instance to forge decalin cores. In addition, the intramolecular variant (IMDA reaction) has become extraordinarily powerful [48]. In consequence a considerable number of total syntheses feature IMDA cyclization as the key step to form the entire molecular skeleton in one step. The advantages of the Diels–Alder cyclization are obvious: besides its efficiency, atom economy, etc. one of the most valuable characteristics of this transformation is the fact that the stereochemical information of the starting material (in the form of geometrically defined olefins) is preserved in the course of the reaction. Therefore, it is possible to use well-established (*E*)- or (*Z*)-selective olefination methodology for the synthesis of diene and dienophile. Subsequently, this high grade of stereoselection can be transferred via the Diels–Alder reaction into the final target molecule. However, as far as the stereoselectivity of the reaction is concerned, there are two aspects which must be addressed: (1) *exo* versus *endo* selectivity; and (2) the facial selectivity. As seen in the presented total syntheses a clever choice of the Diels–Alder precursor often allows control to be gained over these two criteria of selectivity. Moreover, a rather uncommon annulation reaction, namely the [5+2] cycloaddition, proved an elegant approach to the elisapterane skeleton.

Overall, this exciting new family of marine natural products leaves plenty of room for future synthetic efforts, especially when the as yet unexplored biosyntheses of these compounds will be more and more uncovered.

Acknowledgements Our own research in this area has been generously supported by the Fonds der chemischen Industrie (Kekulé fellowship to T.J.H.) and the Fonds zur Förderung der wissenschaftlichen Forschung (Austrian Science Foundation). Thanks are also due to Dr. Martin Green for proofreading of the manuscript.

References

1. Rodríguez AD (1995) Tetrahedron 51:4571 and references cited therein
2. Bayer FM (1961) The shallow-water octocorallia of the West Indian region. Martinus Nijhoff, The Hague
3. Look SA, Fenical W (1987) Tetrahedron 43:3363
4. Ata A, Kerr RG, Moya CE, Jacobs RS (2003) Tetrahedron 59:4215
5. Rodríguez AD, Ramírez C (2001) J Nat Prod 64:100
6. Rodríguez AD, Shi Y-P (2000) Tetrahedron 56:9015
7. Harvis CA, Burch MT, Fenical W (1988) Tetrahedron Lett 29:4361
8. Ata A, Kerr RG (2000) Tetrahedron Lett 41:5821
9. Rodríguez AD, Ramírez C, Rodríguez II, González E (1999) Org Lett 1:527; for stereochemical reassignment see Refs. [5] and [28]

10. Look SA, Fenical W, Matsumoto GK, Clardy J (1986) J Org Chem 51:5145
11. Roussis V, Wu Z, Fenical W (1990) J Org Chem 55:4916
12. Lazerwith SE, Johnson TW, Corey EJ (2000) Org Lett 2:2389
13. Tanaka J-I, Ogawa N, Liang J, Higa T, Gravalos DG (1993) Tetrahedron 49:811; for structural revision see Ref. [12]
14. Rodríguez AD, Ramírez C, Rodríguez II (1999) J Nat Prod 62:997 (see also Ref. [6])
15. Rodríguez II, Rodríguez AD (2003) J Nat Prod 66:855
16. Rodríguez AD, Ramírez C, Medina V, Shi Y-P (2000) Tetrahedron Lett 41:5177
17. Rodríguez AD, González E, Huang SD (1998) J Org Chem 63:7083
18. Rodríguez AD, Ramírez C, Rodríguez II, Barnes CL (2000) J Org Chem 65:1390
19. Shi Y-P, Rodríguez II, Rodríguez AD (2003) Tetrahedron Lett 44:3249
20. Rodríguez AD, Ramírez C, Shi Y-P (2000) J Org Chem 65:6682
21. Rodríguez AD, Ramírez C (2000) Org Lett 2:507
22. Rodríguez AD, Rodríguez II (2002) Tetrahedron Lett 43:5601
23. Ata A, Kerr RG (2000) Heterocycles 53:717
24. Rodríguez AD, Ramírez C, Rodríguez II (1999) Tetrahedron Lett 40:7627
25. Kohl AC, Ata A, Kerr RG (2003) J Ind Microbiol Biotechnol 30:495
26. Look SA, Fenical W, Jacobs R, Clardy J (1986) Proc Natl Acad Sci USA 83:6238
27. Thornton RS, Kerr RG (2002) J Chem Ecol 28:2083
28. Johnson TW, Corey EJ (2001) J Am Chem Soc 123:4475
29. Lazerwith SE, Corey EJ (1998) J Am Chem Soc 120:12777
30. Miyaoka H, Honda D, Mitome H, Yamada Y (2002) Tetrahedron Lett 43:7773
31. (a) Nicolaou KC, Zhong Y-L, Baran PS, Sugita K (2001) Angew Chem Int Ed 40:2145; (b) Nicolaou KC, Sugita K, Baran PS, Zhong Y-L (2002) J Am Chem Soc 124:2221
32. Heckrodt TJ, Mulzer J (2003) J Am Chem Soc 125:4680; for earlier work see: Heckrodt TJ, Mulzer J (2002) Synthesis 1857
33. Heckrodt TJ, Siddiqi SA (2003) Z Naturforsch B Chem Sci 328
34. Kosugi M, Negishi Y, Kameyama M, Migita T (1985) Bull Chem Soc Jpn 58:3383
35. Waizumi N, Stankovic AR, Rawal VH (2003) J Am Chem Soc 125:13022
36. Seyferth D, Marmor RS, Hilbert P (1971) J Org Chem 36:1379
37. (a) Review: Evers M (1986) Chem Scr 26:585; (b) Ahmad R, Saá JM, Cava MP (1977) J Org Chem 42:1228
38. (a) Van Dort HM, Geursen HJ (1967) Recl Trav Chim Pays-Bas 86:520; (b) Parker KA, Petraitis JJ (1981) Tetrahedron Lett 22:397
39. Kim AL, Rychnovsky SD (2003) Angew Chem Int Ed 42:1267
40. Meyers AG, Yang BH, Chen H, McKinstry L, Kopecky DJ, Gleason JL (1997) J Am Chem Soc 119:6496
41. (a) Kowalski CJ, Haque MS (1986) J Am Chem Soc 108:1325; (b) Kowalski CJ, Lal GS (1986) J Am Chem Soc 108:5356; (c) Kowalski CJ, Lal GS (1987) Tetrahedron Lett 28:2463; (d) Kowalski CJ, Reddy RE (1992) J Org Chem 57:7194
42. (a) Nicolaou KC, Vassilikogiannakis G, Mägerlein W, Kranich R (2001) Angew Chem Int Ed 40:2482; (b) Nicolaou KC, Vassilikogiannakis G, Mägerlein W, Kranich R (2001) Chem Eur J 7:5359
43. Mikami K, Motoyama Y, Terada M (1994) J Am Chem Soc 116:2812
44. (a) Tsuji J, Minami I, Shimizu (1983) Tetrahedron Lett 24:1793; (b) Tsuji J, Ohashi I, Minami I (1987) Tetrahedron Lett 28:2397; (c) Paquette LA, Gallou F, Zhao Z, Young DG, Liu J, Yang J, Friedrich D (2000) J Am Chem Soc 122:9610
45. (a) Barton DHR, McCombie SW (1975) J Chem Soc Perkin Trans I 16:1574; (b) Barton DHR, Crich D, Lobberding A, Zard SZ (1986) Tetrahedron 42:2329; (c) for a review see: Crich D, Quintero L (1989) Chem Rev 89:1413

46. Harrowven DC, Tyte MJ (2001) Tetrahedron Lett 42:8709
47. Chaplin JH, Edwards AJ, Flynn BL (2003) Org Biomol Chem 1:1842
48. For a recent review on DA reactions, see Nicolaou KC, Snyder SA, Montagnon T, Vassilikogiannakis GE (2002) Angew Chem Int Ed 41:1668 and references therein. For quinone-based IMDA reactions also see Layton ME, Morales CA, Shair MD (2002) J Am Chem Soc 124:773

Top Curr Chem (2005) 244: 43–76
DOI 10.1007/b96887

Recent Advances in Vinylogous Aldol Reactions and Their Applications in the Syntheses of Natural Products

Markus Kalesse (✉)

Lehrstuhl A, Institut für Organische Chemie, Universität Hannover, Schneiderberg 1B, 30167 Hannover, Germany
Markus.Kalesse@oci.uni-hannover.de

1 **Introduction** . . . 44

2 **Enolate Activation** . . . 46
2.1 Ti-BINOL Catalysis . . . 46
2.2 Tol-BINAP Catalysis . . . 51
2.3 Applications in Total Synthesis . . . 53
2.4 Aldolization Using Nonracemic Chiral Fluoride Sources . . . 58
2.5 γ-Substituted Vinylogous Ketene Acetals in Aldol Reactions . . . 59
2.6 $SiCl_4$-Based Catalysis . . . 61

3 **Aldehyde Activation** . . . 64
3.1 Activation Using Tris(pentafluorophenyl)borane as the Lewis Acid . . . 64
3.2 R_3Si^+ versus TPPB Activation in Aldolizations with γ-Substituted Ketene Acetals 67
3.3 C_2-Symmetrical Copper(II) Complexes as Chiral Lewis Acids . . . 71
3.4 The Total Synthesis of Callipeltoside . . . 72

References . . . 75

Abstract The synthesis of complex natural products still remains as the bottleneck for the biological evaluation of such compounds. In contrast to the biosyntheses, vinylogous aldol reactions can incorporate more than one acetate or propionate building block into the growing polyketide chain. To facilitate such reactions either enolate activation or aldehyde activation is required. For both variations, selective protocols have been put forward and were shown to be efficient substitutes compared to standard aldol reactions. As can be seen in the total syntheses of natural products such as callipeltoside, their use shortens the synthetic route significantly. One of the most challenging transformations in this context is the use of γ-substituted ketene acetals in both aldehyde and enolate activation. In particular, aliphatic aldehydes give poor yields and selectivities. Protocols for the enantioselective variation are given using either the Tol-BINAP system or C_2-symmetrical copper(II) complexes.

Keywords Vinylogous · Aldol reaction · Lewis acid · Callipeltoside

Abbreviations and Symbols

Ac	Acetyl
18-c-6	18-Crown-6
BINAP	2,2′-Bis(diphenylphosphino)-1,1′-binaphthyl
BINOL	Binaphthol
Bn	Benzyl
Bu	Butyl
Cy	cyclohexyl
DAST	Diethylaminosulfur trifluoride
DDQ	2,3-Dichloro-5,6-dicyano-1,4-benzoquinone
DET	Diethyl tartrate
DIBAl-H	Diisobutylaluminum hydride
DMPU	1,3-Dimethyl-3,4,5,6-tetrahydro-2(1H) pyrimidinone
ee	Enantiomeric excess
Et	Ethyl
LDA	Lithium diisopropyl amide
L-selectride	Lithium tri-*sec*-butylborohydride
m-CPBA	*m*-Chloroperoxybenzoic acid
Me	Methyl
MOM	Methoxymethyl
ms	Molecular sieves
Ms	Methanesulfonyl (mesyl)
NMO	4-Methylmorpholine *N*-oxide
Ph	Phenyl
PMB	4-Methoxyphenyl
Pr	Propyl
PPTS	Pyridinium *p*-toluenesulfonate
PyBOP	Benzotriazol-1-yloxytris(pyrrolidino)phosphonium hexafluorophosphate
rt	Room temperature
TBAF	Tetrabutylammonium fluoride
TBS	*tert*-Butyldimethylsilyl
TBAT	Tetrabutylammonium triphenyldifluorosilicate
TBDPS	*tert*-Butyldiphenylsilyl
TBHP	*tert*-Butyl hydroperoxide
TES	Triethylsilyl
Tf	Trifluoromethanesulfonyl (triflyl)
TFA	Trifluoroacetic acid
THF	Tetrahydrofuran
TMS	Trimethylsilyl
Tol	Toluyl
TPAP	Tetra-*n*-propylammonium perruthenate (VII)

1 Introduction

Aldol reactions are among the most prominent and most frequently applied transformations, since they build up polyketide fragments of important biologically active compounds such as antibiotics and antitumor compounds.

They serve not only as tools to generate carbon–carbon bonds, but also chiral centers are established. The most frequently applied methods of aldol reactions often parallel the biosynthesis of polyketide natural products. In the biosynthesis acetate or propionate units are added; subsequently a series of further transformations (reductions, eliminations, hydrogenations) are performed by large polyketide synthetases in order to provide the substrate for the next aldol reaction. The laboratory synthesis mostly follows this modular approach by adding acetate and propionate fragments following reduction and oxidation steps, often coupled with extensive protecting group shuffling and additional transformations such as Wittig olefinations (Scheme 1). Even though one can access almost any thinkable polyketide structure with this kit of established transformations, it might not always be the most economical way to generate natural products. Therefore a substantial number of research groups have focused on the development of more efficient methods for the construction of larger polyketide segments in one step. The focus was naturally put on the development of enantio- or diastereoselective procedures and establishing catalytic systems.

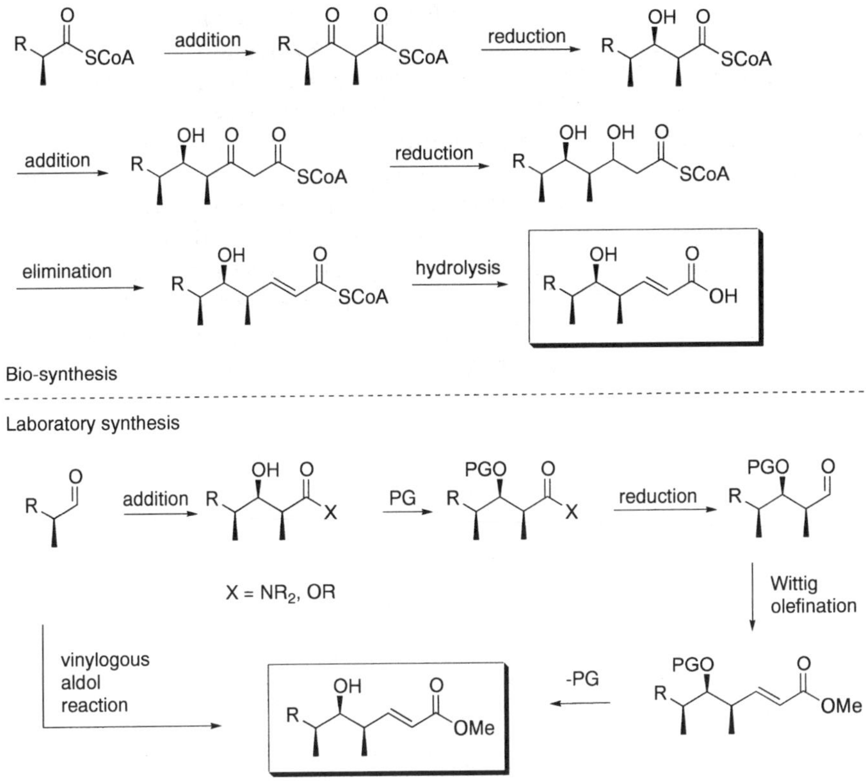

Scheme 1 Comparison of biosynthesis and laboratory synthesis of polyketide fragments

Herein we will focus on the recent development of vinylogous [1] aldol reactions and their application in the synthesis of natural products [2–5]. In particular the synthesis of unsaturated esters through the vinylogous Mukaiyama aldol reaction is of great interest, since it provides rapid access to larger carbon frameworks and allows for a wide variety of transformations of the double bond (dihydroxylation, epoxidation, cuprate addition etc.).

2 Enolate Activation

2.1 Ti-BINOL Catalysis

In 1995, both Sato [6] and Carreira [7–10] independently developed catalytic asymmetric aldol reactions of enolized β-keto esters. They utilized substituted 1,3-dioxan-4-ones as precursors for the synthesis of dienolates and chiral Lewis acids for the generation of the new chiral centers (Scheme 2). The silyl enol ethers can be generated by simple deprotonation and trapping of the enolate with TMSCl. Dienolates **1** and **2** resemble the two acetate equivalents, whereas **3** acts as the combined synthon of one propionate and one acetate group. Sato et al. identified the chiral binaphthol-derived titanium diisopropoxide catalyst **5** to provide higher chemical yields and higher enantioselectivity compared to the corresponding titanium dichloride complex **4** (Scheme 3). The chiral tartaric acid derived acylborane gave ee values ranging from 62 to 76%, albeit 50% of the catalyst had to be used for satisfactory chemical yields.

Table 1 summarizes the yields and selectivities of the addition of dienolate **2** to various aldehydes. It can be seen that the yield and enantioselectivity very much depend on the solvent, with THF being the best tested. The best results were obtained with pentanal (55% yield, 92% ee). On the other hand, when spiro dienolate **1** was used in these aldol reactions, benzaldehyde gave the best

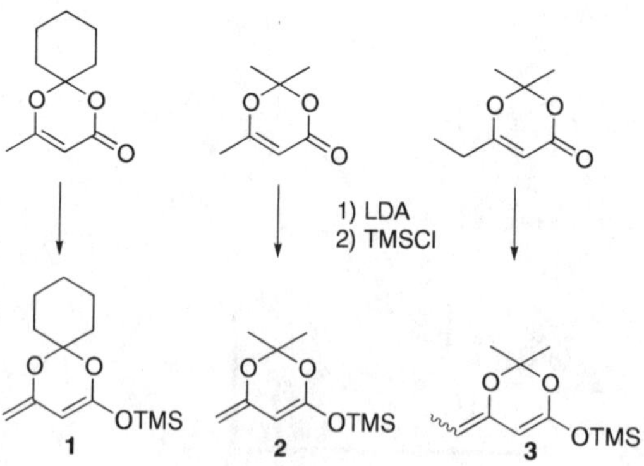

Scheme 2 Generation of dienolates from the corresponding α,β-unsaturated esters

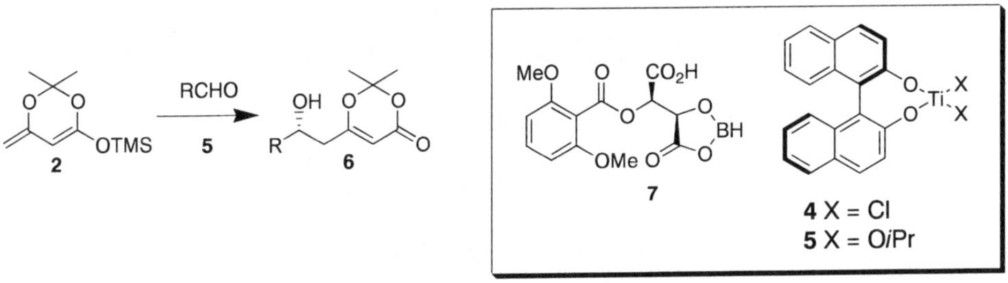

Scheme 3 Vinylogous aldol reaction based on Sato's method

Table 1 Reaction of dienolate **2** with different aldehydes

Entry	Aldehyde	Solvent	Yield (%)	ee (%)
1	PhCHO	THF	38	88
2	PhCHO	EtCN	23	75
3	PhCHO	CH_2Cl_2	10	50
4	*trans*-PhCH=CHCHO	THF	32	33
5	BuCHO	THF	55	92

chemical yield and enantioselectivity (93% yield, 92% ee) (Table 2). This clearly indicates that both the catalyst and the structure of the dienolate have to be adjusted in order to obtain optimal results for a given product.

Since dienolates **1** and **2** represent diacetate synthons, the dienolate derived from 6-ethyl-2,2-dimethyldioxinone can be seen as a propionate–acetate synthon. The synthesis of the corresponding dienolate provides a mixture of the *E* and *Z* enolates in a 3:5 ratio. The reaction with Ti-BINOL complex **5** generates a 5:1 mixture with the *syn* isomer as the major diastereomer. After separation of the diastereomers, the enantiomeric excess of the *syn* isomer was determined to be 100%. The *anti* isomer was formed in 26% ee. The same transformation performed with boron Lewis acid **7** gave the *anti* isomer as the major compound, but only with 63% ee. The minor *syn* isomer was produced with 80% ee. The observed selectivity could be rationalized by an open transition state in which minimization of steric hindrance favors transition state C (Fig. 1). In all three

Table 2 Reaction of dienolate **1** with different aldehydes

Entry	Aldehyde	Solvent	Yield (%)	ee (%)
1	PhCHO	THF	92	92
2	*trans*-PhCH=CHCHO	THF	58	79
3	BuCHO	THF	37	76

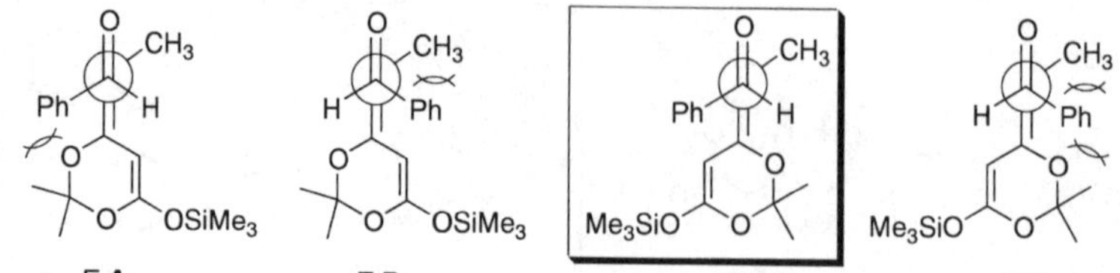

Fig. 1 Possible transition states in the vinylogous aldol reaction of **8** with benzaldehyde

other arrangements (A, B, D) with either the *Z* or *E* dienolate, one or two unfavorable steric interactions disfavor these transition states.

These first examples of the catalytic asymmetric aldol reaction not only provided first results that could be utilized for such transformations but also highlighted the problems that had to be overcome in further elaborations of this general method. It was shown that truly catalytic systems were required to perform an enantioselective and diastereoselective vinylogous aldol reaction, and it became obvious that γ-substituted dienolates that serve as propionate–acetate equivalents provide an additional challenge for diastereoselective additions. To date, the latter problem has only been solved for diastereoselective additions under Lewis acid catalysis (*vide infra*) (Scheme 4, Table 3).

Scettri et al. [11] worked on the improvement of the Sato protocol and investigated the influence of catalyst concentration on chemical yields and selec-

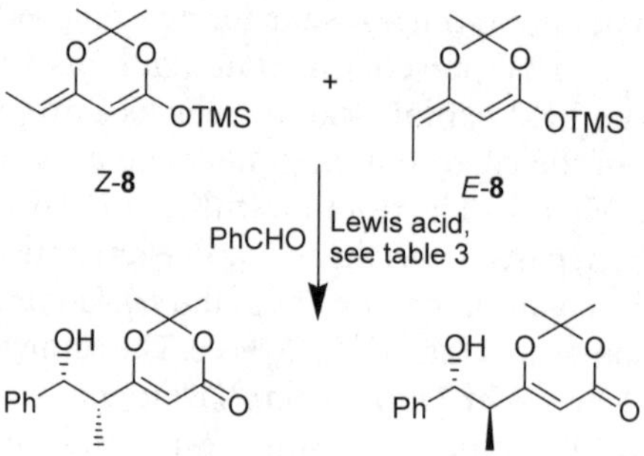

Scheme 4 Vinylogous aldol reaction with dienolate **8**

Table 3 Reaction of dienolate **8** with Lewis acid **7**

Entry	Lewis acid	Yield (%) *syn*-**9**	ee (%) *syn*-**9**	Yield (%) *anti*-**9**	ee (%) *anti*-**9**
1	**5**	51	100	11	26
2	**7**	25	80	75	63

Table 4 Yields and selectivities in reactions with different catalyst loading

Entry	Product	Catalyst **5** (mol%)	Yield (%)	ee (%)
1	**12**	17	42	80
2	**12**	50	65	89
3	**12**	100	81	91
4	**13**	17	59	87
5	**13**	50	89	94

tivities. For this purpose they compared the dioxanone-derived dienolates **1** and **2**. In Table 4 it can be seen that for both dienolates, an increase in catalyst concentration also increases the chemical yield and enantiomeric excess. With the cyclohexanone derivative (**1**) a very good yield of 89% (94% ee) could be achieved at 50 mol% of the catalyst (entry 5), whereas comparable results were observed with dienolate **2** only at stoichiometric concentrations (entry 3).

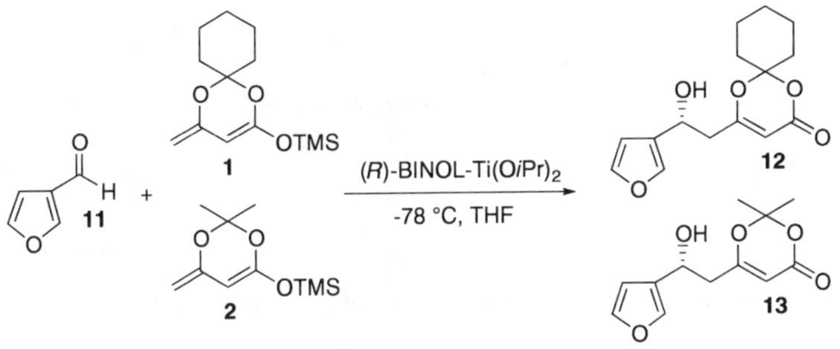

Scheme 5 Reaction of dienolates with 3-formylfuran

This trend observed for 3-formylfuran (**11**) (Scheme 5) could also be confirmed for other aromatic and unsaturated aldehydes (Table 5, Scheme 6). The yields that were achieved using 50 mol% together with dienolate **2** were matched with dienolate **1** using only 17 mol% of the Ti-BINOL system. Interestingly, the reaction of cinnamaldehyde with **2** gave no conversion. When aliphatic aldehydes were used, only notoriously poor yields were observed, indicating the constant problem in this type of transformation that is also encountered by other groups with different dienolates in one way or the other (*vide infra*).

Further investigations from the same group using 1,3-bis(trimethylsilyloxy)-1-methoxy-buta-1,3-diene (Chan's diene, **16**) [12, 13] together with 2–8 mol% of the catalyst derived from $Ti(O\textit{i}Pr)_4$ and (*R*)-BINOL revealed that vinylogous aldol reactions can be achieved in high yields with good enantioselectivities [14] (Scheme 7). The silylated products obtained could be liberated through treatment with 15 eq. TFA in aqueous THF. As can be seen in Table 6, even aliphatic aldehydes can be used to provide good yield and excellent enan-

Table 5 Aldol reactions using the (*R*)-BINOL-Ti(O*i*Pr)$_2$ catalyst

Entry	Product	Aldehydes	Catalyst (mol%)	Yield (%)	ee (%)
1	**15a**	Benzaldehyde	50	63	92
2	**14a**	Benzaldehyde	17	69	92
3	**15a**	Benzaldehyde	20	24	98
4	**15b**	Anisaldehyde	50	59	98
5	**14b**	Anisaldehyde	17	55	95
6	**15c**	*p*-Nitrobenzaldehyde	50	72	75
7	**14c**	*p*-Nitrobenzaldehyde	17	76	90
8	**15d**	*E*-Cinnamaldehyde	50	n.r.	
9	**14d**	*E*-Cinnamaldehyde	17	30	90
10	**15e**	Hexanal	50	39	89
11	**14e**	Hexanal	17	37	36

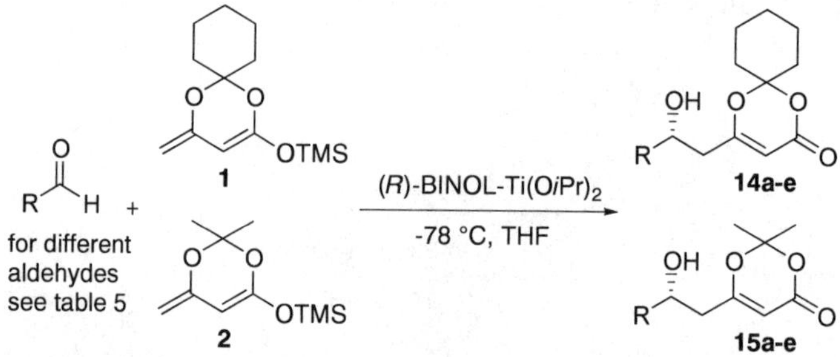

Scheme 6 Vinylogous aldol reactions with different catalyst loading

tiomeric excess (Table 6, entry 4). The authors observed that free alcohols isolated in small quantities showed higher ee values than those derived from acidic deprotection, in which racemization in the deprotection step was accounted for. Another synthetically important feature of this reaction is the fact that the aldol reaction essentially takes place at room temperature. Reactions starting at –78 °C needed to be stirred for an additional 6 h at room temperature in order to provide the reported high yields and selectivities.

Table 6 Reaction of Chan's diene with aldehydes under BINOL-Ti(O*i*Pr)$_4$ catalysis

Entry	Aldehyde	Yield (%)	ee (%)
1	Furfural	82	99
2	Benzaldehyde	94	92
3	Cinnamaldehyde	84	99
4	Hexanal	70	98

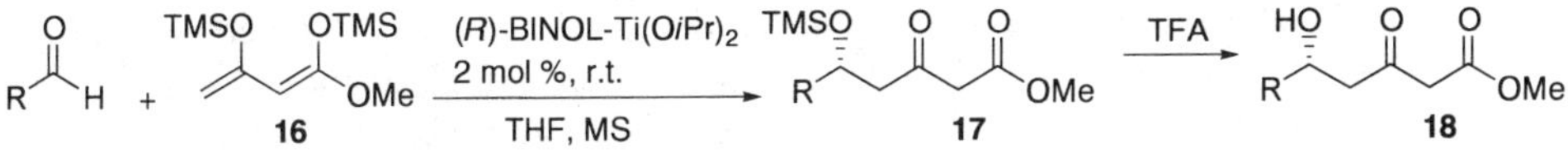

Scheme 7 Reactions using Chan's diene

2.2
Tol-BINAP Catalysis

Carreira et al. used their experience in the addition of simple silyl ketene acetals to aldehydes under Lewis acid catalysis [15]. In these experiments their 2-amino-2′-hydroxy-1,1′binaphthyl-derived catalyst (**19**) was used to provide aldol products with very high enantioselectivity (Scheme 8, Table 7).

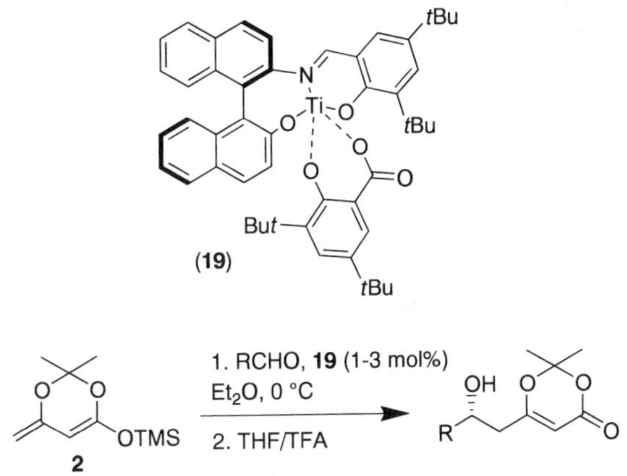

Scheme 8 Vinylogous aldol reaction using Carreira's catalyst

Table 7 Reactions of various aldehydes with Carreira's catalyst (*R*-**19**)

Entry	Aldehyde	Yield (%)	ee (%)
1	TIPS–≡–CHO	38	88
2	TBSO–CH=CH–CHO (Z)	97	94
3	Ph–CH=CH–CHO	88	92
4	Ph–CH=CH–CH=CH–CHO	95	92
5	PhCHO	83	84
6	Ph–CH2CH2–CHO	97	80
7	Bu_3Sn–CH=CH–CHO	79	92

As a development of the existing method, Carreira et al. introduced a chiral metal enolate (Schemes 9 and 10) for the addition of **2** to aldehydes. In contrast to established Mukaiyama aldol reactions, where a Lewis acid activates the aldehyde and by doing so facilitates the addition of the nucleophile, additions occur through enolate activation. The chiral dienolate was generated from (*S*)-Tol-BINAP with $Cu(OTf)_2$ and $(Bu_4N)Ph_3SiF_2$. This soft metal complex allowed desilylation of the silyl dienolate followed by transmetallation. The reaction with various aromatic and unsaturated aldehydes provided the expected aldol product in high yields and enantioselectivities ranging from 65 to 95% ee (Table 8). The best selectivities were observed for aromatic aldehydes. On the other hand, aliphatic aldehydes were shown to give only yields below 40%. This indicates that, at least for enolate activation, aliphatic aldehydes seem to be the more challenging compounds.

In a mechanistic investigation they identified the active species to be the Cu(I) enolate. A catalyst derived from CuOTf together with TBAT and (*S*)-Tol-BINAP was as active as the previously described catalyst that was derived from

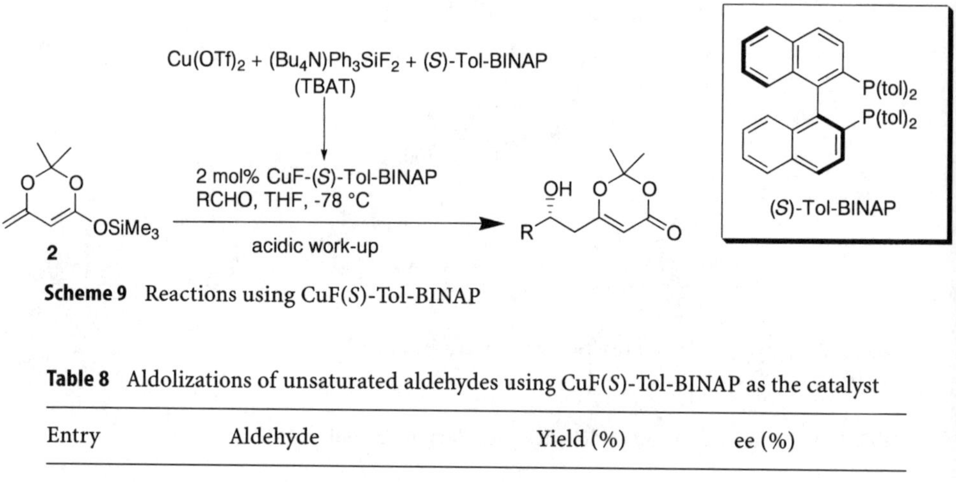

Scheme 9 Reactions using CuF(*S*)-Tol-BINAP

Table 8 Aldolizations of unsaturated aldehydes using CuF(*S*)-Tol-BINAP as the catalyst

Entry	Aldehyde	Yield (%)	ee (%)
1	PhCHO	92	94
2	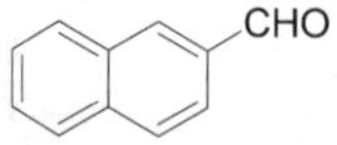	86	93
3	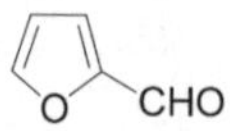	91	94
4	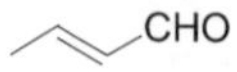	48	91
5	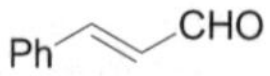	83	85
6	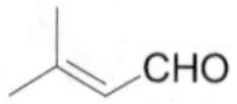	81	83
7	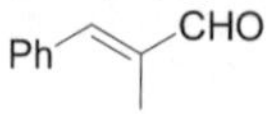	74	65

$Cu(II)OTf_2$. Therefore they rationalized that the Cu(I) complex might be the active species and is generated through a process in which silyl enol ether reduces Cu(II) to Cu(I) [16, 17]. To test this hypothesis, they performed the aldol reaction with a catalyst derived from (*S*)-Tol-BINAP and CuO*t*Bu to obtain the aldol product in approximately the same yield and selectivity. Additional IR-react investigations clearly indicated that enolate **A** (Scheme 10) is the first intermediate that concomitantly adds to benzaldehyde. The generated secondary Cu(I) alcoholate is then trapped as the silyl ether and starts a new catalytic cycle by generating the next Cu enolate (Scheme 10). Consequently, the fluoride is needed only in catalytic amounts and serves just to initiate the catalytic cycle by removing the TMS-protecting group. This process can now be used as an alternative to established aldol reactions under Lewis acid catalysis.

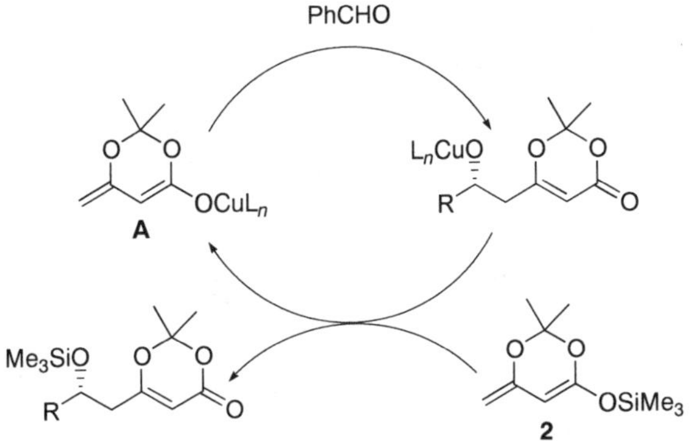

Scheme 10 Proposed catalytic cycle of the CuF-Tol-BINAP catalysis

2.3 Applications in Total Synthesis

In order to demonstrate the efficiency and applicability of their method, Carreira et al. described the synthesis of the C1–C13 polyol segment of amphotericin (Fig. 2). The vinylogous aldol reaction was utilized twice, during the

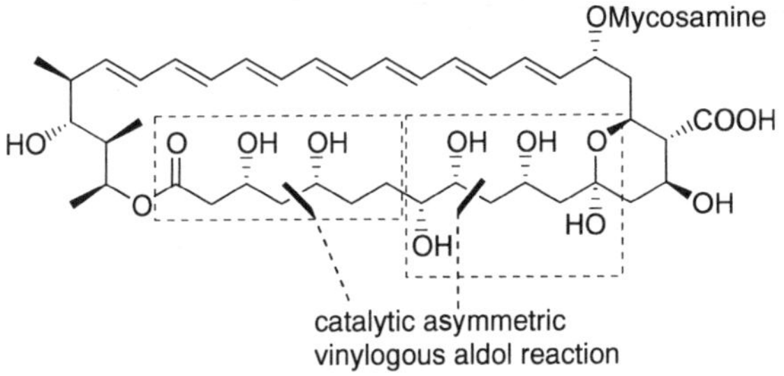

Fig. 2 Retrosynthetic analysis of amphotericin

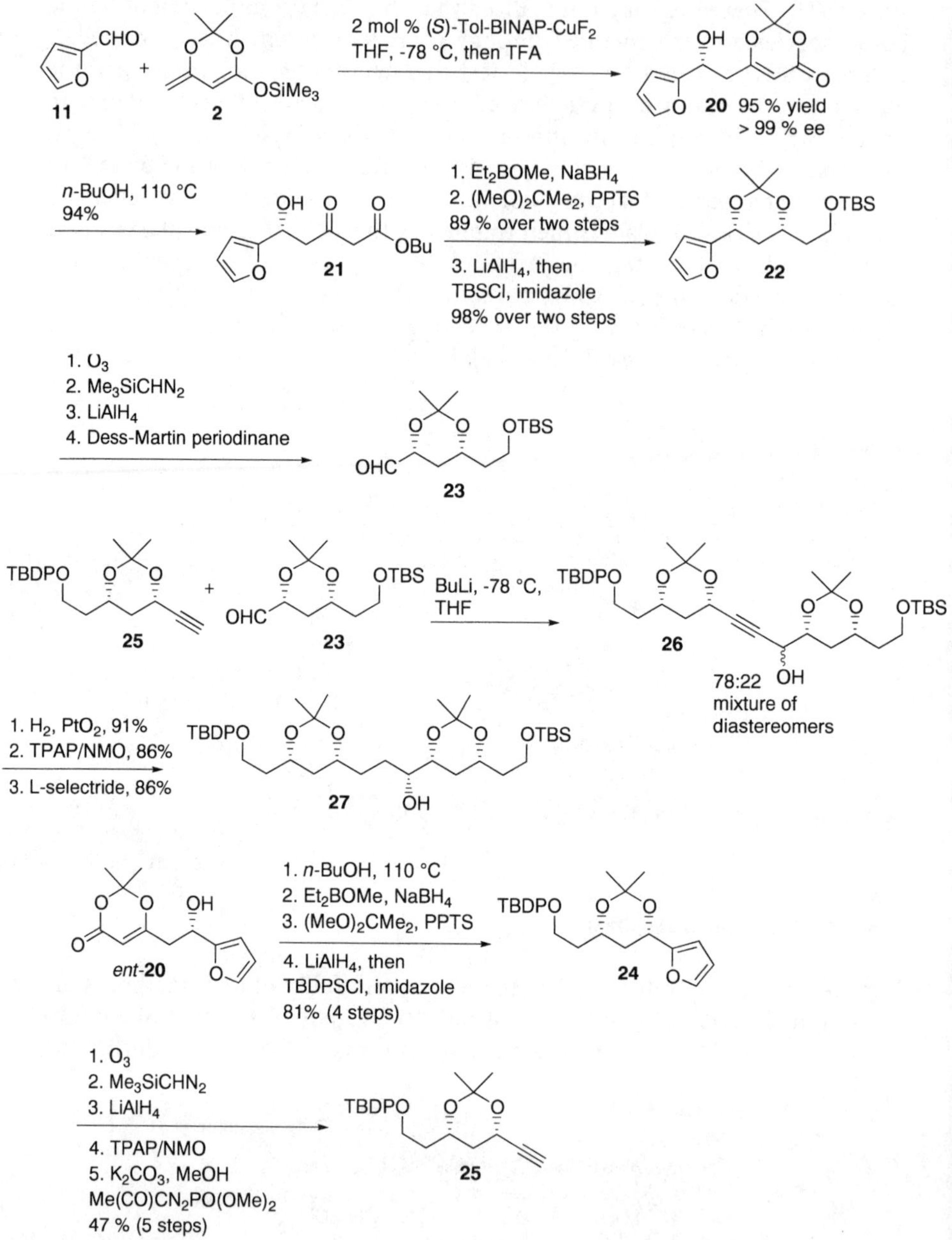

Scheme 11 Synthesis of the C1–C13 segment of amphotericin

C4–C5 bond formation and between C9 and C10. In both cases, the addition of silyl enol ether **2** to furfural was achieved with the Tol-BINAP-CuF_2. For the synthesis of the C1–C7 segment the *S*-catalyst was used and the *R*-catalyst furnished segment **27** (Scheme 11).

Hydrolysis of **20** with the aid of butanol followed by *syn*-selective reduction of *β*-keto ester **21** and protection as the isopropylidene acetal was accomplished in 87% yield. $LiAlH_4$ reduction and TBS protection of the primary alcohol gave **22** in very good yields. In this strategy, the furan residue serves as an aldehyde synthon and ozonolysis followed by esterification gave the corresponding methyl ester. Reduction and consecutive oxidation established aldehyde **23** in 71% yield.

For the synthesis of the C8–C13 segment, aldol reaction with (*R*)-Tol-BINAP·CuF_2 as the catalyst gave *ent*-**20**. Again a sequence of hydrolysis, reduction, and protection as the acetonide generated the 1,3-*syn*-triol. Ozonolysis followed by standard operations established the aldehyde, which was transformed into the corresponding terminal alkyne (**25**) with the ketophosphonate described by Bestmann [18]. These two fragments were coupled through nucleophilic attack of the acetylene moiety at the aldehyde yielding a 78:22 mixture of diastereomers at the so-generated secondary alcohol (**26**). Hydrogenation of the triple bond and oxidation of the alcohol was followed by selective reduction to establish the C1–C13 segment of amphotericin (**27**). In summary, the authors demonstrated how efficient their asymmetric catalysis could be when applied to the fragment synthesis of a complex natural product.

Another example where this particular combination of enolate and catalyst was applied to a total synthesis was given by Campagne et al. in their synthesis of the C9–C23 fragment of streptogramin antibiotics [19]. The retrosynthetic analysis dissects the molecules at the amide linkages. The resulting segment can be further deconvoluted into the linear keto ester **29**, which shows a typical substitution pattern that can be assembled using the vinylogous aldol reaction (Scheme 12). In the synthetic direction dioxane **2** can be added to aldehyde **28** with the (*R*)-Tol-BINAP·CuF_2 catalyst in 80% yield and 81% ee. Hydrolysis of the resulting TMS ether was accomplished using PPTS in methanol. Protection of the alcohol moiety as MOM ether was followed by amidation with protected serine derivative. After liberation of the serine alcohol the stage was set for oxazole formation using DAST followed by dehydrogenation using NiO_2. Applying the vinylogous aldol reaction the C9–C23 segment was assembled in only seven steps including all protecting group manipulations.

In an effort to further extend the scope of the vinylogous enolate chemistry, Bluet and Campagne used Carreira's (*S*)-Tol-BINAP·CuF_2 catalyst in combination with a truly vinylogous enolate [20]. Carreira's and Sato's dioxane represents the dienolate of acetoacetate and provides additionally a donor in the alpha position to the reacting center. The dienolates used by Campagne produce an *α*,*β*-unsaturated ester in contrast to the *β*-keto ester obtained with **2**. In order to demonstrate the advantage over aldehyde activation they compared the

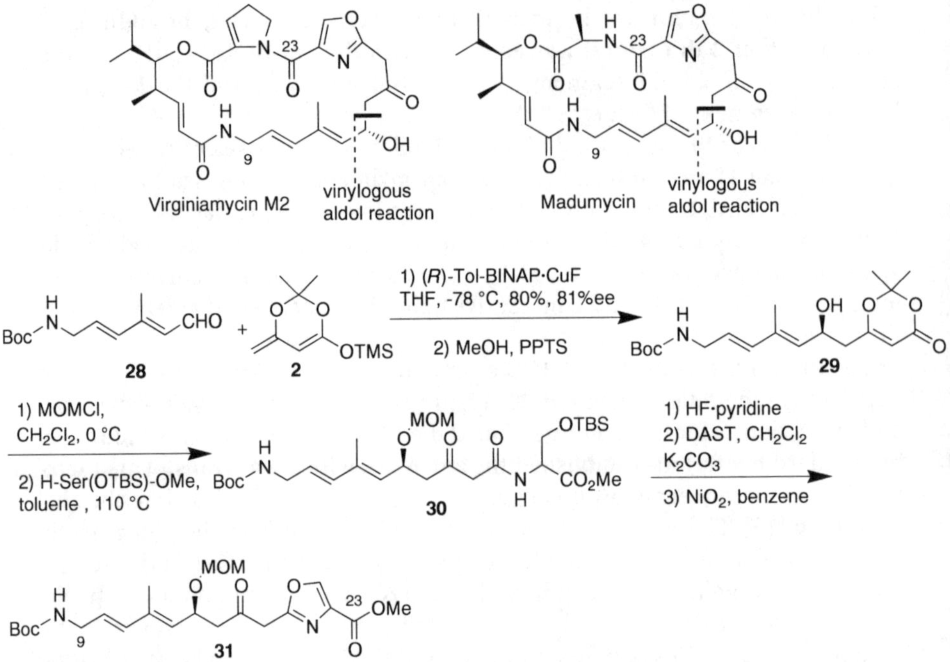

Scheme 12 Synthesis of the C9–C23 segment of streptogramin antibiotics

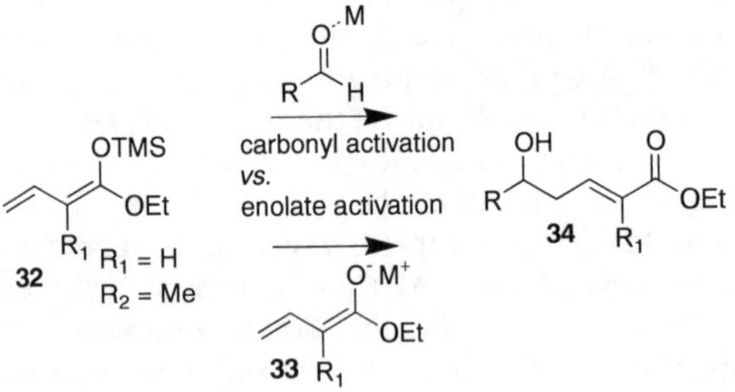

Scheme 13 Enolate activation vs. aldehyde activation

aldol reaction to those under Lewis acid activation (Scheme 13). From the results obtained it became obvious that enolate activation produces the aldol products in significantly higher yields with comparable selectivities (Table 9). It is noteworthy that isobutyraldehyde can be transformed through enolate activation in 68% yield, whereas aldehyde activation produces only 18% of the same product. The yield using cinnamaldehyde was rather low since formation of the 1,4-adduct (Fig. 3) occurred as a side reaction.

An application of this variation was put forward in the total synthesis of octalactin (Scheme 14). In retrosynthetic disconnection the macrolactone sim-

Table 9 Vinylogous aldol reactions using enolate and aldehyde activation

Entry	Aldehyde	Enolate activation		Aldehyde activation	
		Yield (%)	ee (%)	Yield (%)	ee (%)
1	PhCHO	80	70	45	75
2	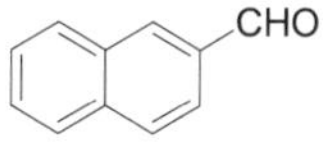	70	48	25	75
3	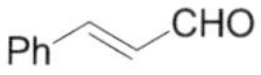	35	56	45	60
4	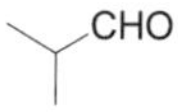	68	77	18	70

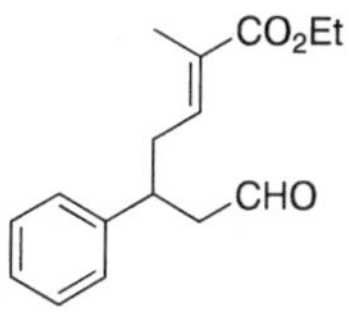

Fig. 3 1,4-adduct from cinnamaldehyde

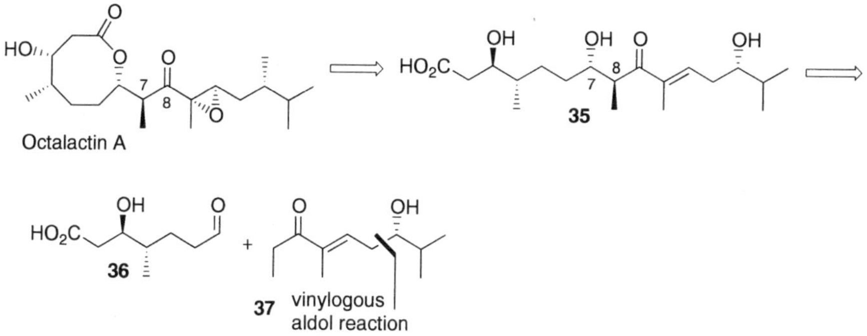

Scheme 14 Retrosynthetic analysis of octalactin

plifies to an open-chain polyketide structure. An aldol coupling between C7 and C8 can be used to assemble both subunits to the macrolactonization precursor. On the other hand, fragment **37** could be synthesized through the vinylogous aldol reaction with isobutyraldehyde (Scheme 15). Aldolization with the (*S*)-Tol-BINAP·CuF catalyst provides the unsaturated ester **38** in 90% yield with 80% ee. TBS protection and sequential elaboration to the Weinreb amide **39** is followed by the addition of EtMgBr to furnish ethyl ketone **37** used for the aldol coupling between both segments.

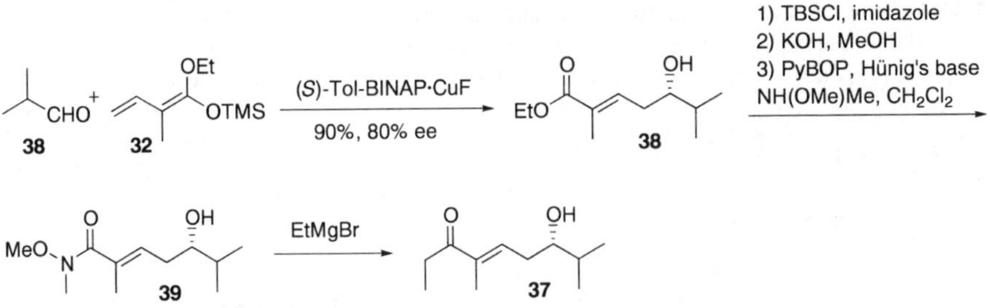

Scheme 15 Synthesis of the vinylogous aldol segment of octalactin

2.4 Aldolization Using Nonracemic Chiral Fluoride Sources

In an attempt to optimize the reaction conditions [21] it was realized that temperature had little influence on the enantioselectivity of the reaction, but the yields were optimum at approximately 0 °C. Also, a small negative nonlinear effect was observed as the relationship between the ee of the ligand and the ee values of the products, implying that aggregation takes place. In addition, the nature of the so-called precatalyst seems to play a significant role in the enantioselective catalytic cycle. This became obvious when Campagne et al.

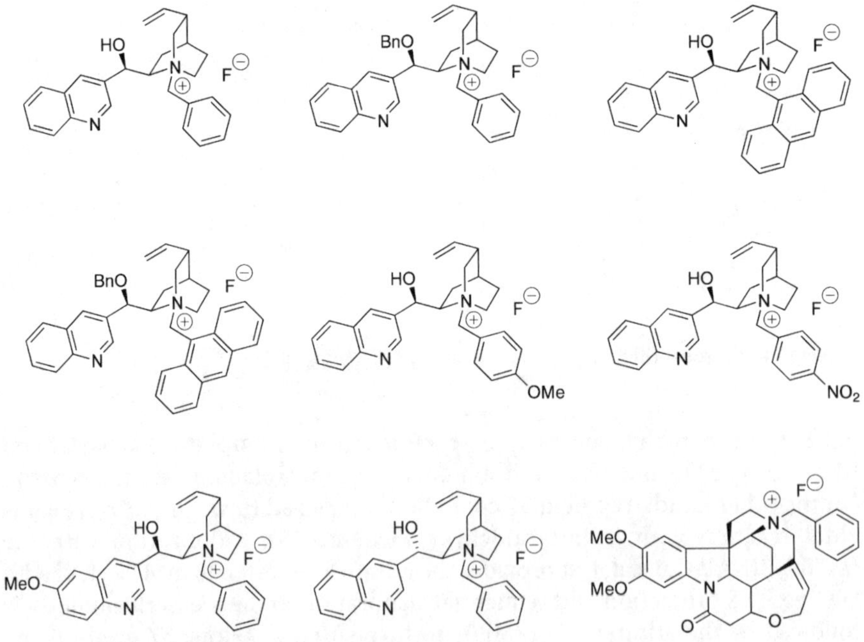

Fig. 4 Ammonium ions derived from cinchona alkaloids

compared the enantioselectivities observed with the catalyst prepared from CuF_2, Tol-BINAP, and TBAT to the aldol reaction with the (*S*)-Tol-BINAP-CuO*t*-But catalyst. They found that the enantiomeric excess significantly dropped by approximately 20%. Consequently, they turned their attention to nonracemic fluorides with counter ions derived from cinchona alkaloids (Fig. 4) [22]. Different vinylogous aldol reactions with "naked" enolates were tested, but even though the γ/α ratio was 100:0, the ee values were only in the range between 11 and 30% (Scheme 16). These findings led to the conclusion that Carreira's catalyst was the most selective one for these systems.

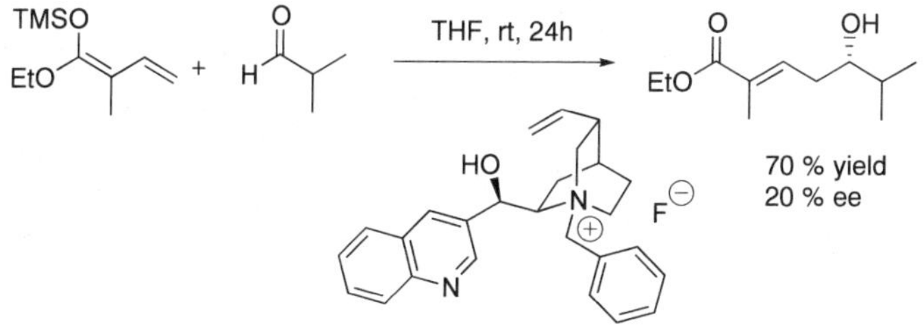

Scheme 16 Aldolization using a nonracemic fluoride

2.5 γ-Substituted Vinylogous Ketene Acetals in Aldol Reactions

As the next challenge they identified the use of γ-substituted vinylogous ketene acetals as nucleophiles [23]. The addition of **40** could not only generate mixtures of α and γ adducts, but the *syn*/*anti* relationship due to the presence of one additional substituent adds to the complexity of this transformation. In first experiments with two different fluoride sources, inspired by the above-described investigations, they isolated the vinylogous aldol product in 45% yield as a 1:1 *syn*/*anti* mixture using TBAT as the fluoride source. In a quest for a fluoride source that could be synthesized and handled without special precautions, the nonracemic ammonium fluorides shown in Fig. 4 were tested. Surprisingly, *N*-benzyl cinchodinium fluoride exclusively gave the α-aldol product **42** in 68% yield, again as a 1:1 *syn*/*anti* mixture. This was in contrast to the previously described γ selectivity for terminal unsubstituted dienolates, and again indicates that the vinylogous aldol reaction is sensitive to subtle differences such as an additional methyl group. Changing back to the Carreira catalyst ((*S*)-Tol-BINAP·CuF) they could isolate the vinylogous aldol product **41** together with its corresponding lactone **43** as a 14:86 mixture (Scheme 17, Table 10).

For applications in total synthesis this method was also thought to be applicable to chiral aldehydes, leading to matched and mismatched situations. Therefore, vinylogous ketene acetal **40** was put to reaction with chiral aldehyde **44** and both enantiomers of Carreira's catalyst. Reaction of aldehyde **44** with the (*S*)-Tol-BINAL·CuF catalyst (matched case) produced only one diastereomeric

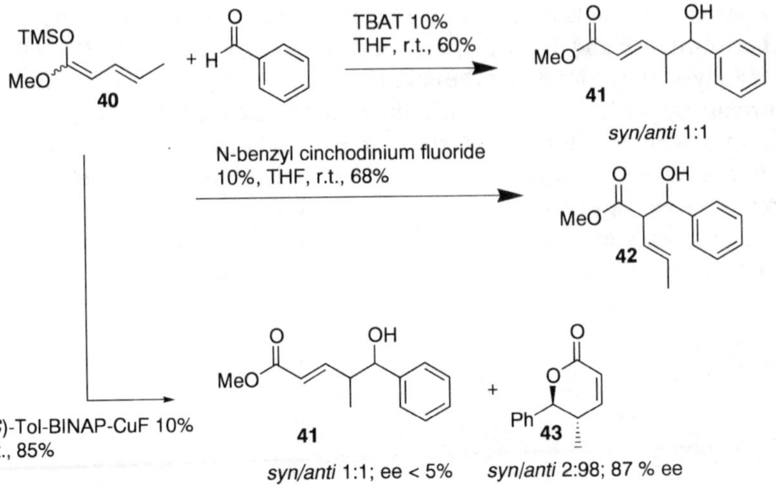

Scheme 17 Vinylogous aldol reaction with γ-substituted ketene acetal **40** using Carreira's catalyst

Table 10 Vinylogous aldol reaction of **40** with different aldehydes

Entry	Aldehyde	Yield (%) (**41**+**43**)	Ratio **43**/**41**	Lactones *Syn*/*anti*	ee
1	PhCHO	85	86/14	>98/2	87
2	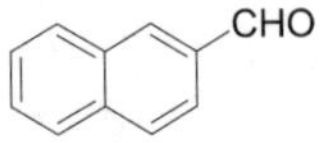	95	80/20	>98/2	85
4	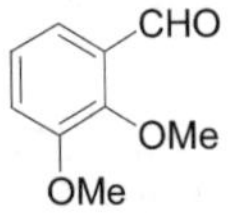	87	81/19	>98/2	91
5	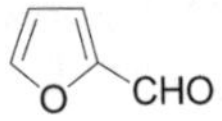	60	50/50	>98/2	86
6	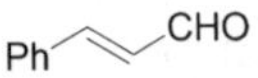	60	70/30	>98/2	82
7	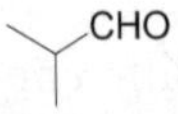	95	64/36	>98/2	91

lactone **45** (60%) and only a small trace (<10%) of the open-chain product could be detected. On the other hand, using the (*R*)-Tol-BINAL·CuF catalyst, both diastereomeric ketones were observed but gratifyingly in a 9:1 ratio for the 4,5-*anti*-5,6-*anti* isomer in 55% yield (Scheme 18).

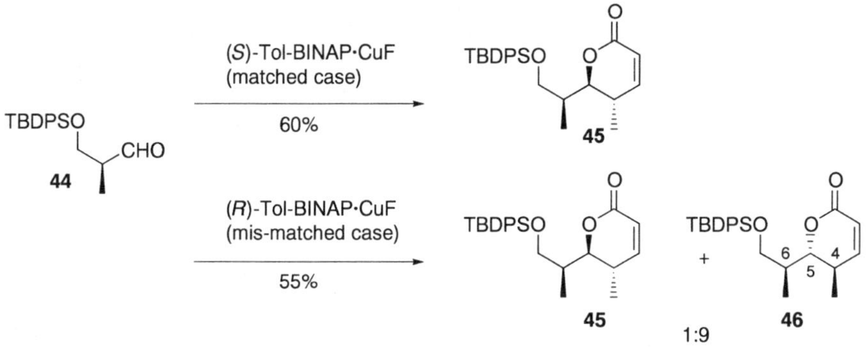

Scheme 18 Vinylogous aldol reactions with chiral aldehyde **44**

2.6 $SiCl_4$-Based Catalysis

Based on the same strategy, Denmark and coworkers developed a vinylogous aldol reaction using enolate activation with a catalyst derived from $SiCl_4$ and dimeric phosphoramide **47** [24,25]. This strategy relies on the observation that not all Lewis acid – Lewis base interactions diminish the Lewis acidity [26–28]. Due to the formation of a pentacoordinated silicon cation (**48**), both the enolate and the substrate can be assembled in a closed transition state, giving rise to the observed high selectivities (Scheme 19) [29,30].

Denmark et al. used all possible combinations of methyl-substituted vinylogous ketene acetals with three different aldehydes under their catalytic condi-

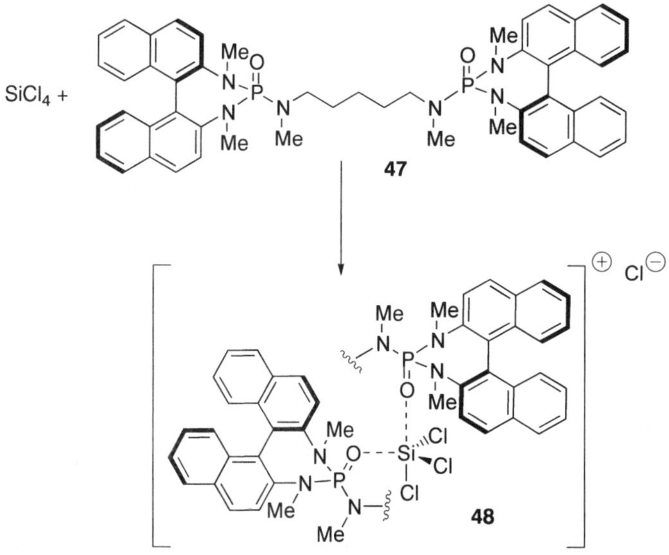

Scheme 19 Pentacoordinated catalyst derived from a dimeric phosphoramide

Table 11 Aldol reactions with chiral catalyst **48**

Entry	Dienolate	R^1	Yield, %	γ/a	d. r.	e. r.
1	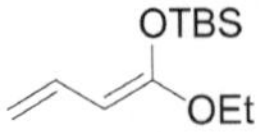	Ph	89	>99:1		99:1
2	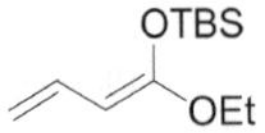	PHCH=CH	84	>99:1		98:1
3	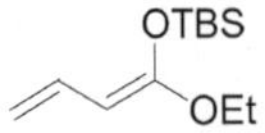	$PhCH_2CH_2$	68	>99:1		95:1
4	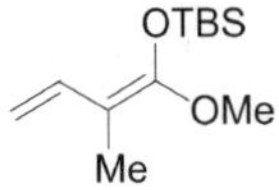	Ph	93	>99:1		99.5:0.5
5	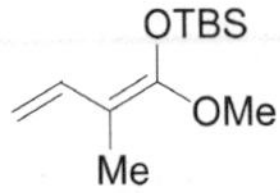	PHCH=CH	88	>99:1		99.5:0.5
6	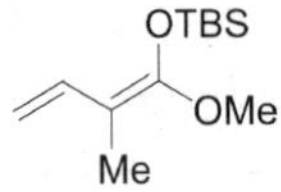	$PhCH_2CH_2$	n.d.	n.d.		n.d.
7	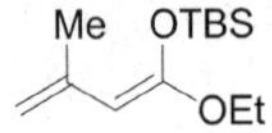	Ph	91	>99:1		96:4
8	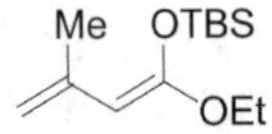	PHCH=CH	97	>99:1		94:6
9	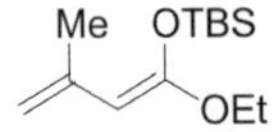	$PhCH_2CH_2$	73	>99:1		97.5:2.5
10	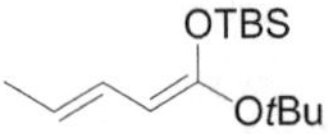	Ph	92	>99:1	>99:1	94.5:5.5
11	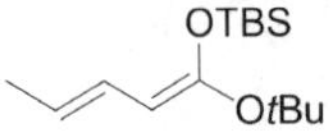	PHCH=CH	71	>99:1	>99:1	91:9
12	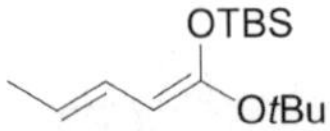	$PhCH_2CH_2$	n.d.	n.d.	n.d.	n.d.

tions (Table 11, Scheme 20) [31]. The catalyst **48** is generated in situ from $SiCl_4$ and the chiral phosphoramide (*R*,*R*)-**47**. With the ketene acetal derived from crotonaldehyde (entries 1–3), the reaction with benzaldehyde (entry 1) required only 1 mol% of the catalyst and exclusively provided the γ product (>99:1) in 89% yield and a remarkable enantiomeric ratio of 99:1. The reaction with cinnamaldehyde gave comparable results (entry 2). The transformation using the saturated aldehyde required 5 mol% of the catalyst and an additional

5 mol% of Hünig's base. Compared to the former aldehydes the yields significantly dropped (68%), indicating that aliphatic aldehydes represent the more challenging substrates. Using a different substitution pattern of the ketene acetal (entries 4–6), very good results were also observed for benzaldehyde and cinnamaldehyde but the reaction with the saturated aldehyde failed (entry 6). The dienolate derived from methyl tiglate gave the expected γ product with all three aldehydes in very good yield and excellent selectivity (entries 7–9). Finally, the most challenging ketene acetal bearing a methyl substituent at the γ position was evaluated under these enolate conditions. In initial studies with ethyl 2-pentenoate, only poor regioselectivity was observed. Nevertheless, increasing the steric bulk of the ester group led to improved selectivity. Consequently, the aldol reactions with the *t*-butyl 2-pentenoate-derived ketene acetal yielding exclusively the *anti* γ product were performed with good yields and selectivities comparable to the above-mentioned examples. However, no reaction could be observed with the saturated aldehyde (entry 12). It should be mentioned that as in Campagne's examples, the *anti* aldol products are formed as the major isomer.

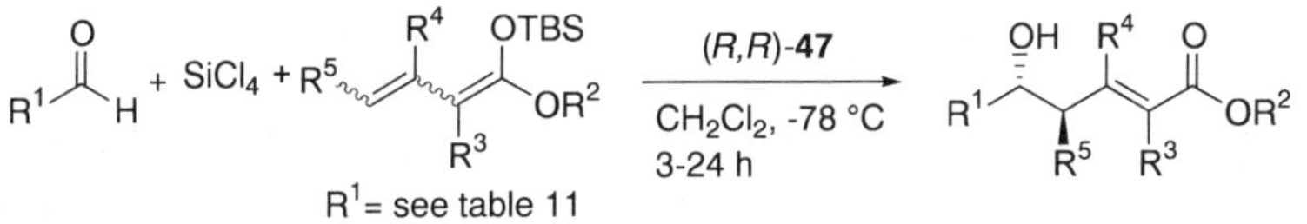

R^1 = see table 11
R^2 = Me, Et, *t*Bu
R^3, R^4 = H, Me

Scheme 20 Vinylogous aldol reactions using catalyst **48**

Having used their catalytic systems with dienolates derived from unsaturated esters, Denmark performed aldol reactions with the dioxanone-derived dienol ether described above in the context of Carreira's and Campagne's vinylogous aldol reactions (Scheme 21). Here, exclusively, the γ product was formed with the nucleophile attacking from the *Re* face. For all three aldehydes, very good yields (83–92%) and selectivities (74–89% ee) were observed with only 0.01–0.05 mol% of the catalyst.

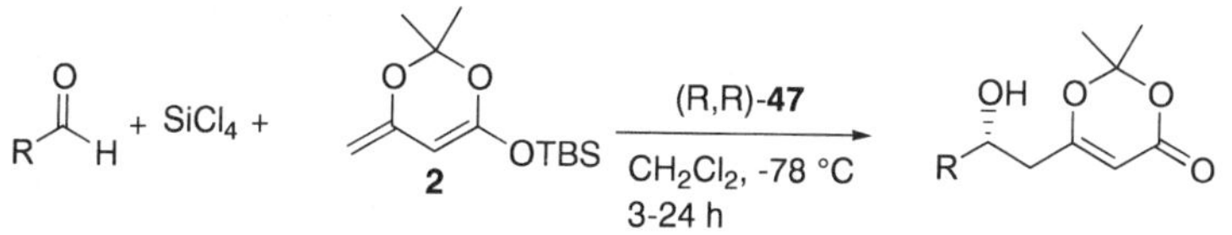

Scheme 21 Vinylogous aldol reaction using dienolate **2** and catalyst **48**

3
Aldehyde Activation

As mentioned above, two principal ways of activation in vinylogous aldol reactions can be envisioned. The research described was dedicated to enolate activation and to developing enantioselective catalytic variations. In another approach, activation of the carbonyl group of aldehydes can be achieved through coordination of Lewis acids to the carbonyl oxygen. Even though various examples were known for quite some time, a selective addition to aldehydes always proved to be a challenge and very often tedious optimizations were required in order to obtain synthetically useful procedures. The aldol addition in the synthesis of swinholide A as described by Paterson et al. [32] provides a picture of such "fine tuning" (Scheme 22).

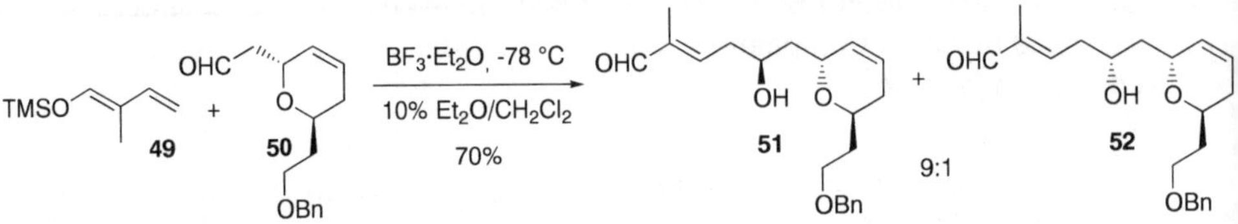

Scheme 22 Vinylogous aldol reaction used in Paterson's synthesis of swinholide A

3.1
Activation Using Tris(pentafluorophenyl)borane as the Lewis Acid

Based on these results, Kalesse et al. applied the vinylogous Mukaiyama aldol reaction in their total synthesis of ratjadone [33, 34]. In the synthesis of the C14–C24 segment (A-fragment), the vinylogous aldol reaction was used together with different Lewis acids to achieve the addition of this diacetate synthon in a diastereoselective manner under Felkin control (Scheme 23).

After optimization of the reaction conditions, the first aldol couplings were achieved with boron trifluoride as the Lewis acid yielding the *syn* (Felkin) product as a 3:1 mixture with the *anti* isomer (Fig. 5). It was rationalized that the poor selectivity arose as a consequence of the small steric hindrance of the vinylogous ketene acetal. Consequently, a bulkier BF_3 analog was sought to overcome this drawback. When 20 mol% of tris(pentafluorophenyl)borane (TPPB) was used, the desired product could be isolated as a single isomer **54** (Schemes 24 and 25). Gratifyingly, the newly generated alcohol was in situ protected as TBS ether [35]. In the course of the reaction, the Lewis acid acts to initiate the vinylogous Mukaiyama aldol reaction and the catalytic cycle can then be carried on utilizing catalysis through a Si^+ species. In contrast to the reaction using BF_3, where it was crucial to use CH_2Cl_2:Et_2O as a 9:1 mixture for obtaining good yields and selectivities, the reaction using $B(C_6F_5)_3$ could be performed in CH_2Cl_2 without any cosolvent (Table 12, entry 11). Additionally,

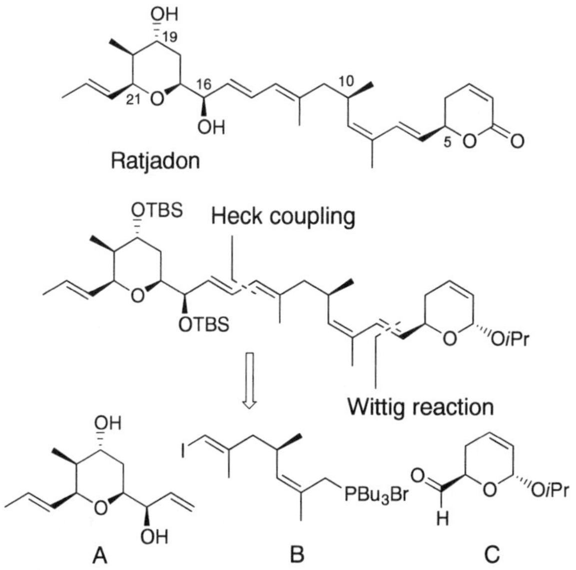

Scheme 23 Retrosynthetic analysis of ratjadone

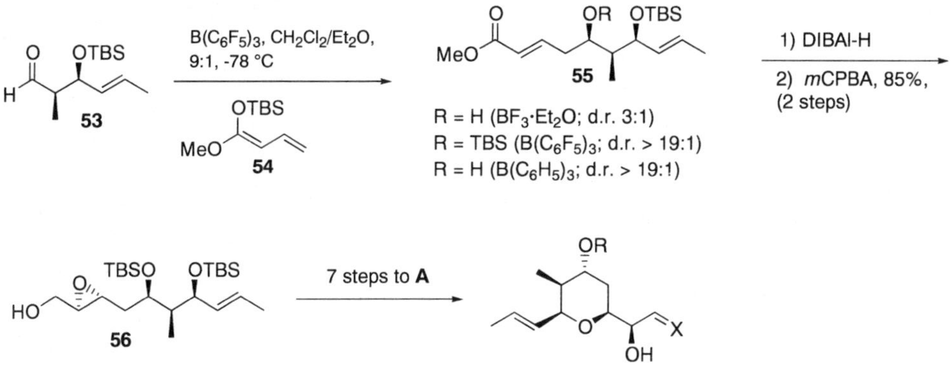

Scheme 24 The Kalesse synthesis of the A-fragment of ratjadone

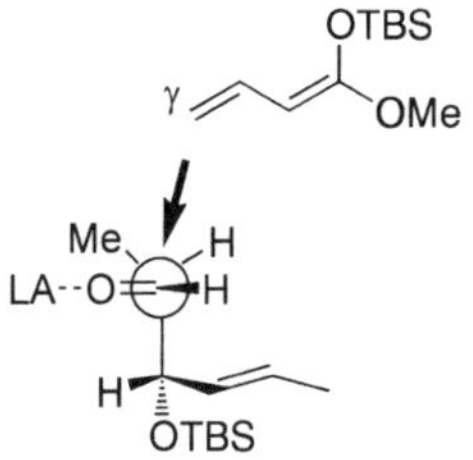

Fig. 5 Felkin–Anh transition state

Table 12 Vinylogous Mukaiyama aldol reaction using different Lewis acids

Entry	Lewis acid	eq.	Solvent	5/6 *syn/anti*	Product	Yield %	Temp. (°C)
1	$BF_3 \cdot OEt_2$	1.5	CH_2Cl_2	3:1	**58**	95	–78
2	$Ti(O\textit{i}Pr)_4$	1.2	CH_2Cl_2		n.r.		0→RT
3	$Ti(O\textit{i}Pr)_4 \cdot$Binol	1.2	CH_2Cl_2		n.r.		0→RT
4	$TiCl_2(O\textit{i}Pr)_2$	2	CH_2Cl_2		n.r.		–78→RT
5	$TiCl_4$	1	CH_2Cl_2		decomp.		–78
6	$B(C_6H_5)_3$	1	CH_2Cl_2/Et_2O (9:1)	>95:5	**57**	85	–78
7	$B(C_6F_5)_3$	1	CH_2Cl_2/Et_2O (9:1)	>95:5	**58**	81	–78
8	$B(C_6F_5)_3$	0.5	CH_2Cl_2/Et_2O (9:1)	>95:5	**58**	78	–78
9	$B(C_6F_5)_3$	0.2	CH_2Cl_2/Et_2O (9:1)	>95:5	**58**	74	–78
10	$B(C_6F_5)_3$	0.1	CH_2Cl_2/Et_2O (9:1)	>95:5	**58**	15	–78
11	$B(C_6F_5)_3$	0.2	CH_2Cl_2	>95:5	**58**	72+**57** (8%)	–78

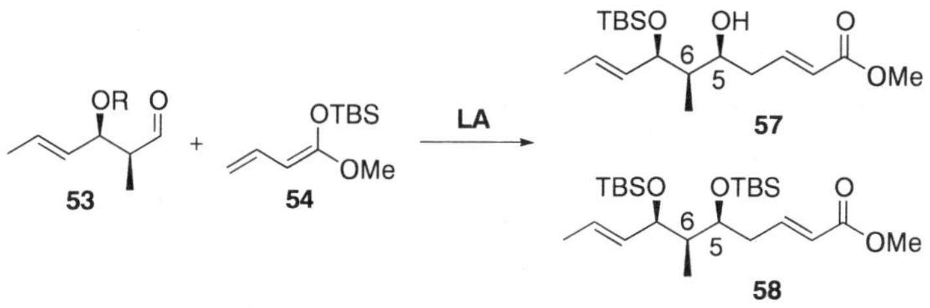

Scheme 25 Vinylogous Mukaiyama aldol reaction of aldehyde **53** and ketene acetal **54** catalyzed by different Lewis acids

$B(C_6F_5)_3$ could be used in substoichiometric concentrations (Table 12, entries 8–10). Only when the reaction was performed with 10% $B(C_6F_5)_3$ did the yield drop significantly (Table 12, entry 10). Remarkably the less expensive triphenyl borane gave the same selectivity (>95%) without TBS transfer (Table 12, entry 6). This was an unexpected result, since it was reported that Mukaiyama aldol reactions are not catalyzed by this Lewis acid [36]. These two reagents can now be used complementarily depending on whether TBS protection of the newly generated hydroxyl group is required or not. Other Lewis acids tested, such as Ti(O*i*Pr)$_4$, $TiCl_2$(O*i*Pr)$_2$, $TiCl_4$, and Ti(O*i*Pr)$_4$·BINOL, gave either no reaction (entries 2–5) or decomposition of the starting materials.

3.2 R_3Si^+ versus TPPB Activation in Aldolizations with γ-Substituted Ketene Acetals

The next focus was set on γ-substituted vinylogous ketene acetals as the most challenging class of nucleophiles resulting in the highest level of complexity during the aldol addition. Transformations in which these ketene acetals could be applied stereoselectively could become of great synthetic value. One obvious question that arose during these efforts was the influence of the C3–C4 double bond geometry for the outcome of the vinylogous Mukaiyama aldol reaction. Both double bond isomers could be synthesized using either the conjugated unsaturated ester (**60**) or the deconjugated isomer (**62**). Both vinylogous ketene acetals were independently reacted with isobutyraldehyde (Scheme 26). The *E*-isomer gave sluggish reaction with TPPB and a mixture of α and γ products with the γ product as a 1:1 mixture of the *syn* and *anti* isomers. With BF_3 on the other hand, the 3,4-*Z*-isomer in Et_2O with isopropanol as the cosolvent gave the γ-alkylation product as the major component with exclusive *syn* selectivity (*rac*-**65**). Isopropanol became essential for obtaining good *syn*/*anti* selectivities since it traps the Si^+ species. From the reaction without any cosolvent it became obvious that two catalytic cycles compete in the transformation. One performed by the Lewis acid TPPB and a second cycle derived from the Si^+ species (Scheme 27). The first one would result in unprotected secondary alcohols during this transformation, whereas the latter one would generate TBS-protected

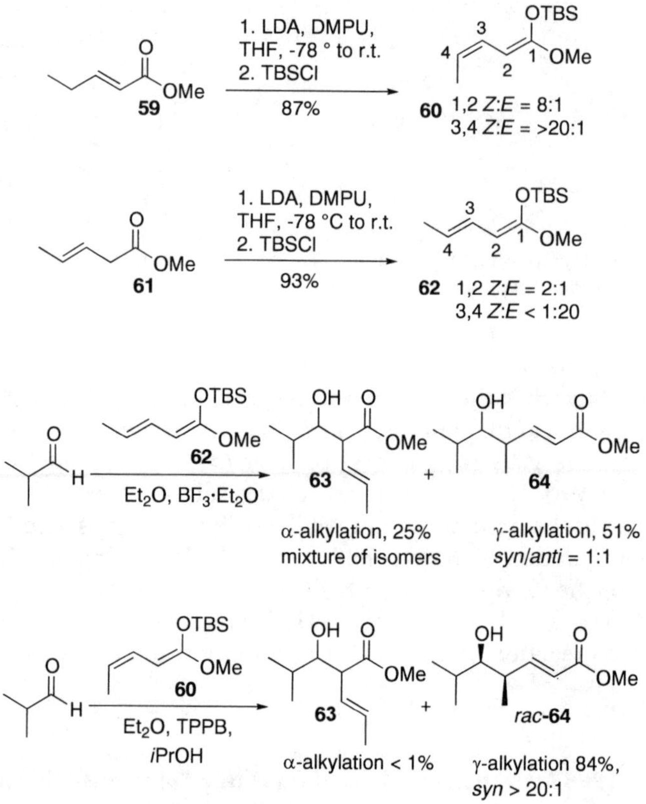

Scheme 26 Vinylogous Mukaiyama aldol reaction using γ-substituted ketene acetals

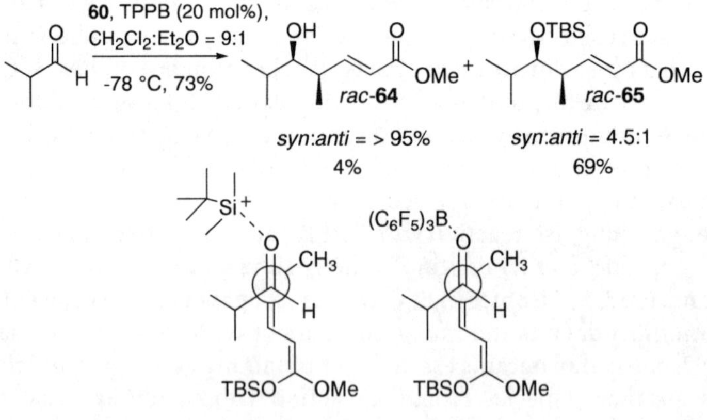

Scheme 27 Si^{+} catalysis versus TPPB catalysis

aldol products, albeit in only 4:1 selectivity. This was exactly the situation for TPPB-catalyzed reactions without isopropanol as cosolvent. The differences in diastereoselectivity can be rationalized by considering that Si^+ catalysis would not form a compact transition state comparable to the one derived from TPPB catalysis with boron as the pivotal element.

With conditions that allow for the diastereoselective addition of γ-substituted vinylogous ketene acetals various aldehydes were tested. Compounds **66** and **68** exhibited very good *syn* (4/5) and Felkin (5/6) selectivity even though the chiral center at C3 would disfavor the Felkin product (Scheme 28).

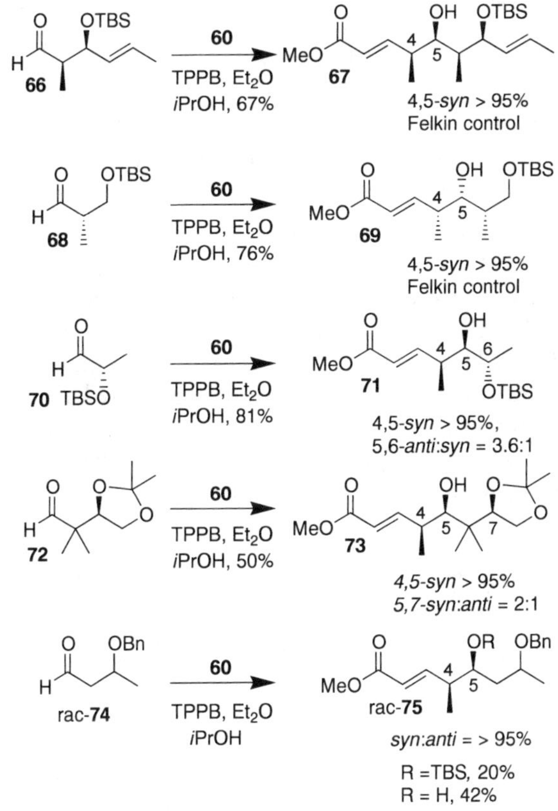

Scheme 28 Vinylogous Mukaiyama aldol reaction using chiral aldehydes

In two studies toward the total synthesis of natural products it could be shown that the α,β-unsaturated esters derived from the vinylogous Mukaiyama aldol reactions can be further functionalized into advanced intermediates. The C1–C7 segment of oleandolide commences with the VMAR of aldehyde **68** derived from the Roche ester. The so-generated stereo-triad was protected as PMB ether and the ester **76** was reduced to the allylic alcohol. Sharpless asym-

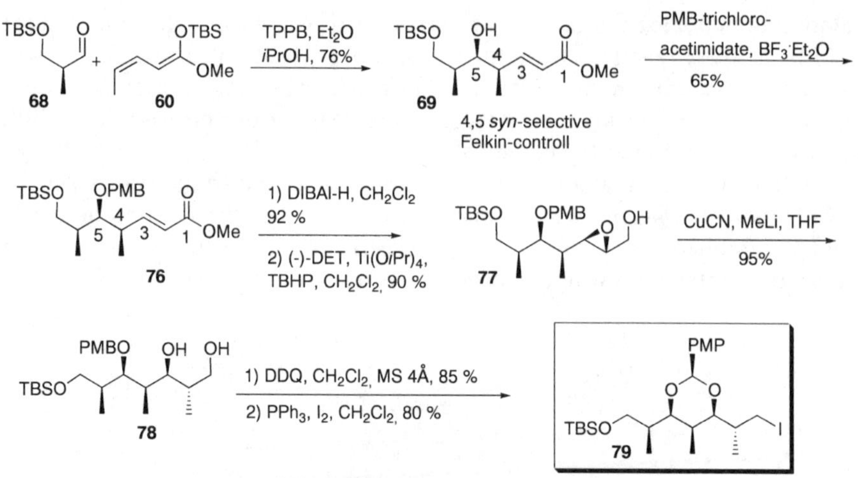

Scheme 29 Synthesis of the C1–C7 segment **79** of oleandolide

metric epoxidation was employed to introduce the appropriate functionality. Cuprate addition to **78** ring-opens the epoxide and establishes the desired *anti* relationship. Two subsequent functional group transformations finish the sequence to the C1–C7 segment **79** [37] (Scheme 29).

In studies directed toward the total synthesis of tedanolide [38], the addition of γ-substituted ketene acetal **60** to aldehyde **84** generated unsaturated ester **85** in 62% yield (Scheme 30). The resulting double bond could be further functionalized by applying Sharpless' asymmetric dihydroxylation conditions. Hence diol **86**, which represents the mismatched case due to unfavorable

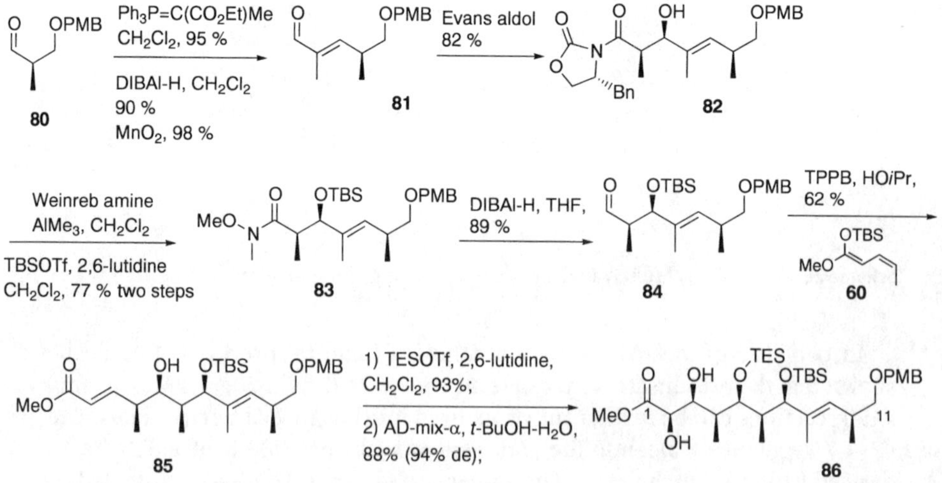

Scheme 30 Synthesis of the C1–C11 segment of tedanolide

interaction with the adjacent methyl group, was isolated in 88% yield with the minor isomer only produced in small quantities (5%).

3.3
C_2-Symmetrical Copper(II) Complexes as Chiral Lewis Acids

Evans and coworkers developed C_2-symmetrical copper(II) complexes as chiral Lewis acids that rely on two-dentate substrates for their catalysis. Consequently,

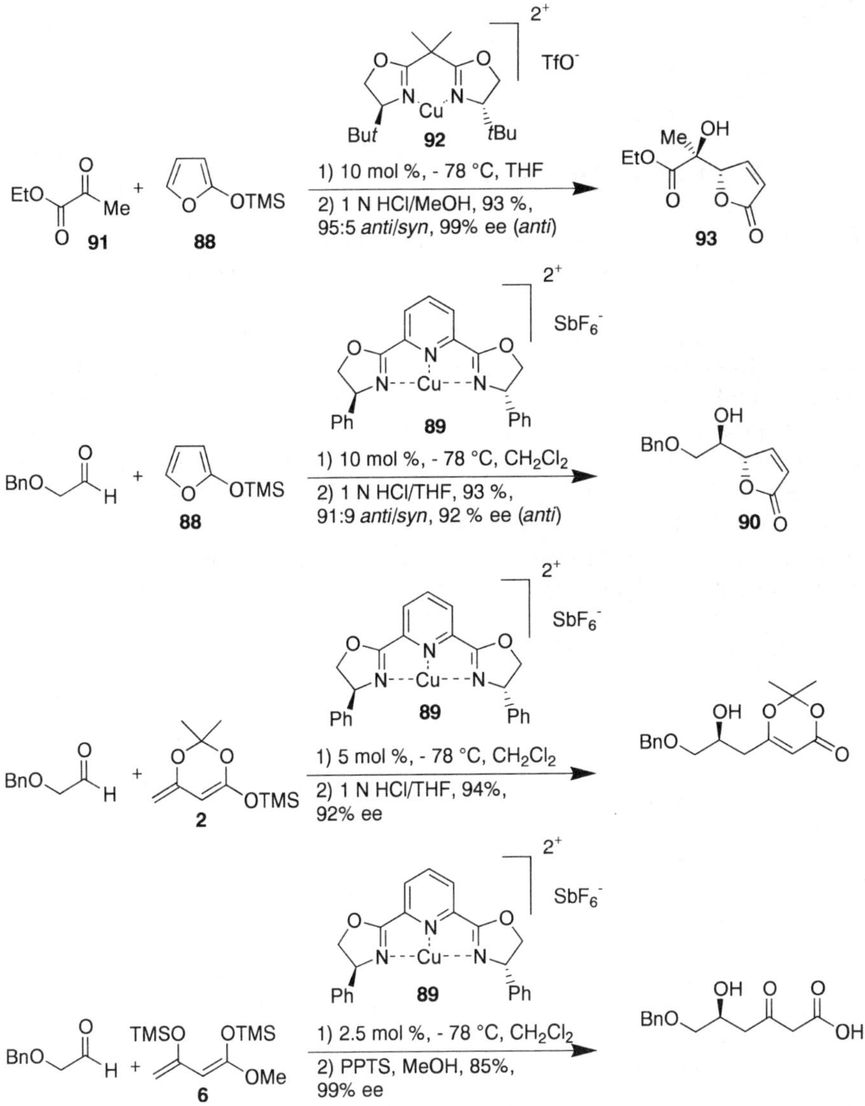

Scheme 31 Vinylogous aldol reactions using Evans' Cu catalysts

additions of ketene acetals to benzyloxyacetaldehyde [39] and pyruvate esters [40] could be achieved in high yield and excellent enantioselectivity (Scheme 31). In these experiments it was shown that cyclic and acyclic dienoxy silanes could be used in the vinylogous aldol reaction with equally good selectivities. As illustrated in Scheme 31, the addition of siloxyfuran **88** to benzyloxyacetaldehyde under catalysis of [Cu((*R*,*R*)PhPyBox)](SbF_6) (**89**) gave aldol product **90** in 93% yield and 92% ee for the *anti* aldol product. Equally high yields and selectivities were observed for the addition using Chan's diene (**6**) and dioxane **2**. The analog Cu(II) box catalyst (**92**) was successfully applied in the aldol transformation of siloxy furanone **88** and pyruvate ester **91** to aldol product **93**. In a first application, this chemistry was applied to the synthesis of the C10–C16 segment of bryostatin [41].

3.4 The Total Synthesis of Callipeltoside

Based on these developments Evans et al. published the synthesis [42] of callipeltoside A (Fig. 6), a cytotoxic natural product isolated from the marine sponge *Callipelt sp.* [43] Their retrosynthetic analysis divided callipeltoside into four equally large segments, of which three were assembled by strategies that allow for the incorporation of larger subunits. Namely, two vinylogous aldol reactions and one *anti*-selective Evans–Metternich variant of the aldol reaction. For the first asymmetric aldol coupling they used the more stable hydrated Cu-PyBox catalyst [Cu((*R*,*R*)PhPyBox)](SbF_6)·$2H_2O$. The vinylogous aldol reaction with 2.5 mol% of catalyst **89** and diene **97** [44] provided the corresponding aldol product (**98**) in 93% yield (Scheme 32). A sequence of TBS protection, reduction, and oxidation to the aldehyde furnished compound **99** used for the Evans–Metternich aldol coupling. Interestingly, only this enantiomer provided

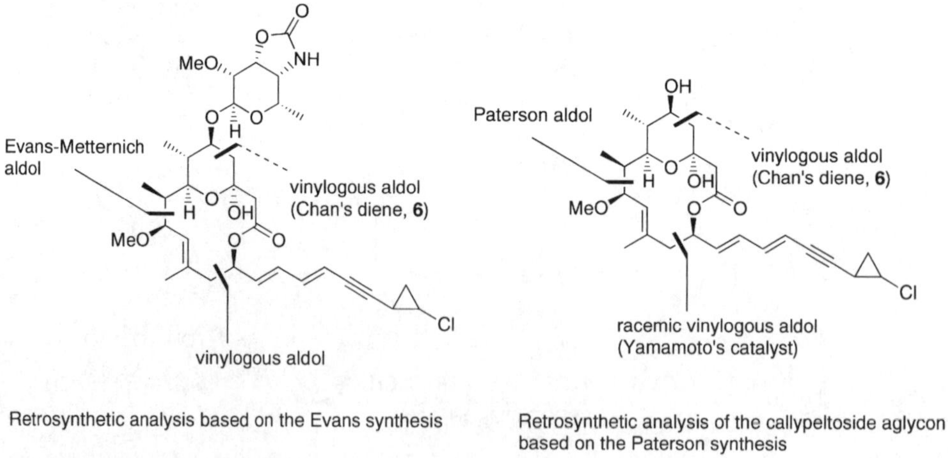

Fig. 6 Retrosynthetic analyses of callipeltoside

satisfactory selectivities in the Evans–Metternich aldol reaction. Next, a sequence of standard transformations generated aldehyde **102** used in the second, diastereoselective vinylogous aldol reaction using $BF_3{\cdot}Et_2O$ as the catalyst and Chan's diene (**6**) at –90 °C. TBS protection of the newly generated secondary alcohol (**103**) and formation of the hemiacetal established compound **104** in 50% yield. Liberation of the secondary alcohol and transformation into the mesylate set the stage for the macrolactonization with inversion of configuration at C13. Hydrolysis of the ester functionality with LiOH was followed by treatment with Cs_2CO_3 and 18-crown-6 in refluxing toluene which led to macrolactonization with 66% yield over two steps (**106**). Further transformations led to the total synthesis of callipeltoside (**107**) (Scheme 32). This synthesis is of

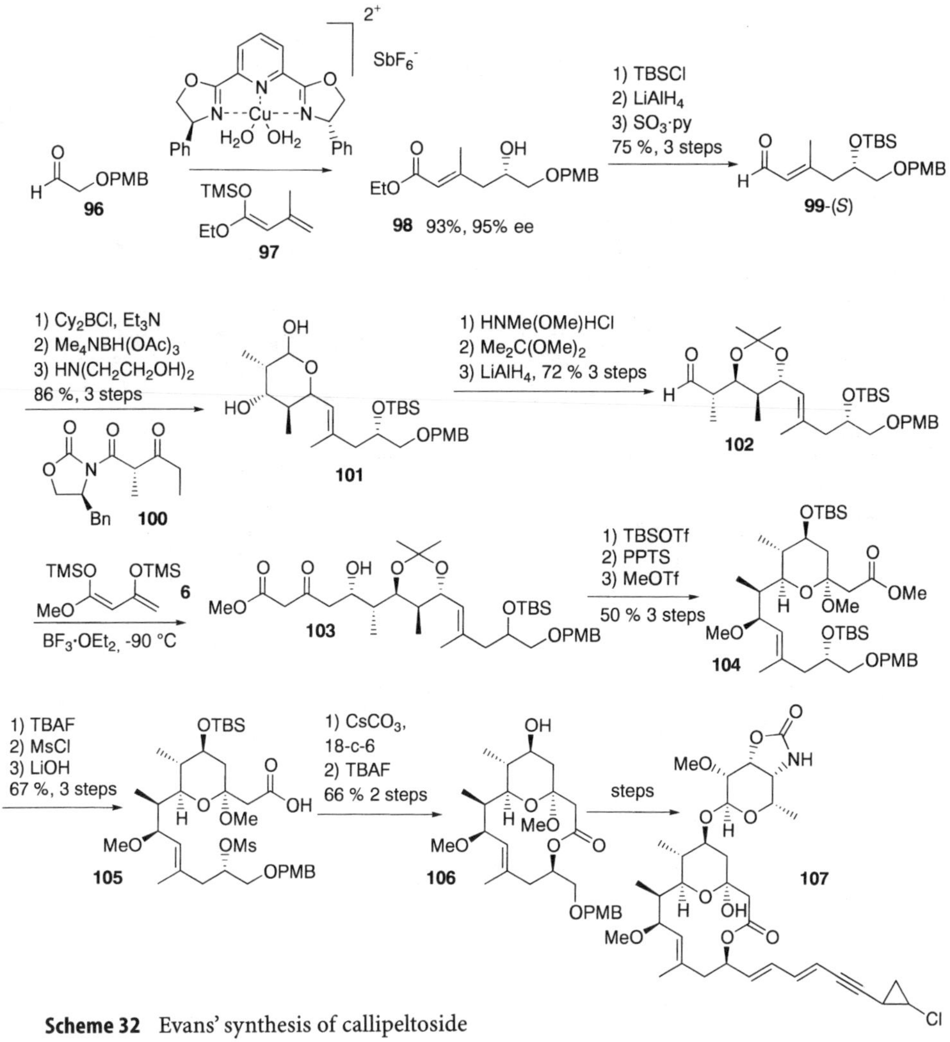

Scheme 32 Evans' synthesis of callipeltoside

special elegance since the construction of the macrolactone involves only large segment couplings, as pointed out above. Both the Evans–Metternich aldol reaction and the vinylogous aldol reaction couple four carbons with respect to the growing polyketide chain. Therefore only three C–C bond formations were necessary in order to construct the macrocycle.

The same natural product was synthesized by Paterson et al. [45] who assembled the carbon skeleton of the macrolide from three larger subunits as well. Instead of the Evans–Metternich variant they used their boron-mediated *anti*-selective aldol strategy which relies as the Evans–Metternich aldol on stereo-induction from the α-chiral center and translates the *E*-enolate geometry, established due to the use of Cy_2BCl, to the *anti* aldol product (Scheme 33).

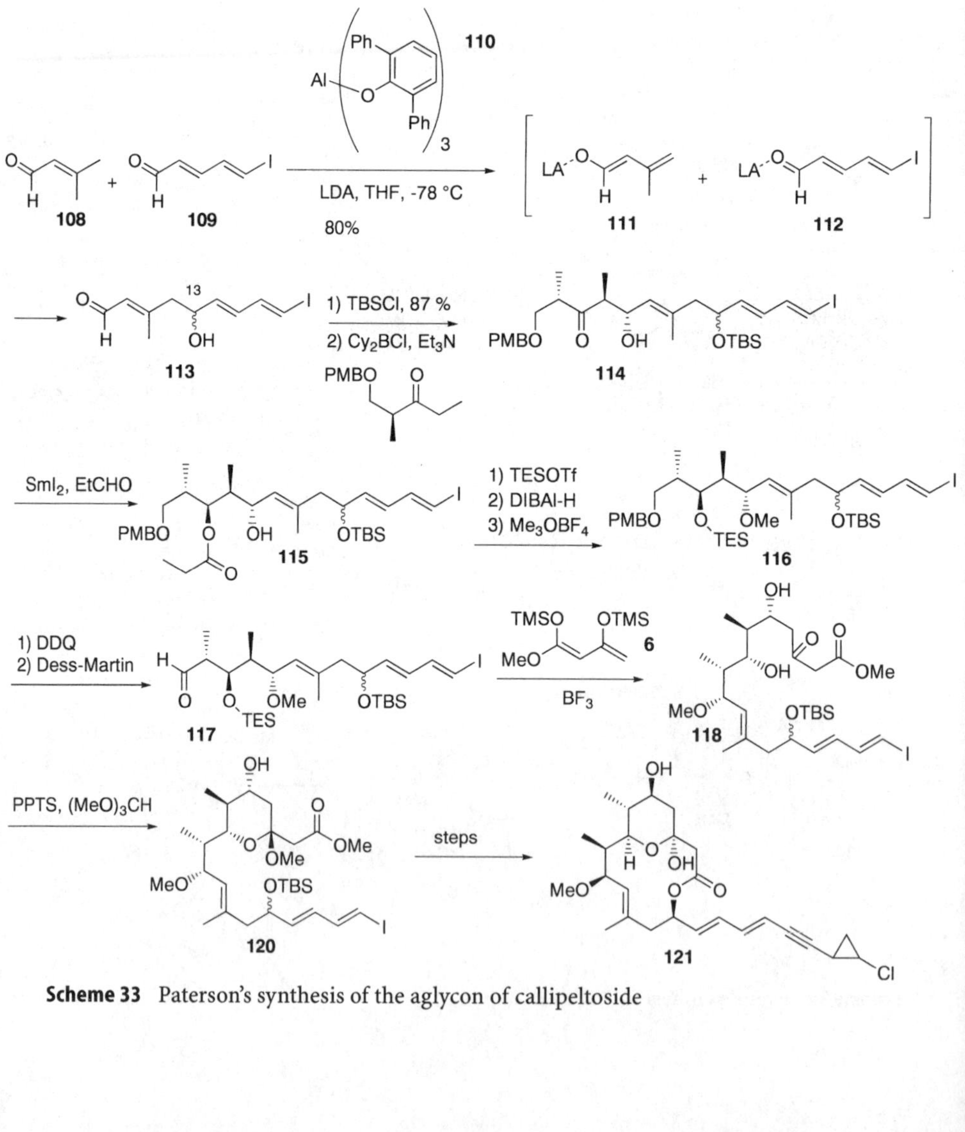

Scheme 33 Paterson's synthesis of the aglycon of callipeltoside

In contrast to the Evans synthesis, the aldol coupling between C12 and C13 achieved was performed such that both configurations at C13 were obtained. Therefore the vinylogous aldol reaction developed by Yamamoto et al. [46, 47] using a sterically demanding Lewis acid was applied, allowing only for electrophilic attack at the γ position. The secondary alcohol was protected and a reduction/oxidation sequence established aldehyde **113**, which was subjected to an aldol reaction based on Paterson's strategy. Evans–Tishchenko reduction with SmI_2 and propionaldehyde produced the 1,3-*anti* diol which was protected, the ester removed by reductive cleavage, and finally transformed into the methyl ether. DDQ deprotection of the PMB group and Dess–Martin oxidation generated aldehyde **117**, which was used in the second vinylogous aldol coupling. For this purpose Chan's diene (**6**) was reacted with BF_3 and produced the desired aldol product (**118**) diastereoselectively. Hemiacetal formation was achieved with PPTS and further transformations gave the aglycon of callipeltoside (**120**).

These two syntheses are prominent examples of advanced synthetic strategies in which polyketides are assembled with only a limited number of C–C bond-forming steps and in which vinylogous aldol reactions play a pivotal role. Besides these syntheses a number of additional examples in which vinylogous aldol reactions were put forward, including Mannich-type reactions, have been published [48–54].

References

1. First review covering the vinylogous principle: Fuson RC (1935) Chem Rev 16:1
2. For reviews covering the development until 2000 see: Casiraghi G, Zanardi F, Appendino G, Rassu G (2000) Chem Rev 100:1929
3. Rassu G, Zanardi F, Battistini L, Casiraghi G. (2000) Chem Soc Rev 29:109
4. For vinylogous Mannich reactions see: Arend M, Westermann B, Risch N (1998) Angew Chem Int Ed 37:1044
5. Bur SK, Martin SF (2001) Tetrahedron 3221
6. Sato M, Sunami S, Sugita Y, Kaneko C (1995) Heterocycles 41:1435
7. Singer RA, Carreira EM (1995) J Am Chem Soc 117:12360
8. Krüger J, Carreira EM (1998) Tetrahedron Lett 7013
9. Pagenkopf BL, Krüger J, Stojanovic A, Carreira EM (1998) Angew Chem 110:3312
10. Krüger J, Carreira EM (1998) J Am Chem Soc 120:837
11. de Rosa M, Soriente A, Scettri A (2000) Tetrahedron Asymmetry 11:3187
12. Chan TH, Brownbridge P (1980) J Am Chem Soc 102:3534
13. For a review on Chan's diene see: Langer P (2002) Synthesis 441
14. Soriente A, de Rosa M, Stanzione M, Villano R, Scettri A (2001) Tetrahedron Asymmetry 12:959
15. Carreira EM, Singer RA, Lee W (1994) J Am Chem Soc 116:8837
16. Kobayashi Y, Taguchi T, Morikawa T, Tokuno E, Sekiguchi S (1980) Chem Pharm Bull 28:262
17. Jardine FH, Rule L, Vohra AG (1970) J Am Chem Soc 238
18. Müller S, Liepold B, Roth GJ, Bestmann HJ (1996) Synlett 521
19. Brennan CJ, Campagne JM (2001) Tetrahedron Lett 42:5195
20. Bluet G, Campagne JM (1999) Tetrahedron Lett 40:5507

21. Bluet G, Campagne JM (2001) J Org Chem 66:4293
22. For enantioselective aldol reactions using chiral quaternary ammonium ions see: Horikawa M, Busch-Petersen J, Corey EJ (1999) Tetrahedron Lett 40:3843
23. Bluet G, Bazán-Tejeda B, Campagne JM (2001) Org Lett 3:3807
24. Denmark SE, Wynn T, Beutner GL (2002) J Am Chem Soc 124:13405
25. Denmark SE, Heemstra JR Jr (2003) Org Lett 5:2303
26. Nelson SG, Wan Z (2000) Org Lett 2:1883
27. Nelson SG, Kim BK, Peelen TJ (2000) J Am Chem Soc 122:9318
28. Denmark SE, Wynn T (2001) J Am Chem Soc 123:6199 and references cited therein
29. Denmark SE, Stavenger RA (2000) Acc Chem Res 33:432
30. Denmark SE, Stavenger RA (2000) J Am Chem Soc 122:8837
31. Denmark SE, Beutner GL (2003) J Am Chem Soc 125:7800
32. Paterson I, Smith JD, Ward RA (1995) Tetrahedron 9413
33. Christmann M, Bhatt U, Quitschalle M, Claus E, Kalesse M (2000) Angew Chem Int Ed 39:4364
34. Christmann M, Bhatt U, Quitschalle M, Claus E, Kalesse M (2001) J Org Chem 1885
35. Christmann M, Kalesse M (2001) Tetrahedron Lett 42:1269
36. Microreview on arylboron compounds in organic synthesis: Ishihara K, Yamamoto H (1999) Eur J Org Chem 527
37. Hassfeld J, Kalesse M (2002) Tetrahedron Lett 43:5093
38. Hassfeld J, Kalesse M (2002) Synlett 2007
39. Evans DA, Kozlowski MC, Murry JA, Burgey CS, Campos KR, Connel BT, Staples RJ (1999) J Am Chem Soc 121:669
40. Evans DA, Burgey CS, Kozlowski MC, Tregay SW (1999) J Am Chem Soc 121:686
41. Evans DA, Carter PH, Carreira EM, Charette AB, Prunet JA, Lautens M (1999) J Am Chem Soc 121:7540
42. Evans DA, Hu E, Burch JD, Jaeschke G (2002) J Am Chem Soc 124:5654
43. Zampella A, D'Auria MV, Minale L, Debitus C, Roussalis C (1996) J Am Chem Soc 118:11085
44. Hoffmann RV, Kim HO (1991) J Org Chem 56:1929
45. Paterson I, Davies RDM, Marquez R (2001) Angew Chem 113:623
46. Saito S, Shiozawa M, Ito M, Yamamoto H (1998) J Am Chem Soc 120:813
47. Saito S, Shiozawa M, Yamamoto H (1999) Angew Chem Int Ed 38:1769
48. Bach T, Kirsch S (2001) Synlett 1974
49. Franck X, Vaz Araujo ME, Jullian JC, Hocquemiller R, Figadère B (2001) Tetrahedron Lett 42:2810
50. Planas L, Pérad-Viret J, Royer J, Sekti M, Thomas A (2002) Synlett 1629
51. Rassu G, Auzzas L, Pinna L, Zambiano V, Zanardi F, Battistini L, Marzocchi L, Acquotti D, Casiraghi G (2002) J Org Chem 67:5338
52. von der Ohe F, Brückner R (2000) New J Chem 9:659
53. Reichelt A, Bur SK, Martin SF (2002) Tetrahedron 58:6323
54. Rassu G, Auzzas L, Zambrano V, Burreddu P, Battistini L, Curti C (2003) Tetrahedron Asymmetry 14:1665

Top Curr Chem(2005) 244: 77–113
DOI 10.1007/b96889

Methanophenazine and Other Natural Biologically Active Phenazines

Uwe Beifuss (✉) · Mario Tietze

Bioorganic Chemistry, Department of Chemistry, University of Hohenheim, Garbenstrasse 30, 70599 Stuttgart, Germany
ubeifuss@uni-hohenheim.de

1 **Introduction** . . . 78

2 **Methanogenesis and Methanophenazine** . . . 80
2.1 Methanogenic Organisms . . . 80
2.2 Energy Metabolism . . . 81
2.3 Methanophenazine – a New Electron Carrier from Archaea . . . 84

3 **Synthesis of Methanophenazine** . . . 85
3.1 Retrosynthesis . . . 85
3.2 Synthesis of Racemic Methanophenazine . . . 86
3.3 Synthesis of Enantiomerically Pure Methanophenazine . . . 89

4 **The Biologic Function of Methanophenazine** . . . 90

5 **Redox Properties of Methanophenazine** . . . 92

6 **New Biologically Active Phenazines from Bacteria** . . . 95

7 **Synthesis of Naturally Occurring Phenazines** . . . 100
7.1 General Strategies . . . 101
7.2 Classical Routes . . . 101
7.3 Diphenylamines via Pd-catalyzed *N*-arylation . . . 106
7.4 Intramolecular Pd-catalyzed *N*-arylation in Phenazine Synthesis . . . 107
7.5 Sequential Inter-/Intramolecular Pd-catalyzed *N*-arylations . . . 109

References . . . 110

Abstract Methanophenazine is a naturally occurring phenazine of nonbacterial origin, which has recently been isolated from the cytoplasmic membrane of *Methanosarcina mazei* Gö1 archaea. It is not only the first and so far the sole phenazine derivative from archaea, but also the first one that is acting as an electron carrier in a respiratory chain – a biologic function equivalent to that of ubiquinones in mitochondria and bacteria. The synthesis of racemic as well as enantiomerically pure methanophenazine is presented and experiments toward the characterization of its biologic function are discussed. The second focus will be on recently discovered phenazines of bacterial origin, many of them exhibiting biologic activities. This review deals with the general methods of phenazine synthesis and its most recent applications in natural product chemistry.

Keywords Phenazine · Methanophenazine · Coenzyme · Natural product synthesis · Pd-catalyzed *N*-arylation

Abbreviations

aliquat	Tricaprylmethylammonium chloride
BINAP	(*R*,*S*)-2,2′-Bis(diphenylphosphino)-1,1′-binaphthyl
dba	Dibenzylideneacetone
CoB-SH	Coenzyme B
CoM-SH	Coenzyme M
dihydro-MP	Reduced form of methanophenazine
F_{420}	Coenzyme F_{420}
$F_{420}H_2$	Reduced form of the coenzyme F_{420}
formyl-H_4MPT	Formyltetrahydromethanopterin
formyl-MF	Formylmethanofuran
HMDE	Hanging mercury drop electrode
H_4MPT	Tetrahydromethanopterin
MF	Methanofuran
methenyl-H_4MPT	Methenyltetrahydromethanopterin
methyl-H_4MPT	Methyltetrahydromethanopterin
methylene-H_4MPT	Methylenetetrahydromethanopterin
methyl-S-CoM	Methyl-coenzyme S
MP	Methanophenazine
MPT	Methanopterin
Ms	*Methanosarcina*
MT1, MT2	Methyl transferases
PG	Protecting group
SHE	Standard hydrogen electrode

1 Introduction

Naturally occurring phenazines from bacteria have been known since the middle of the nineteenth century [1, 2]. The first examples included pyocyanine (**1**),

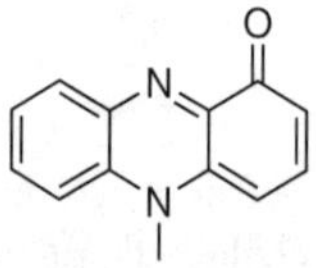

pyocyanine (**1**)

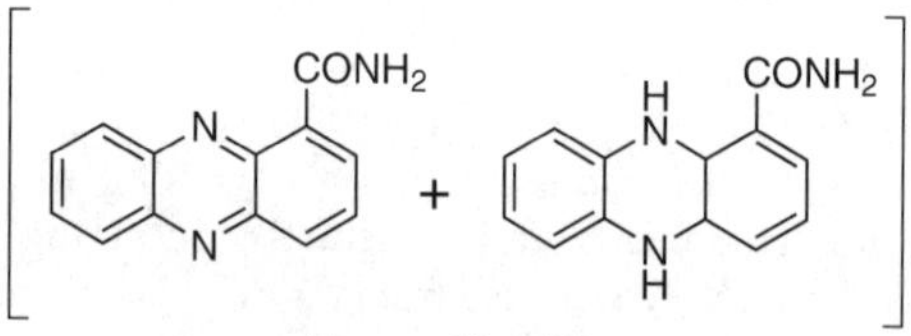

chlororaphine (**2**)

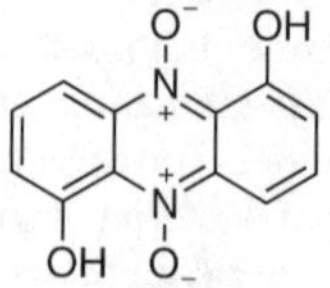

iodinin (**3**)

chlororaphine (**2**), and iodinin (**3**) which were isolated from strains of *Pseudomonas* and *Brevibacterium iodinum* [3–5]. Until now about 100 naturally occurring phenazines have been reported, of which all but one are of bacterial origin. The most important phenazine producers are species of the genera *Pseudomonas* and *Streptomyces*, a minor role being played by *Brevibacterium*, *Microbispora*, and *Sorangium*. Phenazines are known for their significant biologic activity. Mostly they exhibit antibiotic action, with the latter being attributed in model studies to the planar structure of the heteroaromatic phenazines, which allows for DNA intercalation and the resulting inhibition of DNA-dependent RNA synthesis [6].

Especially worth mentioning here are mycomethoxin A (**4**) and B (**5**) exerting effects against various pathogenic mycobacteria, including strains that have been shown to be resistant to the commonly used antibiotics streptomycin and isoniazide [7]. In the search for an effective drug against *Mycobacterium tuberculosis*, *Streptomyces misakiensis* was found to produce the tuberculostatic phenazine-1-carboxylic acid (tubermycin B) (**6**) [8]. Myxin (**7**) isolated from a species of *Sorangium* is a broad-spectrum antibiotic primarily affecting DNA synthesis [9]. A wider application of phenazine antibiotics has been discouraged by a high degree of toxicity of numerous compounds. In recent years the search for novel antibacterial and antiviral compounds has largely increased due to the emergence of antibiotic resistance among pathogens, and with that the interest in new phenazines. On top of this, natural products have recently been isolated that, other than the known biologic activities of phenazines, exhibit antioxidant and/or cell-protective properties. Also, phenazine producers play a role in plant physiology, since some of them may prevent the rhizosphere from being colonized with pathogenic fungi by producing phenazines such as phenazine-1-carboxylic acid (**6**), 2-hydroxyphenazine-1-carboxylic acid (**8**), and 2-hydroxyphenazine (**9**) [10].

Here, we will present a selection of the most attractive naturally occurring phenazines that have been newly discovered. Also, we will deal with the general methods of phenazine synthesis and its most recent applications in natural product chemistry. Until now, the relevance of Pd-catalyzed *N*-arylations to this

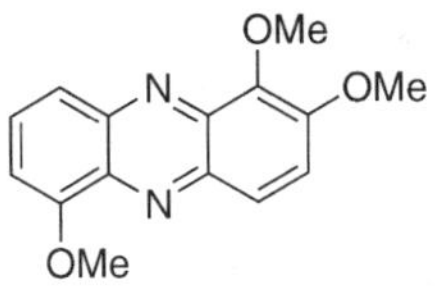

mycomethoxin A (**4**)

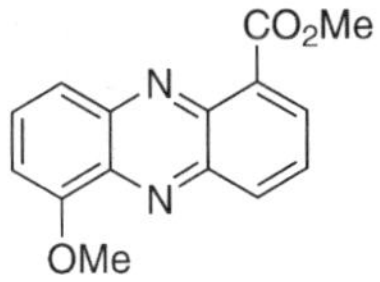

mycomethoxin B (**5**)

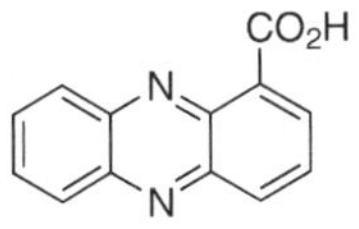

tubermycin B (**6**)

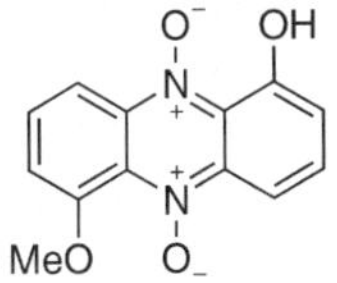

myxin (**7**)

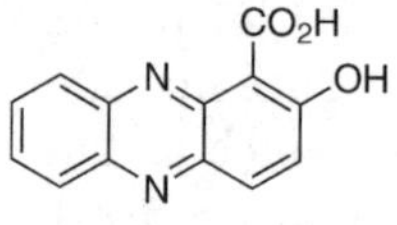

8

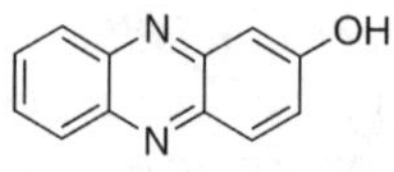

9

field has been negligible. But the first examples employing this new methodology should give some indication of the importance that this reaction type might assume in the very near future.

A naturally occurring phenazine of nonbacterial origin is the methanophenazine (MP) (**10**) which has been isolated from the cytoplasmic membrane of *Methanosarcina (Ms.) mazei* Gö1 archaea. The structure, synthesis, properties, and function of this natural product will be discussed in detail since it is not only the first and so far the sole phenazine derivative from archaea, but also the first one that is acting as an electron carrier in a respiratory chain – a biologic function equivalent to that of ubiquinones in mitochondria and bacteria.

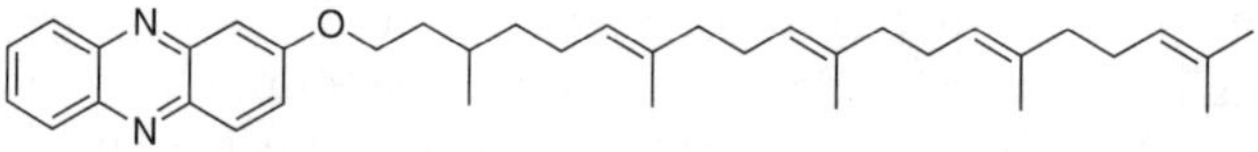

methanophenazine (MP) (**10**)

2
Methanogenesis and Methanophenazine

2.1
Methanogenic Organisms

Strict anaerobic methanogenic archaea inhabit the muds of marshes and tundra regions, the bottoms of lakes, rivers and other waters, as well as the guts of ruminants and the fermenters of sewage treatment plants. With total annual emissions of approx. 10^9 tons, methane formation is one of the most significant biologic processes in the breakdown of organic matter that take place under anaerobic conditions [11]. Hydrolysis of organic polymers to monomers and their transformation by fermentative organisms yield simple compounds such as hydrogen, carbon dioxide, formate, acetate, methanol, and methylamines that all serve as a substrate to methanogenic archaea. These form the end of the anaerobic respiratory chain and convert the substrates mentioned into methane and carbon dioxide, which then reenter the carbon cycle. Due to the influence of civilization the content of methane in the atmosphere, which currently amounts to approx. 1.8 ppm by volume, has dramatically increased over the last few centuries. It is estimated that about 70% of it is generated through the biologic process of methanogenesis [12].

Methanogenic organisms are placed among the archaea, and they differ significantly from the other two domains of life, the eukarya and bacteria. The realm of archaea organisms includes the extreme thermophiles and thermoacidophiles (*Crenarchaeota*), the extreme halophiles and methanogens (*Euryarchaeota*), and the group of *Korarchaeota* [13]. In the phylogenetic tree, archaea establish a distinct evolutionary line. This special position is recognizable from comparative 16S-rRNA analyses that prove their differences with eukaryotes and bacteria. The archaean cell wall does not, as in bacteria, consist of a peptidoglycan skeleton but of pseudomurein, heteropolysaccharides or proteins [14]. And its cytoplasmic membrane contains a number of isoprenoid glycerol ethers in place of the glycerin fatty acid esters of bacteria and eukaryotes [15]. As regards the components participating in the protein biosynthesis, the archaea show some evolutionary proximity to eukarya [16]. But regarding cell division, mode of energy production, and metabolism they share certain characteristics with bacteria.

2.2 Energy Metabolism

The group of methanogens comprises numerous morphologically heterogeneous organisms all of which perform methanogenesis exclusively for energy production [17]. Hydrogenotrophic methanogens are able to activate hydrogen and fix carbon dioxide. Their energy production involves oxidation of hydrogen and reduction of carbon dioxide to methane. They are chemolithoautotrophic, as the only carbon source they use for the production of cell material is carbon dioxide. Other representatives include the methylotrophic methanogens, which employ simple C_1 compounds like formate, methanol, and methylamines to gain energy. The range of substrates of the acetogenotrophic methanogens is limited to acetate, which is converted to methane and carbon dioxide [17].

The transformation of carbon dioxide and hydrogen into methane by methanogenic archaea of the *Methanosarcina* species is represented by Eq. 1:

$$CO_2 + 4H_2 \rightarrow CH_4 + 2H_2O \tag{1}$$

Studies of the cytoplasmic membranes of *Ms. mazei* and *barkeri* demonstrate that membrane-bound enzymes are involved in the corresponding redox reactions [18].

Apart from specific enzyme systems, the carbon dioxide reduction is mediated by a number of unusual coenzymes that can be categorized into two groups, namely C_1 carriers and redox coenzymes [19]. The C_1 carriers transport the C_1 unit from the substrate carbon dioxide to the end-product methane, while the redox coenzymes provide the electrons that are required for the reduction of carbon dioxide to methane. Members of the first group include methanofuran (MF) (**11**), tetrahydromethanopterin (H_4MPT) (**12**), and the coenzyme M (CoM-SH) (**13**), while coenzyme F_{420} (**14**) and coenzyme B (CoB-SH) (**15**) belong to the second group.

Methanogenesis is induced by the activity of the formylmethanofuran dehydrogenase system. Initially, carbon dioxide is fixed by methanofuran (**11**) and

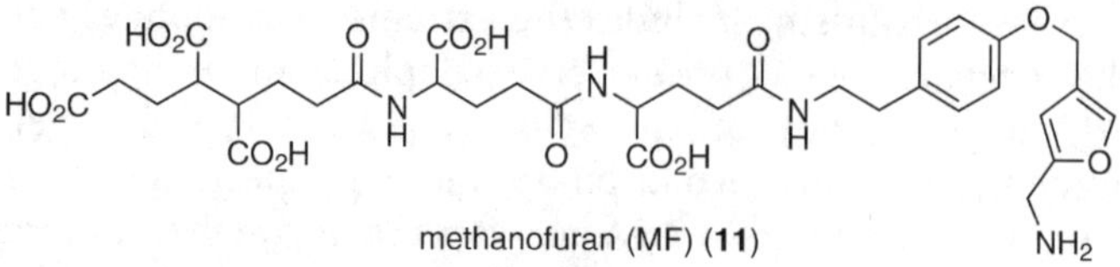

methanofuran (MF) (**11**)

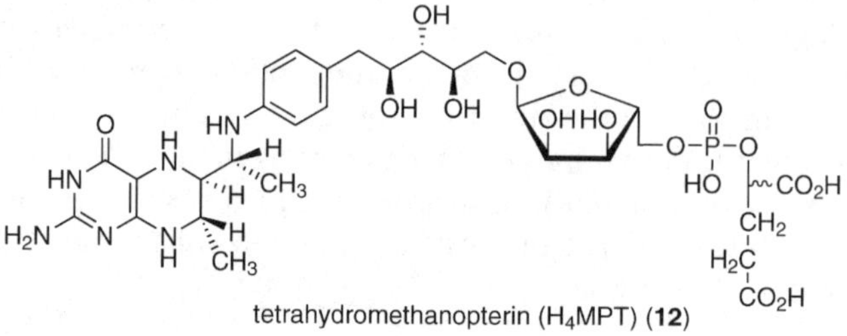

tetrahydromethanopterin (H_4MPT) (**12**)

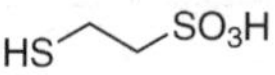

coenzyme M (CoM–SH) (**13**)

then reductively bound to it in the form of formylmethanofuran (formyl-MF) (**16**) (Scheme 1). Subsequent transfer of the formyl group to H_4MPT (**12**) gives formyl-H_4MPT (**17**). The formyl group remains bound to **12** and is then stepwise reduced toward the methyl residue during the following processes (**17**→**20**). Coenzyme F_{420} (**14**) in its reduced form ($F_{420}H_2$) [**14**-H_2], which is provided by the F_{420}-dependent hydrogenase, serves as reducing agent. The methyl group of methyl-H_4MPT (**20**) is then transferred to CoM-SH (**13**), the central intermediate of all methanogenic pathways. Afterwards, methyl-S-CoM (**21**) is reductively demethylated to methane under the catalytic influence of methyl-S-CoM reductase. The two electrons needed are provided by CoB-SH (**15**). This process leads to the formation of the heterodisulfide CoB-S-S-CoM (**22**) from **15** and **13** [12a]. It serves as terminal electron acceptor to the membrane-bound H_2:heterodisulfide oxidoreductase. The two coenzymes **13** and **15** are regenerated by

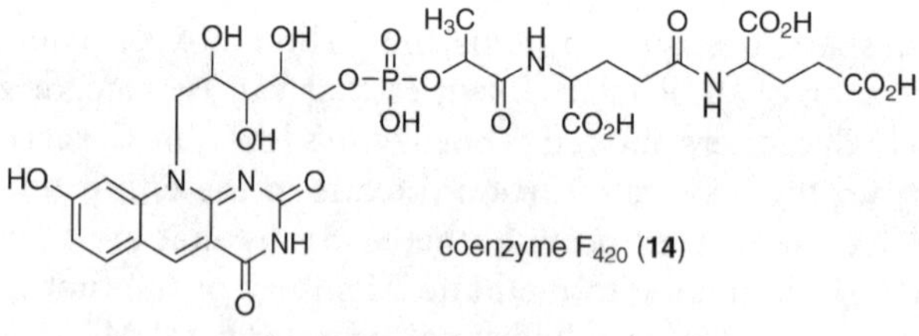

coenzyme F_{420} (**14**)

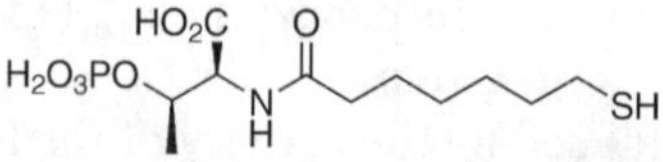

coenzyme B (CoB–SH) (**15**)

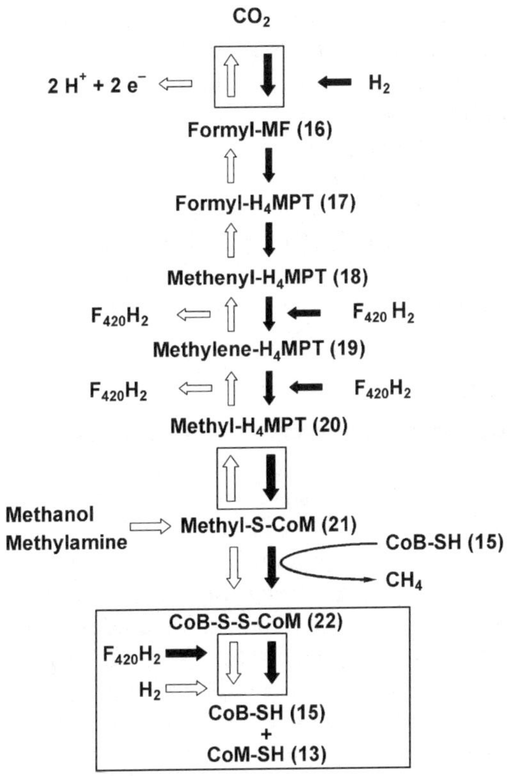

Scheme 1 Methane formation from hydrogen/carbon dioxide, methanol, and methylamine. CoM-SH (**13**)=coenzyme M; F_{420} (**14**)=coenzyme F_{420}; $F_{420}H_2$ (**14**-H_2)=reduced form of the coenzyme F_{420}; CoB-SH (**15**)=coenzyme B; formyl-MF (**16**)=formylmethanofuran; formyl-H_4MPT (**17**)=formyltetrahydromethanopterin; methenyl-H_4MPT (**18**)=methenyltetrahydromethanopterin; methylene-H_4MPT (**19**)=methylenetetrahydromethanopterin; methyl-H_4MPT (**20**)=methyltetrahydromethanopterin; methyl-S-CoM (**21**)=methyl-coenzyme S

the heterodisulfide reductase. The H_2:heterodisulfide oxidoreductase reaction is linked to the generation of a transmembrane proton gradient [20], which is utilized for the ATP synthesis mediated by an ATP synthase activity.

In contrast, disproportioning of methanol occurs with methanogenesis from methanol as given in Eq. 2.

$$4CH_3OH \rightarrow 3CH_4 + CO_2 + 2H_2O \qquad (2)$$

In the first step of the reductive branch of this metabolic pathway three out of four methyl groups are transferred from methanol to CoM-SH (**13**) by methyl transferases, with formation of methyl-S-CoM (**21**) (Scheme 1) [21]. The transformation of **21** and CoB-SH (**15**) into methane and CoB-S-S-CoM (**22**) is catalyzed by the methyl-CoM reductase. Again, reductive cleavage of **22** is mediated by the heterodisulfide reductase [22]. The oxidative part involves oxidation of one out of four methyl groups from methanol to give carbon dioxide. It starts with the formation of **21** and transfer of the methyl group to H_4MPT (**12**) through the membrane integral methyl-H_4MPT:CoM methyl transferase. Following the methyl transfer reaction the successive oxidation of methyl-H_4MPT (**20**) takes place to give formyl-H_4MPT (**17**). Finally, the formylmethanofuran dehydrogenase allows for the oxidation of formyl-MF (**16**) to carbon dioxide. The reducing equivalents accumulating on the oxidation of the methyl residue are transferred to coenzyme F_{420} (**14**) [23]. The oxidation of $F_{420}H_2$ (**14**-H_2) is coupled to the reduction of the heterodisulfide **22**. This exergonic reaction is linked to the generation of an electrochemical proton gradient which is utilized for ATP synthesis.

2.3
Methanophenazine – a New Electron Carrier from Archaea

The strain *Ms. mazei* Gö1 is provided with plenty of enzymes and able to use hydrogen and carbon dioxide as well as methanol, methylamines, and acetate as their substrates. In recent years, studies on this organism have focussed on the two proton-translocating electron transport systems [24]. With the methylotrophic methanogens $F_{420}H_2$:heterodisulfide oxidoreductase plays a major part. It is made up of two components: $F_{420}H_2$ dehydrogenase and heterodisulfide reductase. Some years ago Abken and Deppenmeier succeeded in isolating and purifying $F_{420}H_2$ dehydrogenase from *Ms. mazei* Gö1 [25]. The second component of this system is the heterodisulfide reductase [26]. With the hydrogenotrophic methanogens the H_2:heterodisulfide oxidoreductase system of *Ms. mazei* Gö1 exerts a key function in the electron transport and is made up of two membrane-bound proteins. One part of the system is represented by a membrane-bound hydrogenase [27]. The second component consists of the aforementioned heterodisulfide reductase. Coenzyme F_{420} (**14**) is the central electron carrier in the methanogenic metabolism and its function may be compared to that of coenzyme NAD^+ and $NADP^+$ in eubacteria and eukaryotes, respectively. Among other things, F_{420} serves as a coenzyme of the F_{420}-dependent hydrogenase, the methylene H_4MPT dehydrogenase, and the methylene H_4MPT reductase. Its reduced form $F_{420}H_2$ (**14**-H_2) reacts with $F_{420}H_2$ dehydrogenase.

In contrast, the structure of the electron carrier mediating the electron transport between $F_{420}H_2$ dehydrogenase or membrane-bound hydrogenase and the heterodisulfide reductase was unknown initially. In cooperation with Deppenmeier et al. we have recently been able, though, to isolate and characterize a phenazine ether from the membranes of *Ms. mazei* Gö1, though in small amounts only [28]. Detailed NMR analysis of this natural product shows that its lipophilic side chain, that probably ensures anchorage to the membrane, consists of five isoprenoid units which are linked in a head-to-tail manner. While the C_5 unit that is directly linked to the 2-phenazinyl residue via the ether bridge is saturated, the other four are unsaturated. Three units exhibit (*E*)-configured double bonds. The redoxactive natural product, which was designated methanophenazine (MP) (**10**), is the first phenazine ever isolated from archaea.

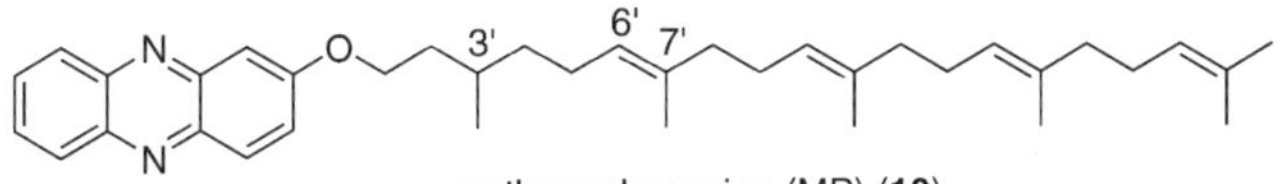

methanophenazine (MP) (**10**)

As methanogens lack the usual quinones it was assumed that **10** functions as an electron carrier in the cytoplasmic membrane and, unlike other naturally occurring phenazines, takes part in the energy-conserving electron transport. Studies on the biologic function of **10** in the membrane-bound electron transport of *Ms. mazei* Gö1 and the elucidation of the absolute configuration at C-3′ required the natural product in larger amounts. As the quantity of **10** made available by extraction from the cytoplasmic membranes of *Ms. mazei* Gö1 was too small and the efforts undertaken to isolate **10** too large, the synthesis of the natural product seemed desirable.

3 Synthesis of Methanophenazine

3.1 Retrosynthesis

The result of the retrosynthetic analysis of *rac*-**10** is 2-hydroxyphenazine (**9**) and the terpenoid unit *rac*-**23**, which may be linked by ether formation [29]. The *rac*-**23** component can be dissected into the alkyl halide *rac*-**24** and the (*E*)-vinyl halide **25**. A Pd(0)-catalyzed sp^3–sp^2 coupling reaction is meant to ensure both the reaction of *rac*-**24** and **25** and the (*E*)-geometry of the C-6′, C-7′double bond. Following Negishi, **25** is accessible via carboalumination from alkyne **27**, which might be traced back to (*E*,*E*)-farnesyl acetone (**28**). The idea was to produce **9** in accordance with one of the methods reported in the literature, and to obtain *rac*-**24** in a few steps from symmetrical 3-methyl-pentane-1,5-diol (**26**) by selective functionalization of either of the two hydroxyl groups.

The advantage of such a prochiral C_6 building block **26** is that both enantiomers (*R*)-**24** and (*S*)-**24**, which are required for the synthesis of the two enantiomeric methanophenazines, are accessible through differentiation of its enantiotopic groups. This stereodivergent synthetic strategy requires a suitable precursor, in this case the half ester of (*R*)-3-methyl glutaric acid [(*R*)-**31**], which needs to be transformed initially into the chiral lactones (*R*)-**30** and (*S*)-**30**. These lactones then were meant to serve as starting material for the two components (*R*)-**29** and (*S*)-**29** constituting the basis of this approach to the two antipodes of **10**.

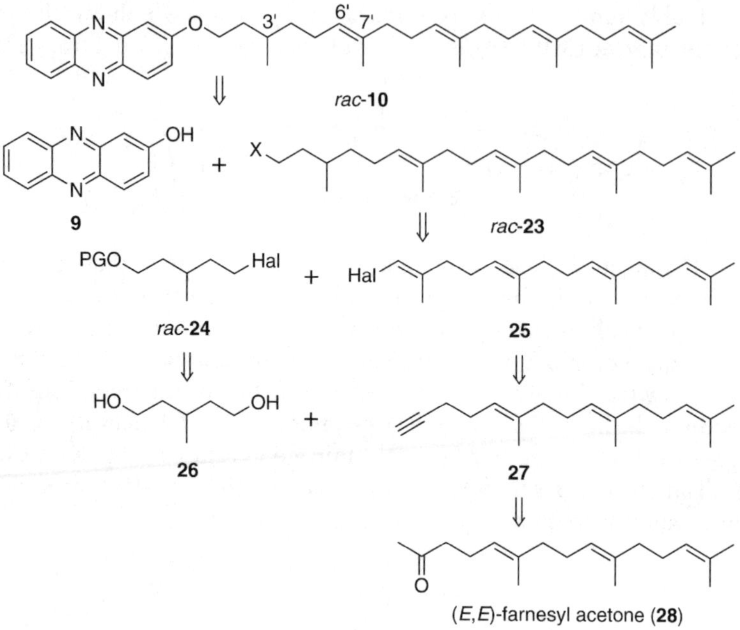

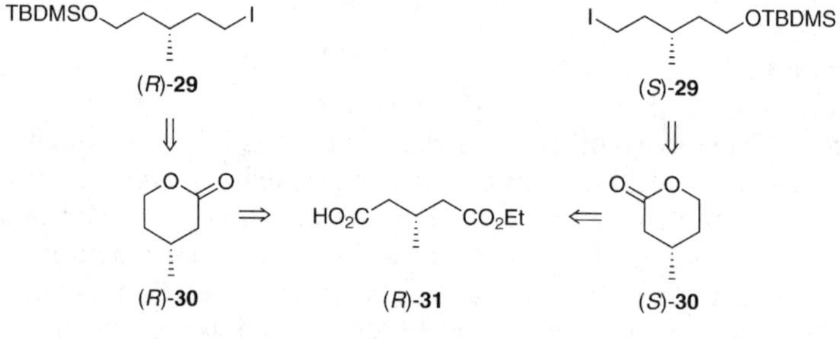

3.2 Synthesis of Racemic Methanophenazine

The synthesis of 2-hydroxyphenazine (**9**) caused unexpected problems, though. The resultant yield of **9** was very low with both the condensation of 2-hydroxy-1,4-benzoquinone (**33**) [30], available from 1,2,4-trihydroxybenzene

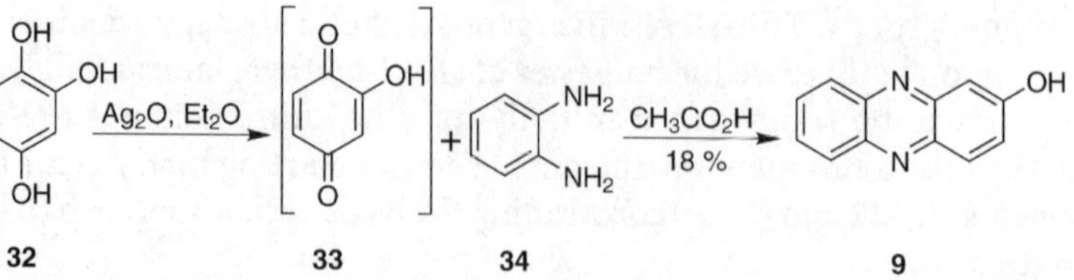

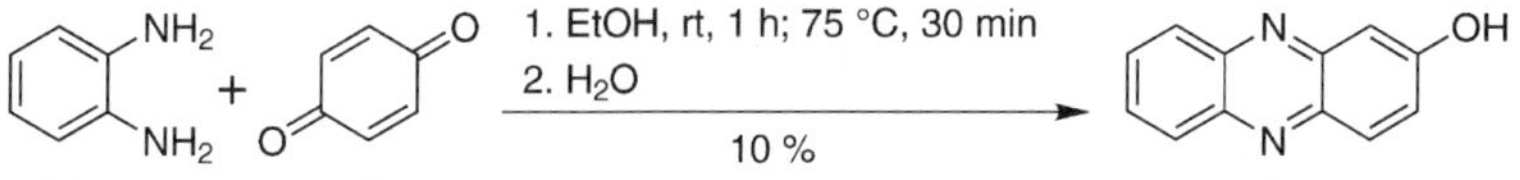

34 35 9

(**32**) [31], with 1,2-phenylenediamine (**34**) according to Kehrmann and Cherpillod [32] and the condensation of 1,4-benzoquinone (**35**) with **34** according to Ott [33]. Eventually, the Beirut reaction proved to be more reliable producing very good yields of **9** by reaction of benzofuroxan **36** and hydroquinone (**37**) to 5,10-dioxide **38** and subsequent reduction [34].

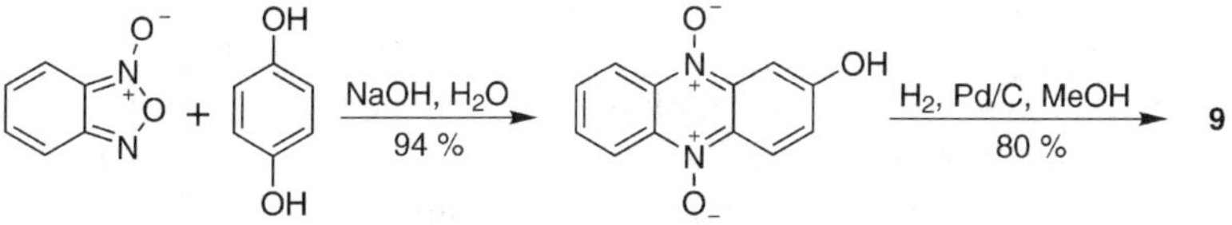

36 37 38

The preparation of vinyl iodide **39** first required the transformation of (*E*,*E*)-farnesyl acetone (**28**), performed according to Negishi, to the terminal alkyne **27** with 75% yield [35]. The latter then gave (*E*)-vinyl iodide **39** in diastereomerically pure form and 74% yield by Zr-catalyzed carboalumination with trimethylaluminum and trapping of the intermediate vinylaluminum species with iodine [36]. The alkyl iodide *rac*-**29** necessary to ensure the coupling with **39** was obtained by selective monofunctionalization of 3-methylpentane-1,5-diol (**26**) in a few steps [37].

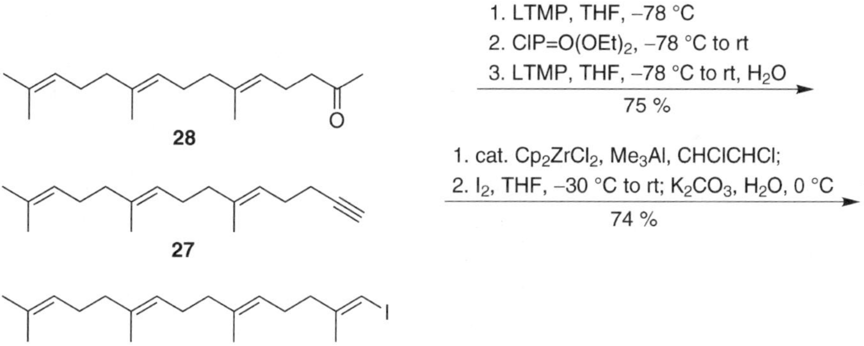

28 27 39

The key step of the synthesis of racemic methanophenazine (*rac*-**10**) implies the coupling of the saturated C_6 component *rac*-**29** to the unsaturated C_{19} component **39**, which is needed for the generation of the terpenoid side chain. To begin with, the sp^2–sp^3 cross-coupling between alkylmetal compounds released from *rac*-**29** in situ and vinylmetal derivatives from **39** posed consid-

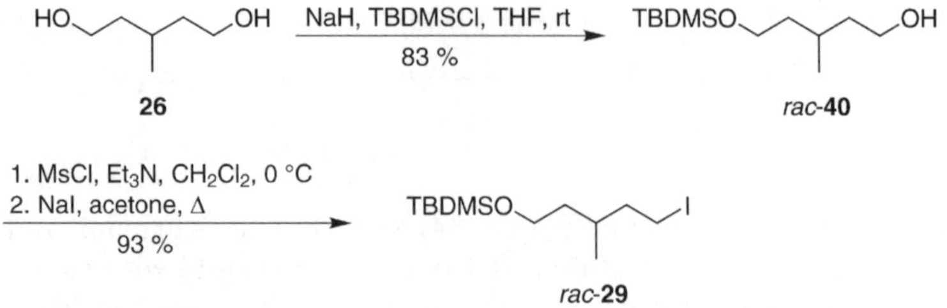

erable difficulties due to the low reactivity of the sp^3 component. In the end, the (*E*)-selective construction of the C-6′, C-7′ double bond [38] of the sesterterpene building block *rac*-**41** was effected by Pd(0)-catalyzed coupling of the organozinc compound obtained in situ from *rac*-**29**, *tert*-butyl lithium, and zinc chloride and the vinyl iodide **39** with 64% yield. Fluoride-induced cleavage of the *tert*-butyldimethylsilyl ether and activation of the resultant alcohol with methanesulfonyl chloride provided the mesylate, which was etherified with 2-hydroxyphenazine (**9**) to give *rac*-**10** [39]. By refraining from isolating or purifying the various intermediate steps, we were able to prepare *rac*-**10** from farnesyl acetone **28** with a total yield of 30%. All spectroscopic data of the synthetic *rac*-**10** corresponded to those of the natural product. Starting from **9**, 3-methyl-pentane-1,5-diol (**26**), and (*E*,*E*)-farnesyl acetone (**28**), this synthesis offers an efficient, convergent approach to methanophenazine (**10**). Also, the possibility of targeted manipulation of the three building blocks opens a simple route to the preparation of derivatives of **10**.

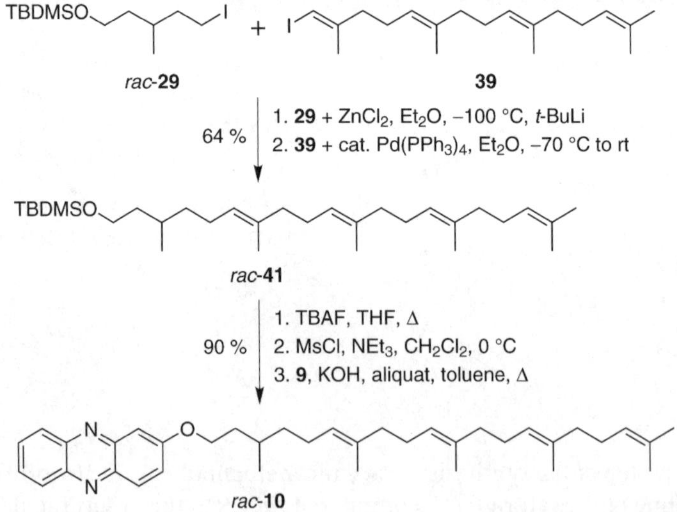

3.3 Synthesis of Enantiomerically Pure Methanophenazine

As mentioned above, the advantage of the prochiral 3-methyl-pentane-1,5-diol (**26**) is that we should be able to derive the two enantiomers (*R*)-**29** and (*S*)-**29** necessary for the synthesis of methanophenazine (**10**) in enantiomerically pure form by differentiating its enantiotopic $(CH_2)_2OH$ groups [40].

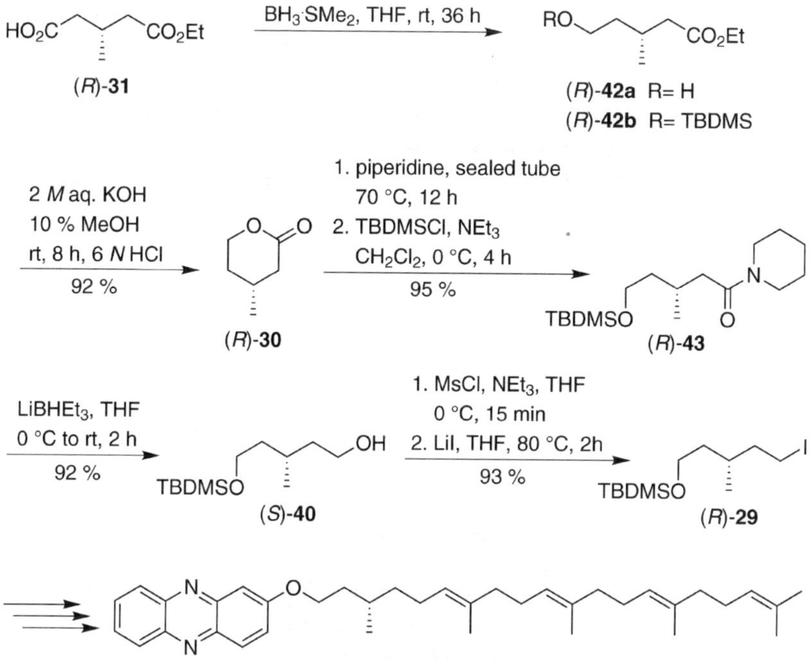

(*S*)-methanophenazine [(*S*)-**10**)]

The monoethyl (*R*)-3-methylglutarate [(*R*)-**31**] available in high optical purity (ee>95%) constitutes a suitable substrate for this stereodivergent synthetic approach. The synthesis of (*S*)-methanophenazine [(*S*)-**10**] requires (*R*)-**29**. The direct conversion of (*R*)-**31** to (*R*)-**42b** via chemoselective reduction of the carboxyl group with $BH_3{\cdot}Me_2S$ and silylation of the resulting hydroxy group could not be accomplished; rather, this approach only gave mixtures of (*R*)-**42b** and the lactone (*R*)-**30** [41]. It turned out, though, that lactone (*R*)-**30** could be obtained with high yields by chemoselective reduction of the carboxylic acid functionality of (*R*)-**31** and subsequent cyclization of the resulting 5-hydroxy ester (*R*)-**42a** under basic conditions. To avoid further difficulties the lactone (*R*)-**30** was opened by reaction with piperidine to give the corresponding hydroxy-amide, which in turn was almost quantitatively transformed to (*R*)-**43** via silylation of the hydroxyl group. The one-step reduction of the tertiary amide (*R*)-**43** with lithium triethyl borohydride [42] provides the primary alcohol (*S*)-**40**, which is then as required converted into the iodide (*R*)-**29** by mesylation and transformation with lithium iodide. As a result, the synthesis of gram amounts of the enantiomerically pure C_6 building block (*R*)-**29** from (*R*)-**31** was accomplished in seven steps with a total yield of 75%. The determination of the enantiomeric excess at the level of the silyloxy alcohol (*S*)-**40** according to Mosher demonstrates that within the course of the synthetic sequence, the integrity of the stereocenter has been left untouched. Corresponding to the synthesis of *rac*-**10** a Pd-catalyzed C,C cross-coupling of the enantiomerically pure building block (*R*)-**29** with (*E*)-vinyl iodide **39**, subsequent desilylation of the coupling product, and final etherification with 2-hydroxyphenazine (**9**) yielded (*S*)-methanophenazine [(*S*)-**10**].

The same starting material can also be employed to synthesize the antipode (*R*)-methanophenazine [(*R*)-**10**], as (*R*)-**31** may easily be transformed into lactone (*S*)-**30** by chemoselective reduction of the ester functionality [43] and subsequent cyclization.

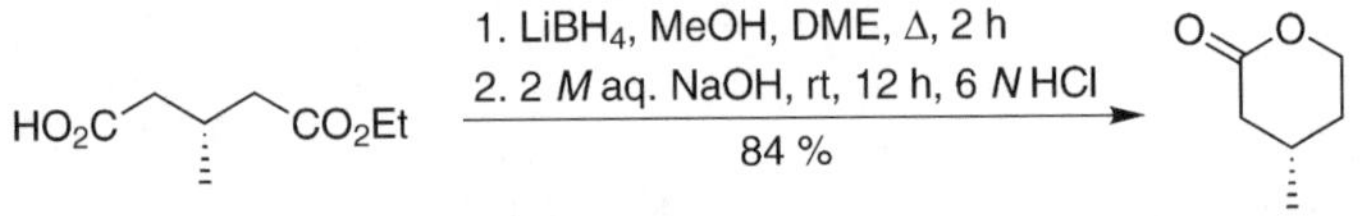

(*R*)-**31** (*S*)-**30**

4
The Biologic Function of Methanophenazine

As already mentioned methanogens lack the usual quinones. This is why it was assumed that in methanogenesis, the redoxactive natural product methanophenazine (MP) (**10**) functions as an electron carrier in the cytoplasmic membrane (Scheme 2). It mediates the transport of electrons between $F_{420}H_2$ dehydrogenase and membrane-bound hydrogenase, respectively, and the heterodisulfide reductase. This would render **10** the first naturally occurring phenazine that is involved in the electron transport of biologic systems. The model compound 2-hydroxyphenazine (**9**) and its reduced form dihydro-**9** were used to perform initial studies with washed cytoplasmic membranes of *Ms. mazei* Gö1. They revealed that all key enzymes – the membrane-bound hydrogenase, $F_{420}H_2$ dehydrogenase, and heterodisulfide reductase – react with the artificial electron carrier (Table 1) [28]. In repeating these experiments with purified enzymes it was possible to functionally reconstitute the $F_{420}H_2$:heterodisulfide oxidoreductase system [44]. This implies that the two partial reactions of this electron transport system can be successfully coupled and successively proceed in the same test tube in terms of a consecutive reaction.

After completion of the total synthesis corresponding experiments could be conducted with methanophenazine (MP) (*rac*-**10**), too. To this end, washed

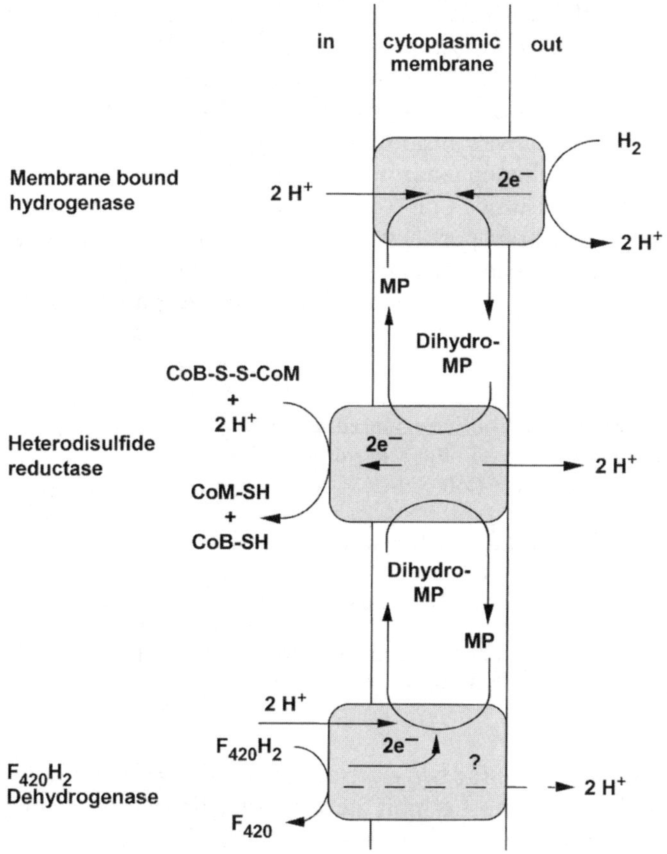

Scheme 2 Model of the membrane-bound electron transfer of *Methanosarcina mazei* Göl according to Ref. [29]. MP=methanophenazine; dihydro-MP=dihydromethanophenazine; CoM-SH=coenzyme M; CoB-SH=coenzyme B; F_{420}=coenzyme F_{420}; $F_{420}H_2$=reduced form of the coenzyme F_{420}

Table 1 Specific activities of the enzymes of the $F_{420}H_2$:heterodisulfide oxidoreductase and H_2:heterodisulfide oxidoreductase systems [29]

Enzyme	Electron donor	Electron acceptor	Spec. activity[a] (U mg protein^{-1})
$F_{420}H_2$ dehydrogenase	$F_{420}H_2$	**9**	0.20
$F_{420}H_2$ dehydrogenase	$F_{420}H_2$	MP	0.15
Membrane-bound hydrogenase	H_2	**9**	2.2
Membrane-bound hydrogenase	H_2	MP	3.2
Heterodisulfide reductase	Dihydro-**9**	CoB-S-S-CoM	2.3
Heterodisulfide reductase	Dihydro-MP	CoB-S-S-CoM	2.6

[a] 1 U=1 µmol substrate converted per minute.

cytoplasmic membranes of *Ms. mazei* Gö1 were reacted with MP and again the activity of the respective enzymes was determined (Table 1). It was shown that the MP served as an electron acceptor to both the membrane-bound hydrogenase and the $F_{420}H_2$ dehydrogenase, when H_2 and $F_{420}H_2$ (**14**-H_2) were added as electron donors, respectively. In addition, the heterodisulfide reductase uses the reduced form of methanophenazine (dihydro-MP) as an electron donor for reducing the heterodisulfide CoB-S-S-CoM (**22**). This means that MP is able to mediate the electron transport between the membrane-bound enzymes [29] (Scheme 2). Reactions mediated/catalyzed by the proton-translocating electron transport systems may thus be subdivided into two partial reactions each (Scheme 3) [24].

$F_{420}H_2$:heterodisulfide oxidoreductase
a) $F_{420}H_2$ + MP → F_{420} + dihydro-MP
b) Dihydro-MP + CoB-S-S-CoM → MP + CoB-SH + CoM-SH

H_2:heterodisulfide oxidoreductase
a) H_2 + MP → dihydro-MP
b) Dihydro-MP + CoB-S-S-CoM → MP + CoB-SH + CoM-SH

Scheme 3 Partial reactions of the electron transport systems $F_{420}H_2$:heterodisulfide oxidoreductase and H_2:heterodisulfide oxidoreductase

As the key enzymes of the two electron transport systems with the electron carriers **9** and *rac*-**10** almost exhibit the same specific activity, we may assume that the phenazine system is wholly responsible for the redox process and that the sole function of the terpenoid side chain in **10** is to anchor the electron carrier in the membrane.

In summary, methanophenazine (**10**) is the first phenazine whose involvement in the electron transport of biologic systems could be established. The experiments indicate that its role in the energy metabolism of methanogens corresponds to that of ubiquinones in mitochondria and bacteria.

5 Redox Properties of Methanophenazine

To further elucidate the function of methanophenazine (**10**) as electron carrier its redox potential was determined [45]. As Scheme 2 reveals, the redox potential of **10**/dihydro-**10** must range somewhere between the values for $F_{420}H_2/F_{420}$ ($E^{0\prime}$=–360 mV) [46] and $H_2/2H^+$ ($E^{0\prime}$=–420 mV), and the redox potential of the heterodisulfide CoB-S-S-CoM (**22**). Whilst the $E^{0\prime}$ values of $F_{420}H_2/F_{420}$ and $H_2/2H^+$ have long been known, the redox potential of **22** remained an unknown quantity. 2-Hydroxyphenazine (**9**) had proved to be a satisfactory model sub-

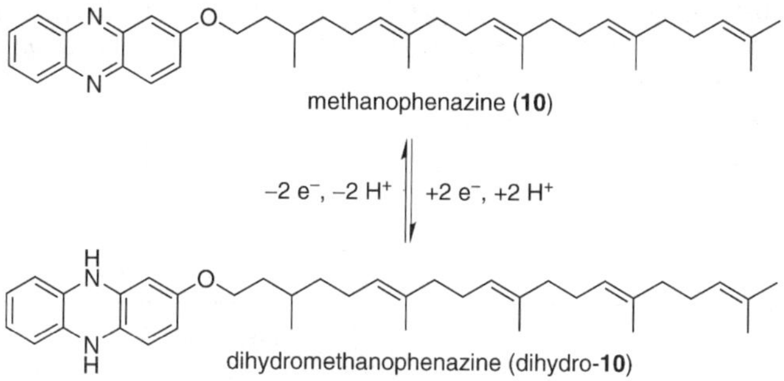

stance for **10** in enzymatic experiments, so the two redox potentials of the two substances were also supposed not to be much different from each other [47].

To ensure the most realistic $E^{0\prime}$ values possible a measuring technique was sought that was suitable for the experiments to be conducted in purely aqueous buffer systems at pH 7. In this case, the measuring technique of choice was cyclic voltammetry using a hanging mercury drop electrode (HMDE) [48]. The conditions described served to determine the redox potential of **9** at $E^{0\prime}$= −191±8 mV vs. standard hydrogen electrode (SHE) in phosphate buffer at pH 7. The same conditions revealed the redox potential of **10** to be $E^{0\prime}$=−165±6 mV vs. SHE (Table 2). The strong dependence of the redox potential on the pH value of the buffer system is particularly noticeable: increase of the pH by one unit led to a shift of the redox potential by −64±0.6 mV. Measurements with the strongly lipophilic **10** could probably only be accomplished because of the strong adsorption of the phenazine at the HMDE.

In addition, the idea of the terpenoid side chain of **10** essentially assisting in anchoring the coenzyme in the cytoplasmic membrane without having any impact on the redox potential was to be explored. To this end, a number of phenazine ethers **44a–g** were synthesized by Williamson ether synthesis and then investigated by electrochemical methods. And indeed, we were able to identify a good match between the redox potentials of the various phenazine ethers, which turned out to be independent of the side chain structure.

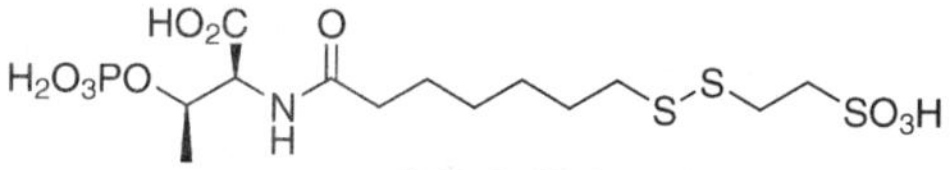

CoB–S–S–CoM (**22**)

Finally, the redox potential of the heterodisulfide CoB-S-S-CoM (**22**), which was synthesized according to known procedures [49], was measured under the same conditions in phosphate buffer at pH 7. An $E^{0\prime}$ value of −143±10 mV vs SHE for **22** exhibits a more positive redox potential than typical disulfides such as glutathione and cysteine ($E^{0\prime}$=−204±6 mV vs SHE and $E^{0\prime}$=−202±3 mV vs. SHE, resp.). That these deviations are probably due to the presence of sulfonate

Table 2 Synthesis of phenazine ethers **10, 44a–g** by etherification of 2-hydroxyphenazine (**9**) with ROMs(–Br) and their redox potentials $E^{0\prime}$ vs. SHE as determined by cyclic voltammetry in phosphate buffer at pH 7 using HMDE

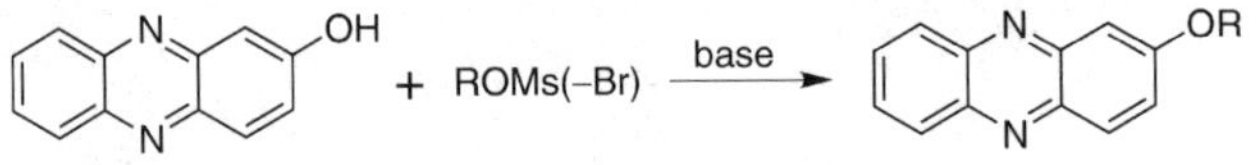

9 **10, 44a-g**

Compound	R	Yield [%]	$E^{0\prime}$
10	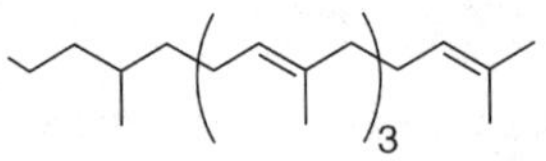	90	–165
44a	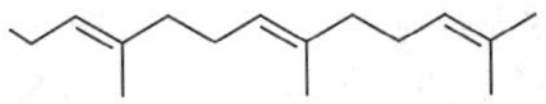	35	–168
44b	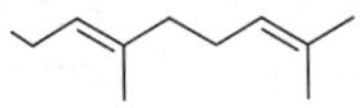	44	–165
44c	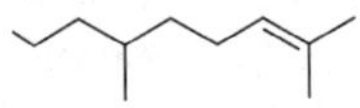	85	–170
44d	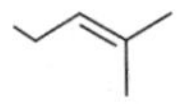	72	–164
44e	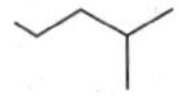	75	–166
44f	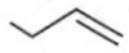	85	–163
44g		74	–169

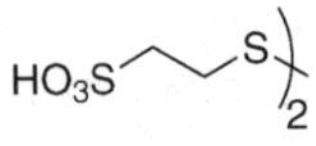

CoM–S–S–CoM (**45**)

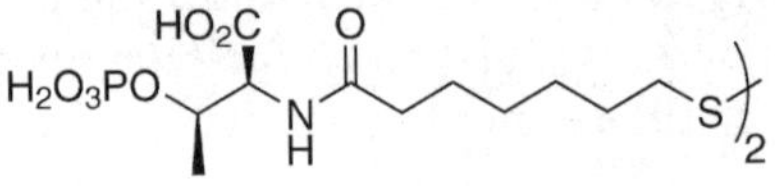

CoB–S–S–CoB (**46**)

and phosphate groups in **22** is indicated by the redox potentials of the homodisulfides CoM-S-S-CoM (**45**) ($E^{0\prime}$=–139±7 mV vs SHE) and CoB-S-S-CoB (**46**) ($E^{0\prime}$=–177±5 mV vs SHE) [50].

What should be emphasized is that the redox potentials measured for **10** and **22** allow for both the reduction of **10** to dihydro-**10** via $F_{420}H_2$ and H_2, and the oxidation of dihydro-**10** to **10** by **22**. This finding, supported by electrochemical experiments, also strongly corroborates the hypothesis that **10** plays a prominent role as an electron carrier in the electron transport system of methanogens.

6
New Biologically Active Phenazines from Bacteria

Since Turner and Messenger's review article [1] covering the occurrence, biochemistry, and physiology of natural phenazines a number of new representatives have been isolated from bacteria that have aroused particular interest either because of their biologic activity or their structural features. Many of these phenazines are 1,6-disubstituted. They include saphenic acid (**47**), saphenic acid methyl ether (**48**), saphenamycin (**49**), and several other esters (**50**) of saphenic acid [51–53]. As regards biologic activity, saphenamycin (**49**), which was isolated from several species including *Streptomyces canarius* [52] and *Streptomyces antibioticus* Tü 2706 [53], is the most interesting. Apart from its antimicrobial activity toward a broad range of bacteria, this compound also exhibits antitumor and antitrichomonal activity as well as larvicidal activity against mosquitoes.

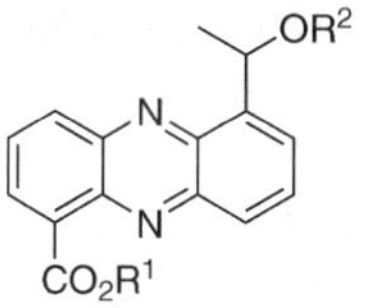

saphenic acid (**47**) $R^1 = R^2 = H$
saphenic acid methyl ether (**48**) $R^1 = H$; $R^2 = CH_3$
saphenamycin (**49**) $R^1 = H$; $R^2 =$ 2-hydroxy-6-methylbenzoyl
saphenyl esters (**50**) $R^1 = H$; $R^2 =$ various fatty acyl residues

The more recently discovered and most unusual structures include the dimeric phenazine derivatives esmeraldin A (**52**) and esmeraldin B (**53**), which are produced by *Streptomyces antibioticus* Tü 2706 together with **49** [54]. Esmeraldins exhibit no antibacterial activity but **53** is effective against tumor cells. Much effort has been directed to the elucidation of the biosynthesis of the esmeraldins [55]. Some rare L-quinovose esters (**55a–d**) of saphenic acid have

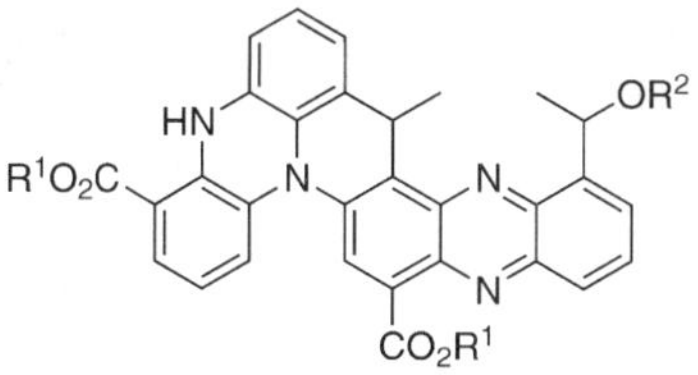

esmeraldic acid (**51**) $R^1 = R^2 = H$
esmeraldin A (**52**) $R^1 = H$; $R^2 =$ various fatty acyl residues
esmeraldin B (**53**) $R^1 = H$; $R^2 =$ 2-hydroxy-6-methylbenzoyl
esmeraldic acid dimethylesters I and II (**54**) $R^1 = Me$; $R^2 = H$
(2 diastereomers)

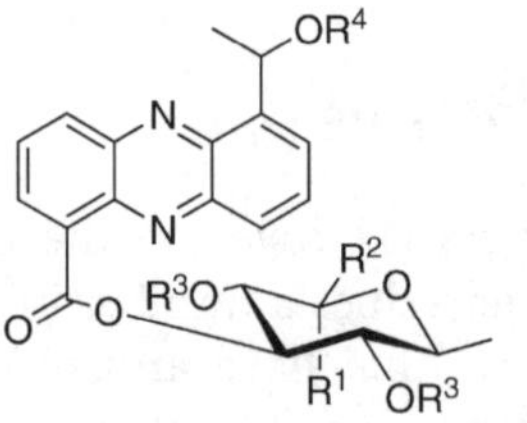

R^1= OH; R^2= R^3= R^4= H (**55a**)
R^1= R^3= R^4= H; R^2= OH (**55b**)

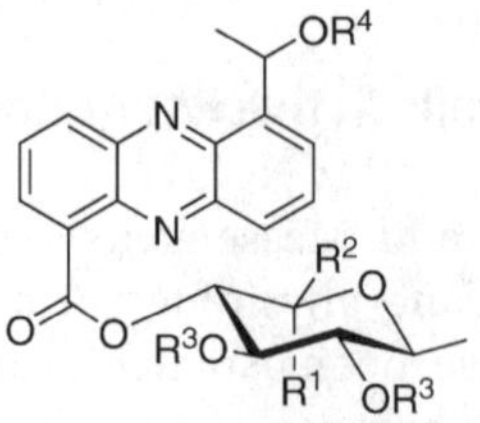

R^1= OH; R^2= R^3= R^4= H (**55c**)
R^1= R^3= R^4= H; R^2= OH (**55d**)

been obtained by Fenical et al. from a marine actinomycete. They exhibit moderate broad-spectrum activity against a number of gram-positive and gram-negative bacteria [56].

Much more exciting biologic properties have been revealed for the equally 1,6-disubstituted phenazoviridin (**56**) isolated from *Streptomyces* sp. HR04. The new free radical scavenger showed strong in vitro inhibitory activity against lipid peroxidation and displayed in vivo antihypoxic activity in mice [57].

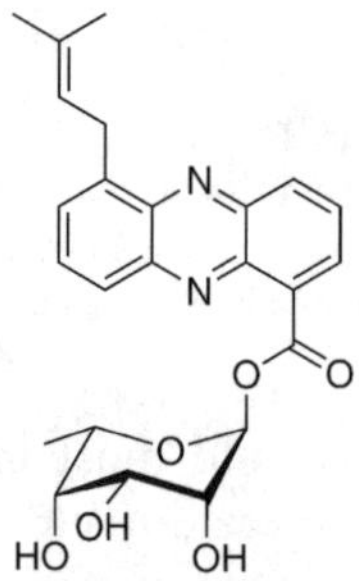

phenazoviridin (**56**)

Phencomycin (**57**) was recovered from *Streptomyces* sp HIL Y-9031725 [58]. A natural product that is structurally very similar was discovered by Laatsch et al. in a marine *Streptomyces* species, interestingly in its reduced form, i.e., the dihydrophencomycin methyl ester (**58**) [59]. Both show weak antibiotic activity.

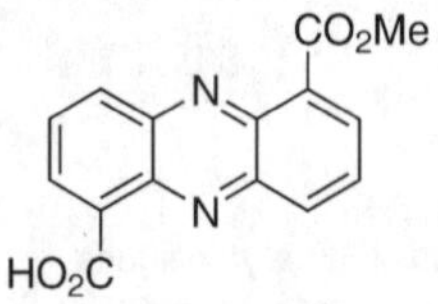

phencomycin (**57**)

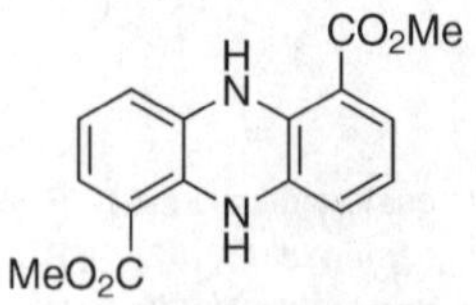

dihydrophencomycin methyl ester (**58**)

Some time ago pelagiomicin A (LL-14I352α) (**59**), B (**60**), C (**61**), and LL-14I352β (griseoluteic acid) (**62**) were found in marine bacteria [60, 61]. Pelagiomicin A (**59**) is not only effective against gram-positive and gram-negative

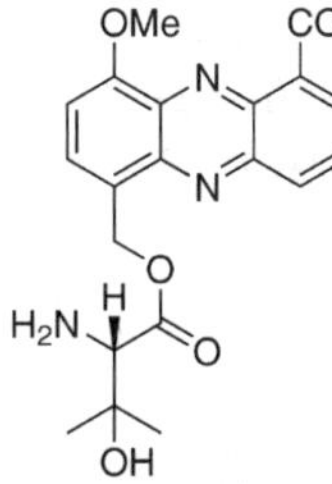

pelagiomicin A (**59**)
(LL-14I352α)

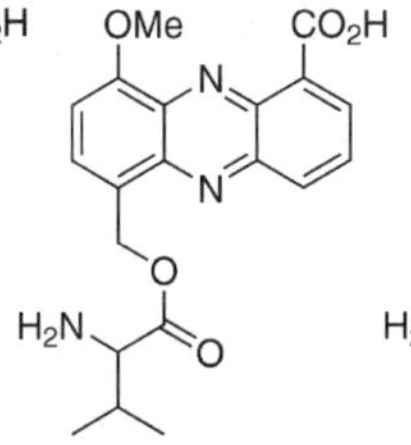

pelagiomicin B (**60**)

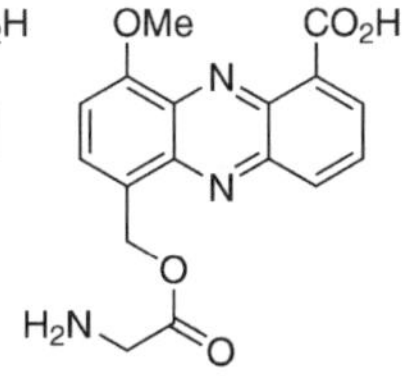

pelagiomicin C (**61**)

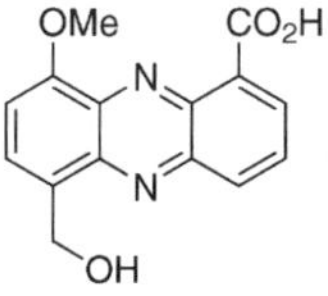

LL-14I352β (**62**)

bacteria, but also possesses in vitro and in vivo antitumor activity. Unusual properties are revealed by tri- and tetrasubstituted phenazines SB 212021 (**63**) and SB 212305 (**64**), inhibiting the zinc-dependent metallo-β-lactamases of several bacteria by chelation of the active site metal ion [62].

Phenazines divide into a larger group without and a smaller group with an *N*-alkyl substituent. The simplest compound of the latter group is the long-known pyocyanine (**1**) that has an *N*-methyl group. Also, other phenazines have been isolated over recent years, some of which have more complex terpenoid side chains. Some possess remarkable biologic activities. This group comprises lavanducyanin or WS-9659 A (**65**) and WS-9659 B (**66**) which have been produced by different species of *Streptomyces* [63, 64]. Compound **65** was shown

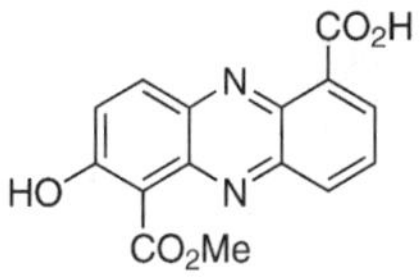

SB 212021 (**63**)

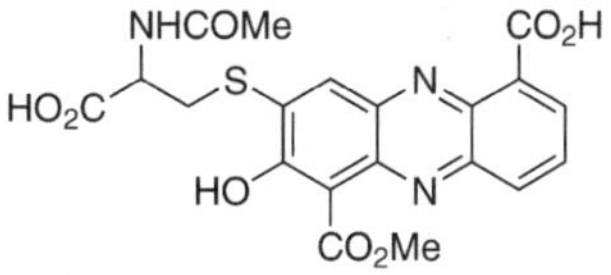

SB 212305 (**64**)

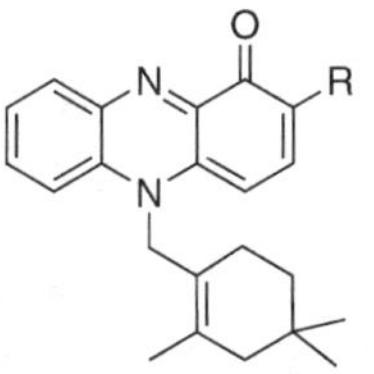

lavanducyanin (WS-9659 A) (**65**) R = H
WS-9659 B (**66**) R = Cl

to not only display interesting antitumor activity [63] but also enzyme (testosterone 5α-reductase)-inhibiting activity [64–66]. In addition, experiments with several mammalian cell lines have demonstrated that **65** represents a new type of cell growth stimulating substance with low molecular weight [67, 68].

A compound of a very similar structure is phenazinomycin (**67**), which has been isolated from mycelial extracts of *Streptomyces* sp. WK-2057 by Omura et al. [69]. This compound possesses in vivo antitumor activity against experimental murine tumor cells and cytotoxic activity against adriamycin-resistant

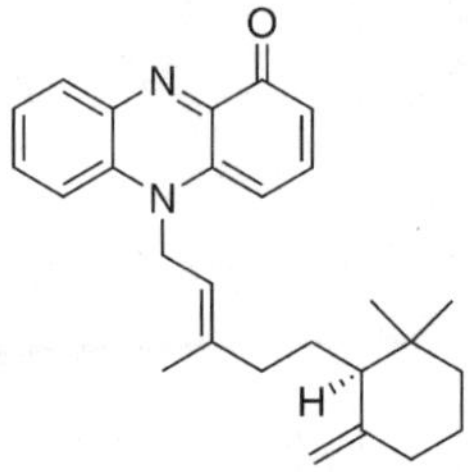

phenazinomycin (**67**)

P388 leukemia cells [69, 70]. Benthocyanins A (**68**), B (**69**), and C (**70**) from *Streptomyces prunicolor* are new potent radical scavengers, and these compounds are more effective than vitamin E in preventing, alleviating or overcoming a variety of diseases that have been proven to be caused by oxygen-derived free radicals [71]. Compounds **68** and **69** have a unique structure that is characterized by a highly conjugated furophenazine system and an *N*-geranyl

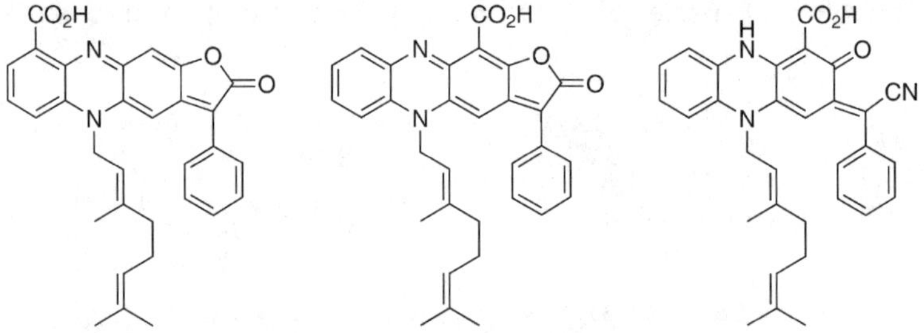

benthocyanin A (**68**) benthocyanin B (**69**) benthocyanin C (**70**)

substituent. The neuroprotective effect of aestivophoenins A (**71**), B (**72**), and C (**73**) produced by *Streptomyces purpeofuscus* 2887-SVS2 seems to be dependent on its antioxidative activity [72].

As part of a screening program for bioactive compounds from endosymbiontic microorganisms, Zeeck et al. successfully isolated a group of new, structurally related phenazine antibiotics, the so called endophenazines A (**74**), B (**75**), C (**76**), and D (**77**) from the arthropod-associated endosymbiont *Strepto-*

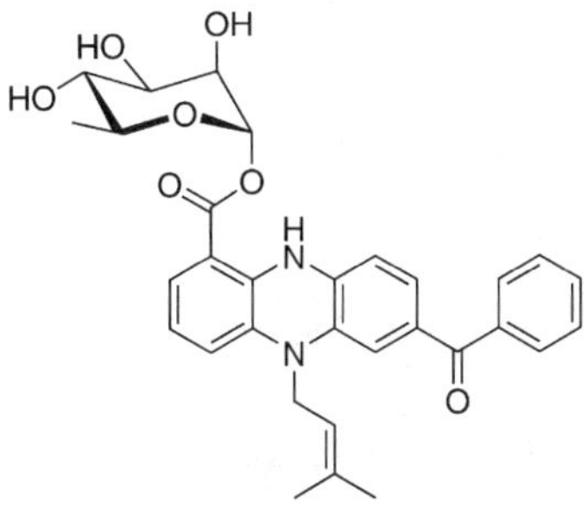

aestivophoenin A (**71**)

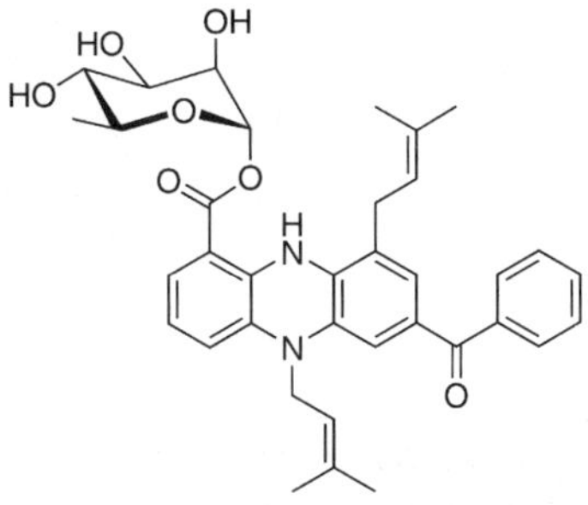

aestivophoenin B (**72**)

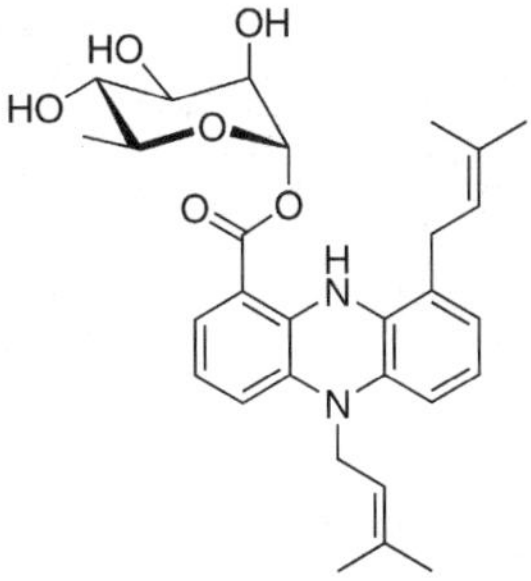

aestivophoenin C (**73**)

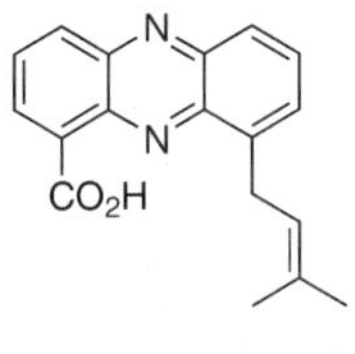

endophenazine A (**74**)

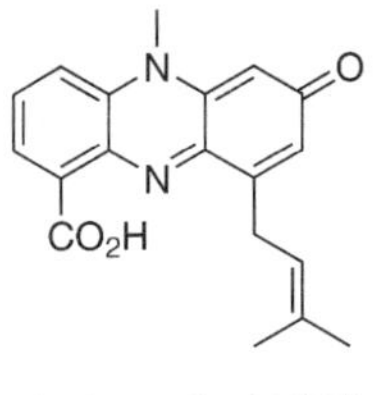

endophenazine B (**75**)

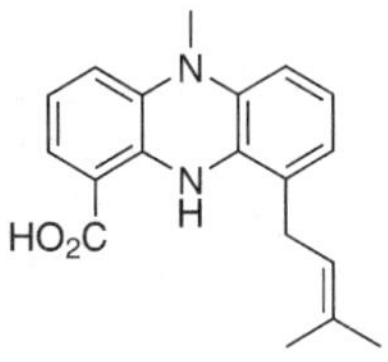

endophenazine C (**76**)

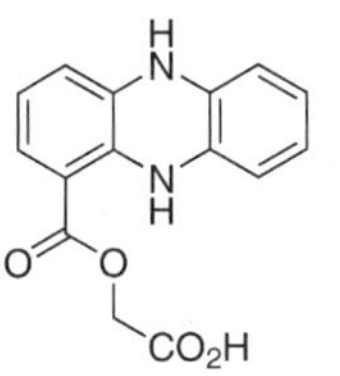

endophenazine D (**77**)

myces anulatus [73]. Also, structurally new types of compounds that are characterized by two phenazine substructural units have been isolated and their structures elucidated: diphenazithionin (**78**), an inhibitor of lipid peroxidation from *Streptomyces griseus* ISP 5236 [74], and phenazostatins A (**79**), B (**80**), and

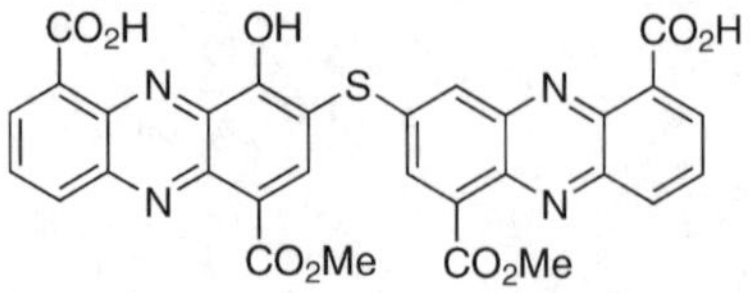

diphenazithionin (**78**)

C (**81**) isolated from *Streptomyces* sp. 833 [75–77] as neuronal cell protecting substances. The stereochemistry of these compounds remains to be established.

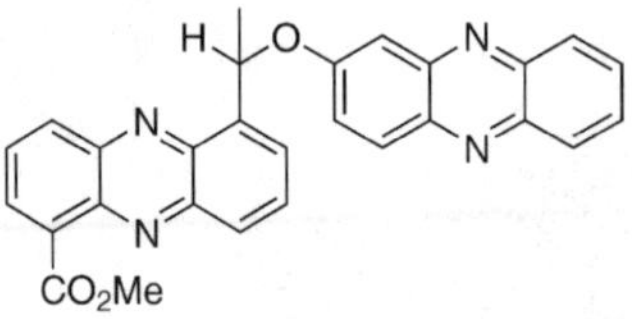

phenazostatin A (**79**)

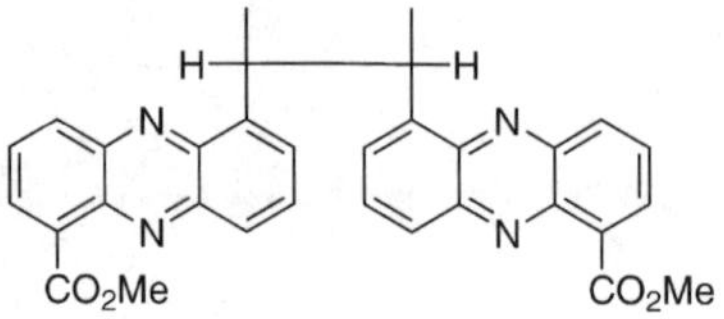

phenazostatin B (**80**)

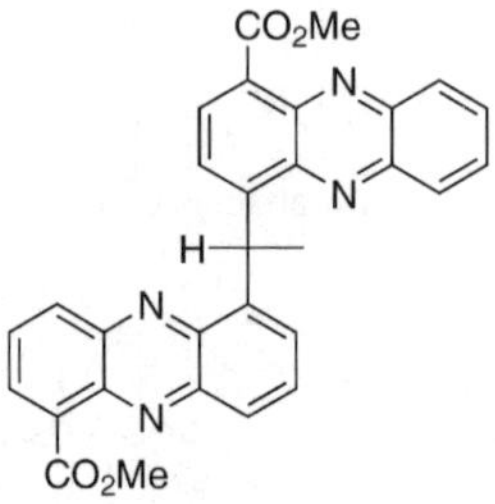

phenazostatin C (**81**)

7 Synthesis of Naturally Occurring Phenazines

In the synthesis of methanophenazine (**10**) and its analogs, the reliable preparation of 2-hydroxyphenazine (**9**) in larger amounts was least expected to pose any problems. But even though **9** was produced by Beirut reaction in good yields, we dealt with the problems inherent in the selective preparation of singly and multiply substituted phenazines in greater detail. The remainder is dedicated to introducing the general methods for the construction of phenazines and to discussing current examples of natural product synthesis. In addition to the more traditional methods, Pd-catalyzed reactions will be addressed in particular because they not only serve to prepare diphenylamines, but are also suitable for the cyclization of substituted diphenylamines to give phenazines. And finally, a new Pd-catalyzed one-pot synthesis of phenazines will be presented.

7.1 General Strategies

In the majority of cases, the most important step in the synthesis of naturally occurring phenazines involves the generation of their basic skeleton by constructing the central heterocyclic ring, for which, apart from the reaction of phenylenediamines with quinoids and the utilization of the Beirut reaction, the cyclization of substituted diphenylamines remains of utmost importance [78]. Mostly, the diphenylamines required are made accessible via nucleophilic aromatic substitution. Amazingly, the Pd-catalyzed arylation of anilines discovered by Buchwald and Hartwig [79] has so far been of minor relevance to the generation of the phenazine skeleton. In the near future, however, this situation can be expected to change dramatically. Syntheses of phenazines that are based on the modification of the phenazine skeleton have never assumed any greater significance. The reason is that phenazines, because of their deactivation due to the electron-withdrawing effects of the two nitrogen atoms, do not easily undergo electrophilic aromatic substitution reactions [78].

7.2 Classical Routes

A more direct, one-step approach to phenazines **84** is possible via condensation of *o*-phenylenediamines **82** with *o*-benzoquinones **83** that can be generated in situ from oxidation of the specific catechols. As reactions of substituted *o*-phenylenediamines with substituted *o*-benzoquinones inevitably suffer from regioselectivity problems, this method is bound to have a limited range of

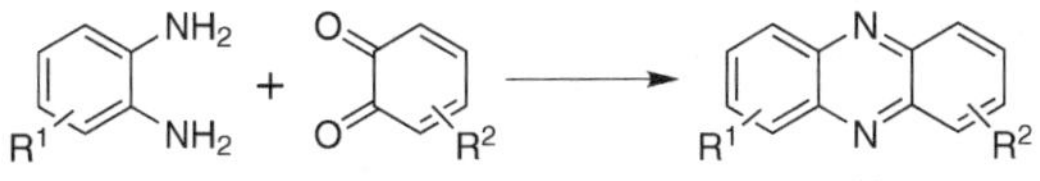

82 **83** **84**

applicability [78]. As expected, the reaction of 1,2-phenylenediamine (**34**) with 2-hydroxy-1,4-benzoquinone (**33**) or 1,4-benzoquinone (**35**) that are generated in situ gives the 2-hydroxyphenazine (**9**) the synthesis of methanophenazine (**10**). But as the yields of pure product in our own experiments were too low, following this approach any further did not seem worthwhile.

Another one-step method of similarly limited value for the synthesis of multiply substituted phenazines **84** is the Beirut reaction [80]. It involves the transformation of the easily accessible benzofuroxans **85** together with, in

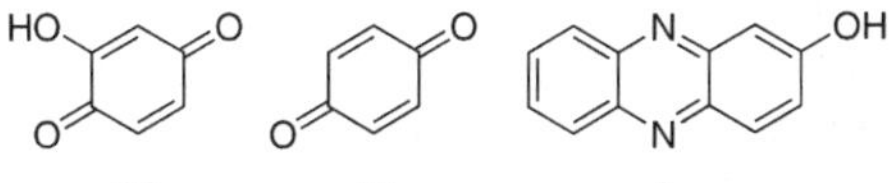

33 **35** **9**

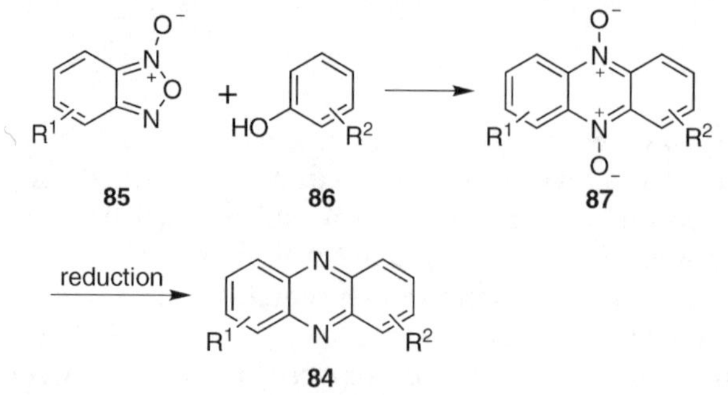

particular, phenols **86** to give the respective phenazine 5,10-dioxides **87** in a single step. The di-*N*-oxides can easily be reduced to the phenazines **84**.

Most syntheses of naturally occurring phenazines, though, are based on a two-step elaboration of the central heterocycle of the phenazine [78]. The first key step involves the generation of *ortho*-monosubstituted **88** or *ortho*, *ortho'*-disubstituted diphenylamines **89–91** via nucleophilic aromatic substitution. Ring formation is then achieved by means of reductive or oxidative cyclization, for which a number of efficient methods are available. The main flaw of this approach is the synthesis of the substituted diphenylamines via nucleophilic aromatic substitution, as this reaction often can only be performed under strongly basic reaction conditions and at high temperatures. In addition, the diphenylamines required may only be achieved with certain substitution patterns with high yields.

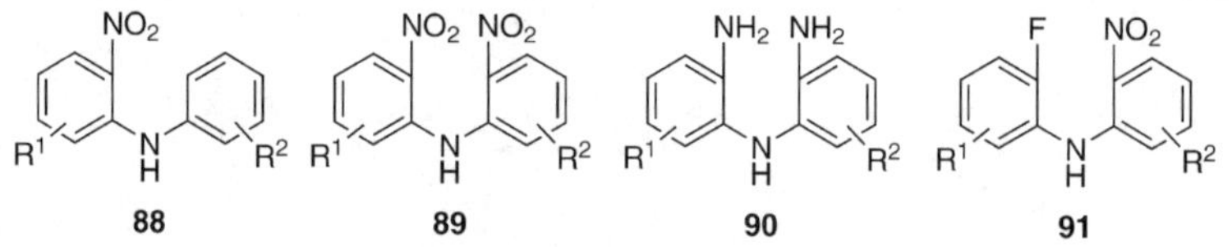

An example of the method described is the synthesis of saphenic acid (**47**) that has recently been reported by Nielsen et al. [81]. Starting from properly substituted aromatic precursors **92** and **93**, the naturally occurring 1,6-disubstituted phenazine was synthesized in racemic form. Here, the first major step involves an intermolecular nucleophilic aromatic substitution that, due to the substitution pattern, has proved to be relatively unproblematic and after hydrolysis of the acetal yields the *o*-nitrodiphenylamine **94**. Much more difficult is the ring formation leading to the final phenazine, which can best be achieved through a high excess of $NaBH_4$, accompanied by reduction of the methyl ketone. But at 32%, the yield is still rather poor.

So the saphenic acid (**47**), which has only weak biologic activity, was available as a template for the preparation of both saphenamycin (**49**) and saphenamycin analogs [82]. Reaction of the allylic ester **95** of the saphenic acid with the acid

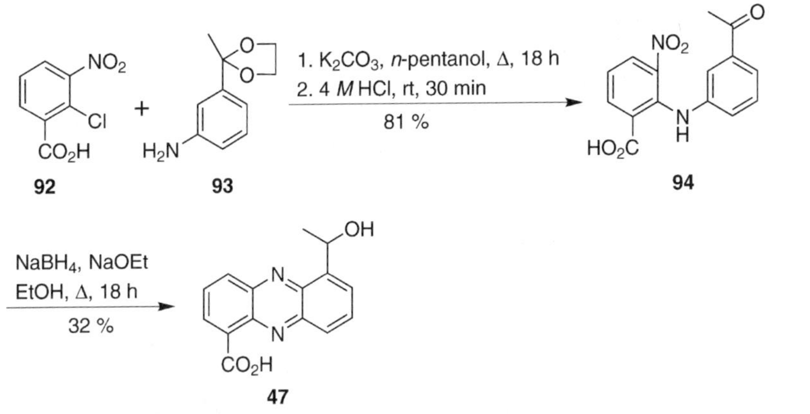

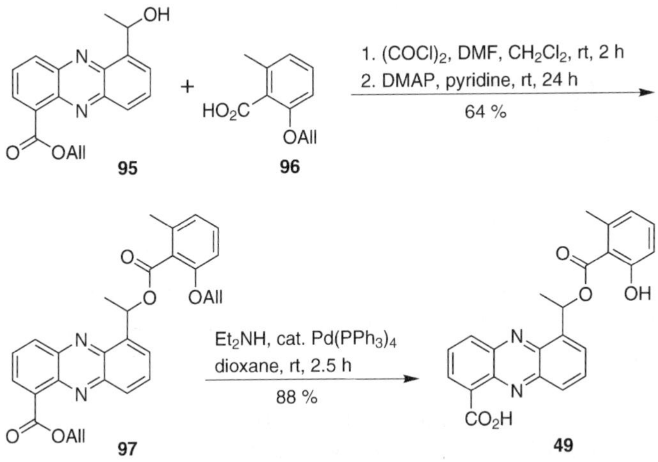

chloride of 6-methyl salicylic acid **96** generated in situ produced the ester **97** in 64% yield. The deprotection gave **49** in racemic form. The same reaction sequence could be performed with enantiomerically pure **47** that was produced in the traditional way by separation of the respective (–)-brucine diastereomeric salts. Interesting to note: no significant differences in the antibiotic activity of the enantiomers of **49** could be observed. Using a solid support

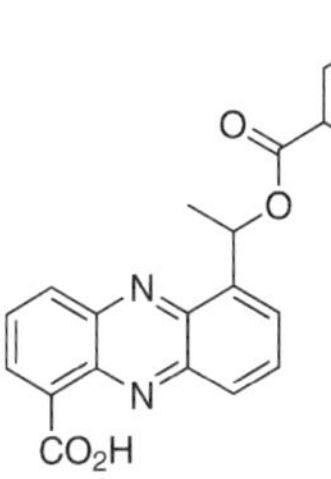

98

approach, Nielsen et al. have also synthesized a small library of saphenamycin analogs **98** modified at the benzoate moiety [83].

Quite recently, the structure of pelagiomicins B (**60**) and C (**61**) was confirmed by a synthesis starting from griseoluteic acid (**62**) [60]. In the first synthesis of **62**, Holliman et al. employed the reductive cyclization of *o*-nitrodiphenylamines using $NaBH_4$ to yield phenazines [84]. To this end, 3-amino-4-methoxybenzyl alcohol (**99**) was treated with methyl 2-bromo-3-nitrobenzoate (**100**) to yield the *o*-nitrodiphenylamine **101**, which by reductive cyclization

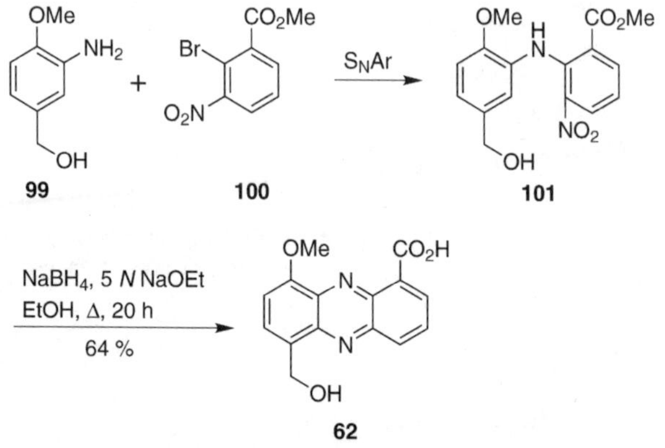

gave griseoluteic acid (**62**) in 64% yield. For the synthesis of **60** and **61** the griseoluteic acid derivative **102** was reacted with *N*-*t*-BOC-L-Val **103** to yield **104**, which in turn was deprotected to give **60**. The structure of **61** was established in a similar way [60].

Kitahara et al. decided on a classical approach for generating the benthocyanin A skeleton [85]. Following Hollimann [86] they started their synthesis with the one-step preparation of the phenazine **106** by reaction of **100** and

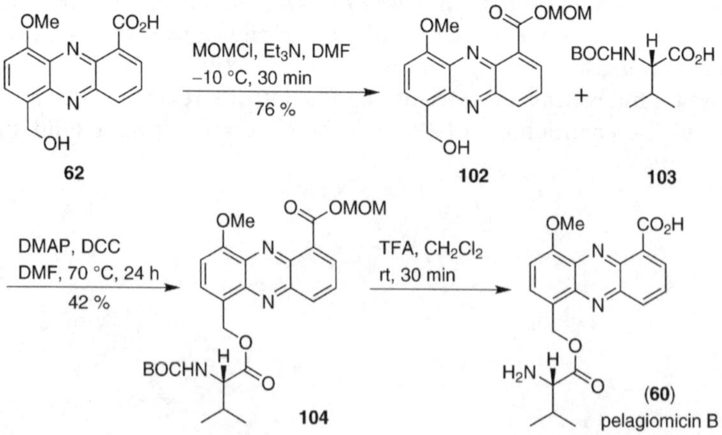

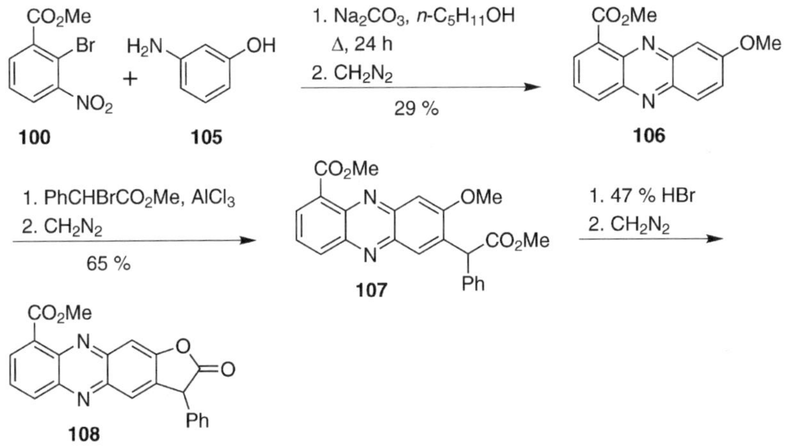

105 – with a yield of 29% only. It is assumed that reaction of the electron-poor methyl 2-bromo-3-nitrobenzoate (**100**) and *m*-aminophenol (**105**) first gives the product of the nucleophilic aromatic substitution, which is then cyclized (with the phenolate anion being probably involved) to provide the respective phenazine-*N*-oxide. At last, this directly leads to the phenazine **106** after disproportioning and methylation. Friedel–Crafts alkylation with methyl phenyl bromoacetate and CH_2N_2 treatment produces the diester **107**, which after hydrolysis with HBr provides the basic skeleton **108** of benthocyanin A. It is true that the completion of the total synthesis of benthocyanin A now only requires the introduction of the terpenoid component at the ring nitrogen, but it is well known that such *N*-allylations of phenazines cannot be forced that easily and with low yields only.

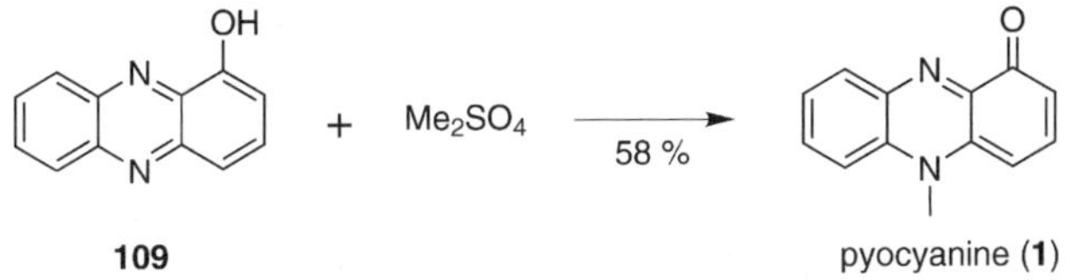

The same problem occurs with the synthesis of lavanducyanin (**65**) [85, 87] and phenazinomycin (**67**) [88]. Here, Kitahara et al. used the same approach in that they first generated the phenazine skeleton and performed the *N*-allylation as the last step. While the *N*-methylation of the unprotected 1-hydroxyphenazine (**109**) proceeds without any major problems and provides pyocyanin (**1**) in acceptable yields [89], model experiments already indicate that the allylation of **110** could not be so readily accomplished. As expected, the yields were poor. The only way to bring about the synthesis of **65** is under high-pressure conditions by reaction of **110** with **111**, and even so the yield of the natural product remains quite low. Similar problems are encountered with the synthesis of **67** by allylation of **110** with the terpenoid component **112**.

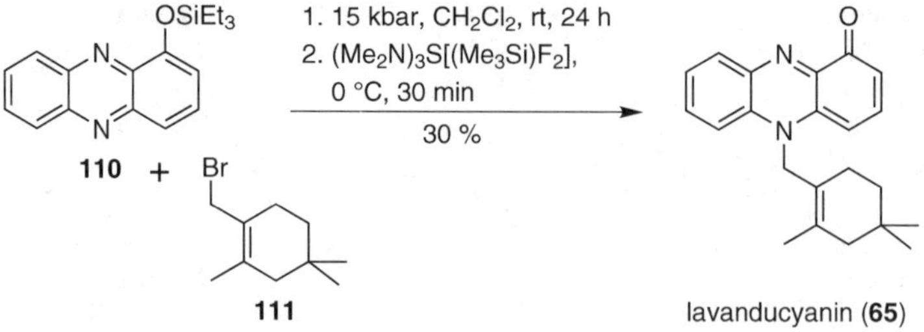

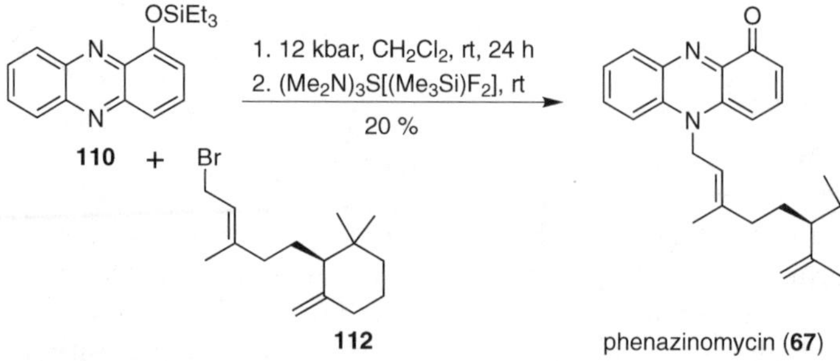

7.3
Diphenylamines via Pd-catalyzed *N*-arylation

The access to substituted diphenylamines has been significantly improved through the development of the Pd-catalyzed *N*-arylation of anilines by Buchwald and Hartwig. The cyclization of the substituted diphenylamines to give the corresponding phenazines may then be conducted according to standard methods [78].

For instance, in the context of our investigations of the synthesis of 2-hydroxyphenazine (**9**), we have succeeded in generating the *o,o′*-dinitrodiphenyl-

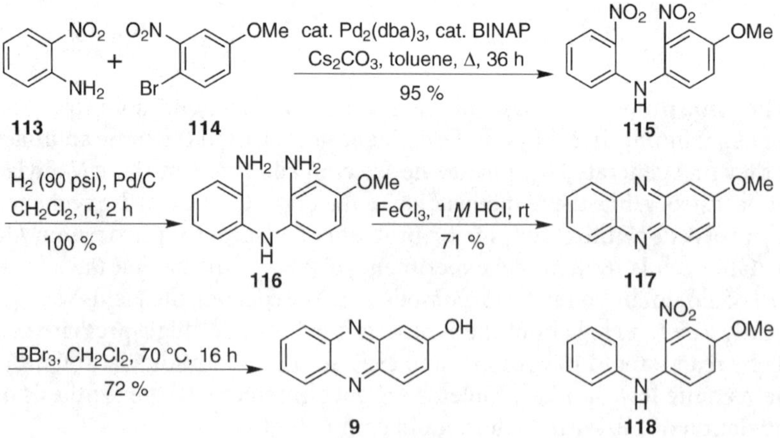

amine **115** by Pd-catalyzed arylation from **113** and **114** with high yields [90]. The reduction of the two nitro groups then gave the *o,o′*-diaminodiphenylamine **116**, which cyclized to provide the phenazine **117** by $FeCl_3$-mediated reaction. In a similar way, we were able to produce and convert the 2-nitrodiphenylamine **118** with $NaBH_4$ into the 2-methoxyphenazine (**117**).

7.4 Intramolecular Pd-catalyzed *N*-arylation in Phenazine Synthesis

Needless to say, the Buchwald–Hartwig reaction can also be usefully employed in ways other than the efficient preparation of diphenylamines. Given the respective substitution, it should be possible to bring about the phenazine skeleton by Pd-catalyzed ring formation as well. There are two ways to proceed: either the substituent pattern required by the intramolecular Buchwald–Hartwig reaction is elaborated after the formation of the diphenylamine (**121**→**124**), or the starting material already contains the substituents necessary for the two *N*-arylations. A reasonable starting point is the intermolecular *N*-arylation of an *o*-haloaniline

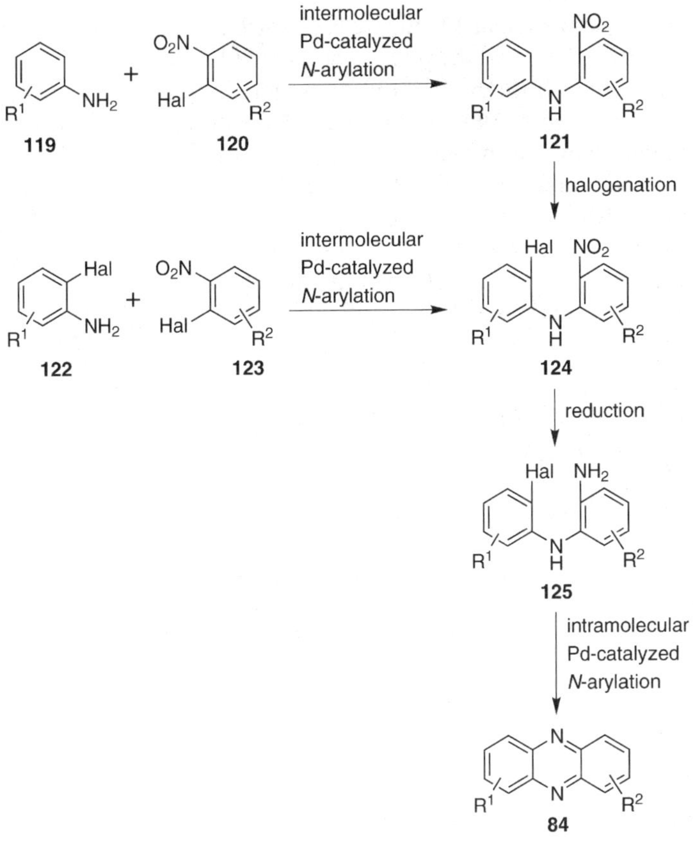

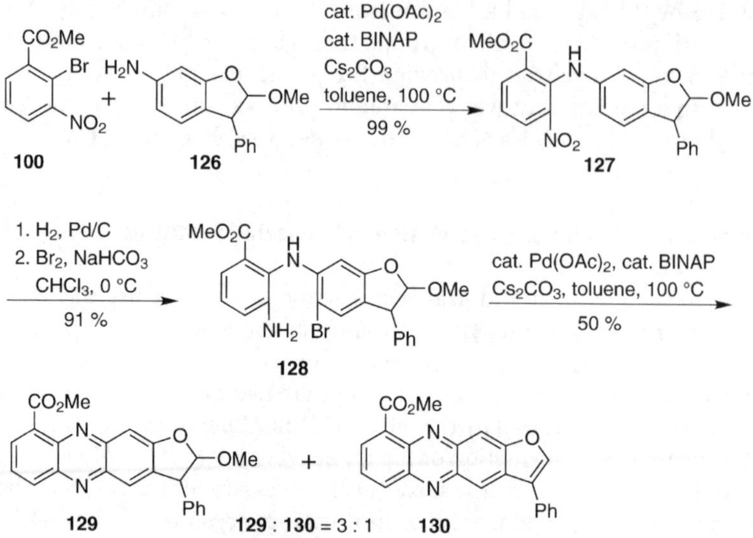

122 with an *o*-halonitroarene **123** to give **124**, which is then further reduced to **125** and finally cyclized to yield the phenazine **84** by Pd catalysis.

Kamikawa et al. chose the first option to generate the benthocyanin skeleton [91]. To begin with, **100** and aniline **126** are transformed into *o*-nitrodiphenylamine **127** by intermolecular *N*-arylation. Reduction of the nitro group and selective bromination produces **128**, and this time an intramolecular Buchwald–Hartwig reaction is used to derive a mixture of the desired phenazine **129** and the elimination product **130**. The fundamental problem with this approach relates to the selective introduction of the halogen substituent that is required for the intramolecular *N*-arylation.

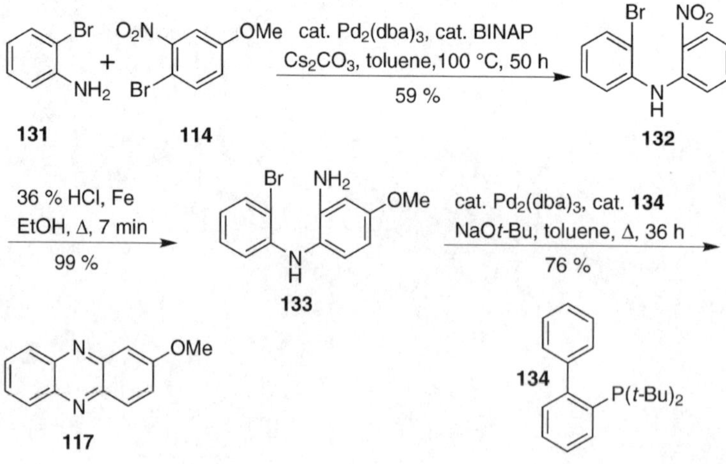

That the second option can also be successfully used has recently been revealed by our synthesis of 2-methoxyphenazine (**117**) [90]. The reduction of *o*-bromo-*o'*-nitrodiphenylamine **132** accessible via intermolecular Pd-catalyzed *N*-arylation provides the *o*-amino-*o'*-bromodiphenylamine **133**, which can then be cyclized to give **117** in a Pd-catalyzed intramolecular *N*-arylation by employing $Pd_2(dba)_3$ as the Pd complex and **134** as the phosphine ligand. It should be noted that the outcome of both the intermolecular and the intramolecular *N*-arylations heavily depends on the appropriate choice of the Pd complex as well as the phosphine. Ether cleavage leads to 2-hydroxyphenazine (**9**).

7.5 Sequential Inter-/Intramolecular Pd-catalyzed *N*-arylations

Even more attractive is the prospect of preparing phenazines in a single synthetic operation by sequential inter-/intramolecular *N*-arylation. This would require the transformation of either two *o*-haloanilines **122** and **135** or of an *o*-phenylenediamine **82** with an *o*-dihalobenzene **136**. Until recently, not even a single example could be found for this one-step synthesis of heterocycles via double Pd-catalyzed *N*-arylation. Numerous experiments conducted in a variety of different combinations of various Pd catalysts, bases, and phosphine ligands in our laboratory failed to realize this novel synthetic principle.

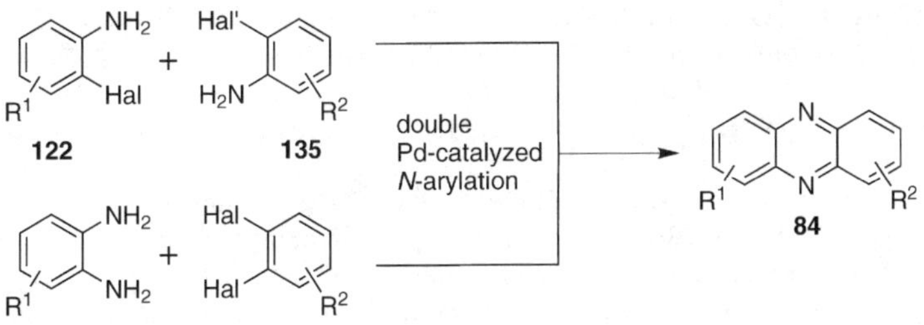

122 **135** **82** **136** **84**

Some breakthrough was finally achieved by using $Pd_2(dba)_3$ as the Pd complex, tris(*tert*-butyl)phosphine as the ligand, and sodium *tert*-butoxide as the base [90]. This combination of reagents proved to bring about the synthesis of unsubstituted phenazine (**137**) by reaction of two molecules of 2-bromoaniline (**131**). Remarkably, no reaction takes place under the conditions of the **133**→**117** transformation. Currently the scope and limitations of this new sequential inter-/intramolecular *N*-arylation for the synthesis of *N*-heterocycles

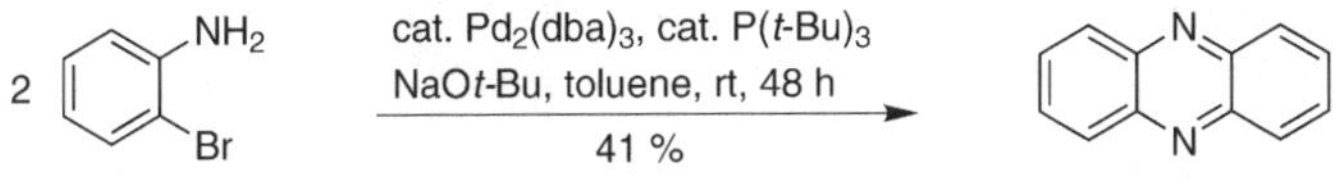

131 **137**

are being investigated. To start with, both the optimization of reaction conditions and the range of applicability of the phenazine synthesis will have to be addressed. Later we will examine whether this reaction proves suitable for generating unsymmetrical phenazines, e.g., by employing solid-phase methodology, and which other heterocycles may be generated by means of sequential inter-/intramolecular *N*-arylations.

References

1. Turner JM, Messenger AJ (1986) Adv Microb Physiol 27:211
2. Ingram JM, Blackwood AC (1970) Adv Appl Microbiol 13:267
3. Fordos J (1859) Recl Trav Soc Emulation Siences Pharm 3:30
4. Guignard L, Sauvageau M (1894) C R Seances Soc Biol 46:841
5. Clemo GR, McIlwain H (1938) J Chem Soc 479
6. Hollstein U, Van Gemert RJ Jr (1971) Biochemistry 10:497; Hollstein U, Butler PL (1972) Biochemistry 11:1345; Hassan HM, Fridovich I (1980) J Bacteriol 141:156
7. Yamanaka S (1972) J Chiba Med Soc 48:63 [Chem. Abstr. (1972) 162986]
8. Isono K, Anzai K, Suzuki S (1958) J Antibiot A 11:264
9. Gräfe U (1992) Biochemie der Antibiotika, Spektrum, Heidelberg, p 66f
10. Thomashow LS, Weller DM (1988) J Bacteriol 170:3499; Mazzola M, Cook RJ, Thomashow LS, Weller DM, Pierson III LS (1992) Appl Environ Microbiol 58:2616; Pierson LS III, Pierson EA (1996) FEMS Microbiol Lett 136:101
11. Friedmann HC, Klein A, Thauer RK (1990) FEMS Microbiol Rev 87:339
12. (a) Thauer RK (1998) Microbiology 144:2377; (b) Cammack R (1997) Nature 390:443
13. Woese CR, Kandler O, Wheelis ML (1990) Proc Natl Acad Sci USA 87:4576; Brown JR, Doolittle WF (1997) Mikrobiol Mol Biol Rev 61:456
14. Kandler O, Hippe H (1977) Arch Microbiol 113:57; Kandler O, König H (1978) Arch Microbiol 118:141
15. Tornabene TG, Langworthy TA (1979) Science 203:51; Gambacorta A, Trincone A, Nicolaus B, Lama L, De Rosa M (1994) Syst Appl Microbiol 16:518
16. Zillig W, Stetter KO, Schnabel R, Madon J, Gierl A (1982) Zbl Bakt Hyg 1 Abt Orig C3:218
17. Boone DR, Whitman WB, Rouvière P (1993) Diversity and taxonomy of methanogens. In: Ferry JG (ed) Methanogenesis. Chapman & Hall, New York, p 35
18. Deppenmeier U, Blaut M, Gottschalk G (1989) Eur J Biochem 186:317
19. Wolfe RS (1985) Trends Biochem Sci 10:396; DiMarco AA, Bobik TA, Wolfe RS (1990) Annu Rev Biochem 59:355
20. Deppenmeier U, Blaut M, Gottschalk G (1991) Arch Microbiol 155:272
21. Harms U, Thauer RK (1996) Eur J Biochem 235:653
22. Mahlmann A, Deppenmeier U, Gottschalk G (1989) FEMS Microbiol Lett 61:115; Harms U, Thauer RK (1996) Eur J Biochem 241:149
23. Ma K, Thauer RK (1990) FEMS Microbiol Lett 70:119; te Brömmelstroet BWJ, Geerts WJ, Keltjens JT, van der Drift C, Vogels GD (1991) Biochim Biophys Acta 1079:293
24. Deppenmeier U, Lienard T, Gottschalk G (1999) FEBS Lett 457:291
25. Abken HJ, Deppenmeier U (1997) FEMS Microbiol Lett 154:231
26. Heiden S, Hedderich R, Setzke E, Thauer RK (1994) Eur J Biochem 221:855; Bäumer S, Ide T, Jacobi C, Johann A, Gottschalk G, Deppenmeier U (2000) J Biol Chem 275: 17968
27. Deppenmeier U, Blaut M, Schmidt B, Gottschalk G (1992) Arch Microbiol 157:505

28. Abken H-J, Tietze M, Brodersen J, Bäumer S, Beifuss U, Deppenmeier U (1998) J Bacteriol 180:2027
29. Beifuss U, Tietze M, Bäumer S, Deppenmeier U (2000) Angew Chem Int Ed 39:2470
30. Willstätter R, Müller F (1911) Ber Dtsch Chem Ges 44:2171
31. Thiele J (1898) Ber Dtsch Chem Ges 1:1247
32. Kehrmann F, Cherpillod F (1924) Helv Chim Acta 7:973
33. Ott R (1959) Monatsh Chem 90:827
34. Ley K, Seng F, Eholzer U, Nast R, Schubart R (1969) Angew Chem Int Ed 8:596
35. Negishi E, King AO, Klima WL (1980) J Org Chem 45:2526
36. Ley SV, Armstrong A, Diez-Martin D, Ford MJ, Grice P, Knight JG, Kolb HC, Madin A, Marby CA, Mukherjee S, Shaw AN, Slawin AMZ, Vile S, White AD, Williams DJ, Woods M (1991) J Chem Soc Perkin Trans I 667; Negishi E, Takahashi T (1988) Synthesis 1; Negishi E, Van Horn DE, King AO, Okukado N (1979) Synthesis 501; Van Horn DE, Negishi E (1978) J Am Chem Soc 100:2252
37. McDougal PG, Rico JG, Oh Y-I, Condon BD (1986) J Org Chem 51:3388
38. Smith AB III, Qiu Y, Jones DR, Kobayashi K (1995) J Am Chem Soc 117:12011; Negishi E, Liu F (1998) Palladium- or nickel-catalyzed cross-coupling with organometals containing zinc, magnesium, aluminium, and zirconium. In: Diederich F, Stang PJ (eds) Metal-catalyzed cross-coupling reactions. Wiley-VCH, Weinheim, p 1
39. Loupy A, Sansoulet J, Vaziri-Zand F (1987) Bull Soc Chim Fr 1027
40. Beifuss U, Tietze M (2000) Tetrahedron Lett 41:9759
41. Mohr P, Tori M, Grossen P, Herold P, Tamm C (1982) Helv Chim Acta 65:1412; Herold P, Mohr P, Tamm C (1983) Helv Chim Acta 66:744
42. Brown HC, Kim SC (1977) Synthesis 635; Brown HC, Kim SC, Krishnamurthy S (1980) J Org Chem 45:1
43. Chen L-Y, Zaks A, Chackalamannil S, Dugar S (1996) J Org Chem 61:8341
44. Bäumer S, Murakami E, Brodersen J, Gottschalk G, Ragsdale SW, Deppenmeier U (1998) FEBS Lett 428:295
45. Tietze M, Beuchle A, Lamla I, Orth N, Dehler M, Greiner G, Beifuss U (2003) Chembiochem 4:333
46. Walsh C (1986) Acc Chem Res 19:216
47. Mann S (1970) Arch Mikrobiol 71:304
48. (a) Heyrovský M, Mader P, Veselá V, Fedurco M (1994) J Electroanal Chem 369:53; (b) Vavřička S, Heyrovský M (1994) J Electroanal Chem 375:371
49. Noll KM, Donnelly MI, Wolfe RS (1987) J Biol Chem 262:513; Ellermann J, Hedderich R, Böcher R, Thauer RK (1988) Eur J Biochem 172:669
50. Hauska G (1988) Trends Biochem Sci 13:415; Thauer RK, Jungermann K, Decker K (1977) Bacteriol Rev 41:100
51. Michel KH, Hoehn MM (1982) US Patent 4,316,959
52. Kitahara M, Nakamura H, Matsuda Y, Hamada M, Naganawa H, Maeda K, Umezawa H, Iitaka Y (1982) J Antibiot 35:1412
53. Geiger A, Keller-Schierlein W, Brandl M, Zähner H (1988) J Antibiot 41:1542
54. Keller-Schierlein W, Geiger A, Zähner H, Brandl M (1988) Helv Chim Acta 71:2058
55. Van't Land CW, Mocek U, Floss HG (1993) J Org Chem 58:6576; McDonald M, Wilkinson B, Van't Land CW, Mocek U, Lee S, Floss HG (1999) J Am Chem Soc 121:5619
56. Pathirana C, Jensen PR, Dwight R, Fenical W (1992) J Org Chem 57:740
57. Kato S, Shindo K, Yamagishi Y, Matsuoka M, Kawai H, Mochizuki J (1993) J Antibiot 46:1485
58. Chatterjee S, Vijayakumar EKS, Franco CMM, Maurya R, Blumbach J, Ganguli BN (1995) J Antibiot 48:1353

59. Pusecker K, Laatsch H, Helmke E, Weyland H (1997) J Antibiot 50:479
60. Imamura N, Nishijima M, Takadera T, Adachi K, Sakai M, Sano H (1997) J Antibiot 50:8
61. Singh MP, Menendez AT, Petersen PJ, Ding W-D, Maiese WM, Greenstein M (1997) J Antibiot 50:785
62. Gilpin ML, Fulston M, Payne D, Cramp R, Hood I (1995) J Antibiot 48:1081
63. Imai S, Furihata K, Hayakawa Y, Noguchi T, Seto H (1989) J Antibiot 42:1196
64. Nakayama O, Yagi M, Tanaka M, Kiyoto S, Okuhara M, Kohsaka M (1989) J Antibiot 42:1221
65. Nakayama O, Shigematsu N, Katayama A, Takase S, Kiyoto S, Hashimoto M, Kohsaka M (1989) J Antibiot 42:1230
66. Nakayama O, Arakawa H, Yagi M, Tanaka M, Kiyoto S, Okuhara M, Kohsaka M (1989) J Antibiot 42:1235
67. Matsumoto M, Seto H (1991) J Antibiot 44:1471
68. Imai S, Noguchi T, Seto H (1993) J Antibiot 46:1232
69. Ōmura S, Eda S, Funayama S, Komiyama K, Takahashi Y, Woodruff HB (1989) J Antibiot 42:1037
70. Funayama S, Eda S, Komiyama K, Ōmura S (1989) Tetrahedron Lett 30:3151
71. Shin-ya K, Furihata K, Hayakawa Y, Seto H (1991) Tetrahedron Lett 32:943; Shin-ya K, Furihata K, Teshima Y, Hayakawa Y, Seto H (1993) J Org Chem 58:4170
72. Kunigami T, Shin-ya K, Furihata K, Furihata K, Hayakawa Y, Seto H (1998) J Antibiot 51:880; Shin-ya K, Shimizu S, Kunigami T, Furihata K, Hayakawa Y, Seto H (1995) J Antibiot 48:1378
73. Gebhardt K, Schimana J, Krastel P, Dettner K, Rheinheimer J, Zeeck A, Fiedler H-P (2002) J Antibiot 55:794; Krastel P, Zeeck A, Gebhardt K, Fiedler H-P, Rheinheimer J (2002) J Antibiot 55:801
74. Hosoya Y, Adachi H, Nakamura H, Nishimura Y, Naganawa H, Okami Y, Takeuchi T (1996) Tetrahedron Lett 37:9227
75. Yun B-S, Ryoo I-J, Kim W-G, Kim J-P, Koshino H, Seto H, Yoo I-D (1996) Tetrahedron Lett 37:8529
76. Kim W-G, Ryoo I-J, Yun B-S, Shin-ya K, Seto H, Yoo I-D (1997) J Antibiot 50:715
77. Kim W-G, Ryoo I-J, Yun B-S, Shin-ya K, Seto H, Yoo I-D (1999) J Antibiot 52:758
78. Urleb U (1998) Phenazines. In: Schaumann E (ed) Methods of organic chemistry (Houben-Weyl), 4th edn, vol E9b/part 2. Thieme, Stuttgart, p 266; McCullough KJ (1989) Pyrazines and related ring structures. In: Coffey S, Ansell MF (eds) Rodd's chemistry of carbon compounds, 2nd edn, vol IV/part IJ. Elsevier, Amsterdam, p 241; McCullough KJ (1995) Pyrazines and related ring structures. In: Ansell MF (ed) Rodd's chemistry of carbon compounds, Suppl 2nd edn, vol IV/part IJ. Elsevier, Amsterdam, p 93; McCullough KJ (2000) Pyrazines and related structures. In: Sainsbury M (ed) Rodd's chemistry of carbon compounds. 2nd suppl 2nd edn, vol IV/part IJ, Elsevier, Amsterdam, p 99
79. Hartwig JF (1998) Angew Chem Int Ed 37:2046; Wolfe JP, Wagaw S, Marcoux J-F, Buchwald SL (1998) Acc Chem Res 31:805; Hartwig JF (1998) Acc Chem Res 31:852; Yang BH, Buchwald SL (1999) J Organomet Chem 576:125; Hartwig JF (2000) Palladium-catalyzed amination of aryl halides and sulfonates. In: Ricci A (ed) Modern amination methods. Wiley-VCH, Weinheim, p 195
80. Ley K, Seng F (1975) Synthesis 415; Gasco A, Boulton AJ (1981) Adv Heterocycl Chem 29:251; Haddadin MJ, Issidorides CH (1993) Heterocycles 35:1503
81. Petersen L, Jensen KJ, Nielsen J (1999) Synthesis 1763
82. Laursen JB, Jørgensen CG, Nielsen J (2003) Bioorg Med Chem 11:723
83. Laursen JB, de Visser PC, Nielsen HK, Jensen KJ, Nielsen J (2002) Bioorg Med Chem Lett 12:171

84. Challand, SR, Herbert RB, Holliman FG (1970) J Chem Soc Chem Commun 1423
85. Kitahara T, Kinoshita Y, Aono S, Miyake M, Hasegawa T, Watanabe H, Mori K (1994) Pure Appl Chem 66:2083
86. Brooke PK, Challand SR, Flood ME, Herbert RB, Holliman FG, Ibberson PN (1976) J Chem Soc Perkin Trans I 2248
87. Kinoshita Y, Watanabe H, Kitahara T, Mori K (1995) Synlett 186
88. Kinoshita Y, Kitahara T (1997) Tetrahedron Lett 38:4993
89. Surrey AR (1946) Org Synth 86
90. Tietze M, Iglesias A, Beifuss U (in preparation)
91. Emoto T, Kubosaki N, Yamagiwa Y, Kamikawa T (2000) Tetrahedron Lett 41:355

Note added in proof: Meanwhite a new review on phenacine natural products has been published: Laursen JB, Nielsen J (2004) Chem Rev 104:1663

Top Curr Chem (2005) 244: 115–148
DOI 10.1007/b96890

Occurrence, Biological Activity, and Convergent Organometallic Synthesis of Carbazole Alkaloids

Hans-Joachim Knölker (✉)

Institut für Organische Chemie, Technische Universität Dresden, Bergstrasse 66, 01069 Dresden, Germany
hans-joachim.knoelker@chemie.tu-dresden.de

1 Introduction . . . 116
1.1 Occurrence, Classification, and Biological Activity of Carbazole Alkaloids . . . 116
1.2 Convergent Synthesis of Carbazoles by Oxidative Coupling of Arylamines . . . 121

2 Iron-Mediated Synthesis of Carbazoles . . . 122
2.1 Arylamine Cyclization . . . 123
2.2 Quinone Imine Cyclization . . . 128
2.3 Oxidative Cyclization by Air . . . 130

3 Molybdenum-Mediated Synthesis of Carbazoles . . . 132

4 Palladium-Catalyzed Synthesis of Carbazoles . . . 135
4.1 Cyclization of *N,N*-Diarylamines . . . 135
4.2 Cyclization of Arylamino-1,4-benzoquinones . . . 140
4.3 Cyclization of Arylamino-1,2-benzoquinones . . . 142

5 Conclusion . . . 144

References . . . 144

Abstract A brief overview of the occurrence and biological activity of carbazole alkaloids is provided. Recent progress in the development of novel methodologies for the synthesis of substituted biologically active carbazole alkaloids is summarized. The described methods are based on the transition metal-mediated oxidative coupling of arylamines with unsaturated six-membered hydrocarbons, thus leading to highly convergent total syntheses of the envisaged natural products. The advantages of this strategy, as well as the scope and limitations of the different procedures, are discussed.

1 Introduction

1.1 Occurrence, Classification, and Biological Activity of Carbazole Alkaloids

A variety of structurally diverse carbazole alkaloids has been isolated from different natural sources over the past four decades [1–8]. The majority of carbazole alkaloids was obtained from higher plants of the genus *Murraya*, *Clausena*, and *Glycosmis* belonging to the family *Rutaceae*. The genus *Murraya*, trees growing in southern Asia, represents the richest source of carbazole alkaloids from terrestrial plants, particularly for 1-oxygenated and 2-oxygenated tricyclic carbazole alkaloids (Scheme 1).

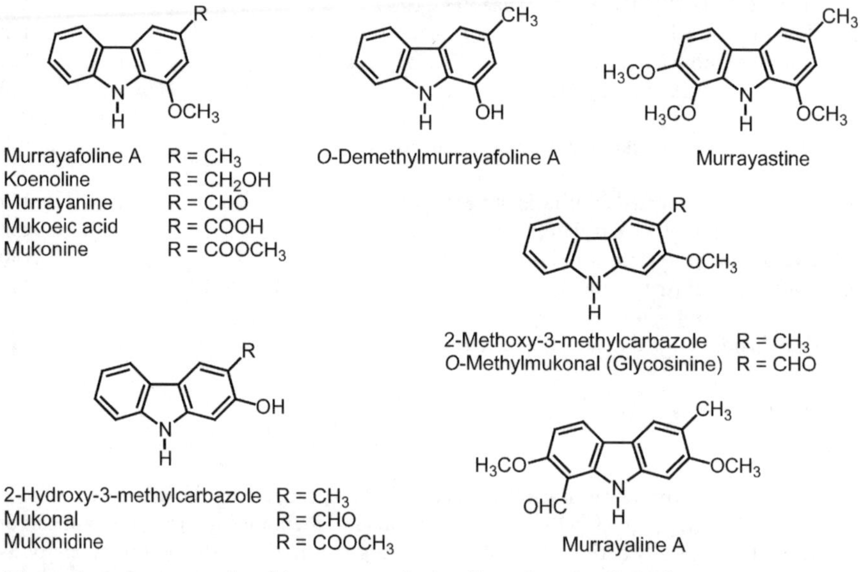

Scheme 1 1-Oxygenated and 2-oxygenated tricyclic carbazole alkaloids

For the 1-oxygenated and 2-oxygenated tricyclic carbazole alkaloids 3-methylcarbazole appears to be the common biogenetic precursor [1], which is subsequently oxygenated at different positions and further oxidized at the methyl group [2]. The classification of carbazole alkaloids used in this article is based on the functionalization of ring A bearing the methyl substituent [7]. The first carbazole alkaloid isolated from natural sources was murrayanine, which was obtained by Chakraborty in 1965 from the stem bark of *Murraya koenigii* [9]. Subsequently, murrayanine was also found in *Clausena heptaphylla* [10]. Murrayanine was shown to exhibit antibiotic acitivity [11]. *Murraya koenigii* (family *Rutaceae*) is an Indian medicinal plant commonly known as curry-leaf

Fig. 1 Blooming plant of *Murraya koenigii* (curry-leaf tree). (Courtesy of Calvin Lemke, Department of Botany and Microbiology, University of Oklahoma)

(Hindi: *kari patta*) tree (Fig. 1). The leaves of this small tree are used as a spice for the preparation of curry. Extracts from various parts of the plant have a strong antibacterial and antifungal activity and therefore are applied in folk medicine [12].

On closer investigation of different parts of *Murraya koenigii* over the following years a broad range of 1- and 2-oxygenated tricyclic carbazole alkaloids was discovered. Among the 1-oxygenated carbazole alkaloids mukoeic acid [13], its methyl ester, mukonine [2], and *O*-demethylmurrayafoline A [14] were all isolated from the stem bark, whereas the cytotoxic carbazole alkaloid koenoline [15] was obtained from the roots. Murrayafoline A [16] was isolated from the root bark and murrayastine [17] from the stem bark of *Murraya euchrestifolia* Hayata collected in Taiwan. The 2-oxygenated carbazole alkaloids 2-hydroxy-3-methylcarbazole [18], mukonal [19], mukonidine [20], and 2-methoxy-3-methylcarbazole [21] were extracted from different parts of *Murraya koenigii*. Mukonidine was also obtained from the root bark of *Clausena excavata* by Wu [22], who reported spectral data different from those described by Chakraborty [19, 20]. In China *Clausena excavata* is used in traditional folk medicine for the treatment of various infectious diseases and poisonous snakebites. *O*-Methylmukonal (glycosinine) was isolated from both *Murraya siamensis* [23] and from *Glycosmis pentaphylla* [24]. Murrayaline A was found in the stem bark of *Murraya euchrestifolia* [17, 25]. The structures of murrayastine and murrayaline A demonstrate that the 1- and 2-oxygenated tricyclic carbazole alkaloids may be further functionalized in ring C.

In contrast to the 1- and 2-oxygenated tricyclic carbazole alkaloids, most of the 3-oxygenated and the 3,4-dioxygenated tricyclic carbazole alkaloids were isolated from *Streptomyces* (Scheme 2). Their common structural feature is a 2-methylcarbazole, although biosynthetic studies on carbazomycin B have shown that tryptophan is the precursor of the carbazole nucleus [26]. 4-De-

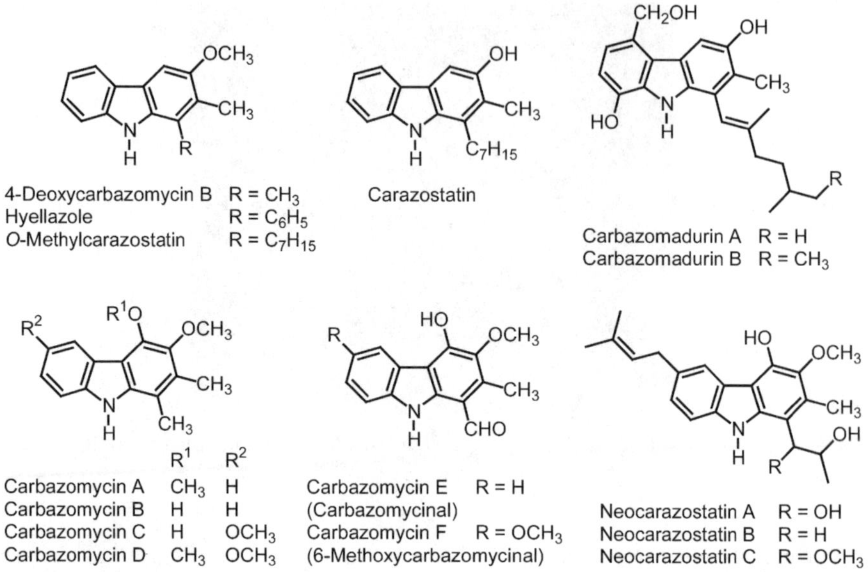

Scheme 2 3-Oxygenated and 3,4-dioxygenated tricyclic carbazole alkaloids

oxycarbazomycin B, originally obtained by degradation of natural carbazomycin B [27], exhibits a significant inhibitory activity against various Gram-positive and Gram-negative bacteria [28]. Over the years, 4-deoxycarbazomycin B became something of a benchmark for synthetic approaches to these alkaloids [7]. Moore isolated hyellazole, the first carbazole alkaloid from marine sources, from the blue-green alga *Hyella caespitosa* [29]. Carazostatin was isolated from *Streptomyces chromofuscus* and shows a strong inhibition of lipid peroxidation induced by free radicals [30]. Electrochemical studies confirmed that both carazostatin and its synthetic derivative *O*-methylcarazostatin are useful antioxidants [31]. In liposomal membranes carazostatin has a stronger antioxidant activity than α-tocopherol [32]. The neuronal cell protecting alkaloids carbazomadurin A and B were obtained by Seto from *Actinomadura madurae* 2808-SV1 [33]. The biological data suggest that the mode of action of the carbazomadurins is based on their antioxidative activity. Nakamura described the isolation of the carbazomycins A–F from *Streptoverticillium ehimense* H 1051-MY 10 [27, 34]. Carbazomycin E (carbazomycinal) and carbazomycin F (6-methoxycarbazomycinal) were also obtained from a different *Streptoverticillium* species of the strain KCC U-0166 [35]. The carbazomycins A and B inhibit the growth of phytopathogenic fungi and exhibit antibacterial and antiyeast activities. Carbazomycin C and D show only very weak antimicrobial activity. Carbazomycin E and F inhibit selectively the formation of aerial mycelia without any effect on the growth of substrate mycelia. Moreover, carbazomycin B and C were found to be inhibitors of 5-lipoxygenase [36]. Kato isolated the neocarazostatins A–C from the culture of *Streptomyces* sp. strain

GP 38 [37]. The neocarazostatins show a strong inhibitory activity against lipid peroxidation induced by free radicals.

The carbazole-1,4-quinones represent an important subgroup of the carbazole alkaloids and were mainly isolated from plants of the genus *Murraya* (Scheme 3) [7, 38]. Murrayaquinone A, isolated by Furukawa from the root bark of *Murraya euchrestifolia* [16], exhibits cardiotonic activity [39]. Koeniginequinone A and B were obtained by Chowdhury from the stem bark of *Murraya koenigii* [40]. Wu isolated clausenaquinone A from the stem bark of *Clausena excavata* [41]. Clausenaquinone A inhibits the growth of several tumor cell lines and moreover shows potent inhibition of the rabbit platelet aggregation induced by arachidonic acid. The isomeric pyrayaquinones A and B have an annulated dimethylpyran ring as the characteristic structural feature and were found in the root bark of *Murraya euchrestifolia* [42]. The carbazomycins G and H, isolated from *Streptoverticillium ehimense*, are structurally unique because of the carbazole-1,4-quinol moiety [43]. Carbazomycin G exhibits antifungal activity.

The carbazole-3,4-quinone alkaloids were all isolated by Seto from different *Streptomyces* (Scheme 4). Carbazoquinocin C was obtained from *Streptomyces*

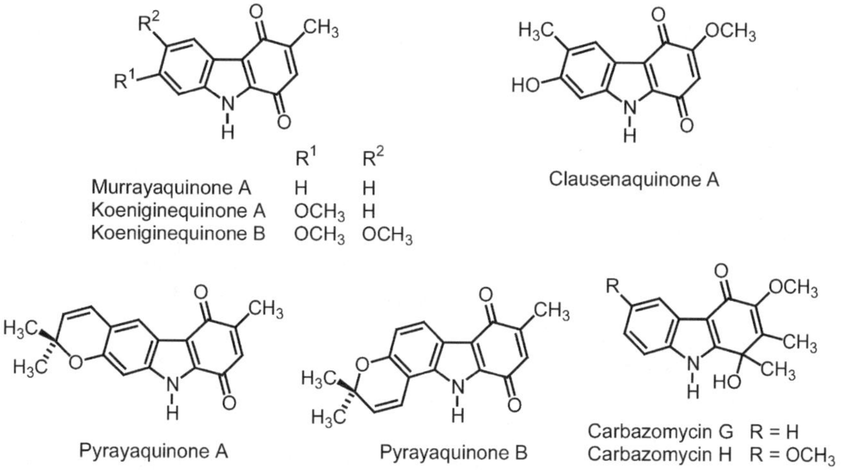

Scheme 3 Carbazole-1,4-quinone and carbazole-1,4-quinol alkaloids

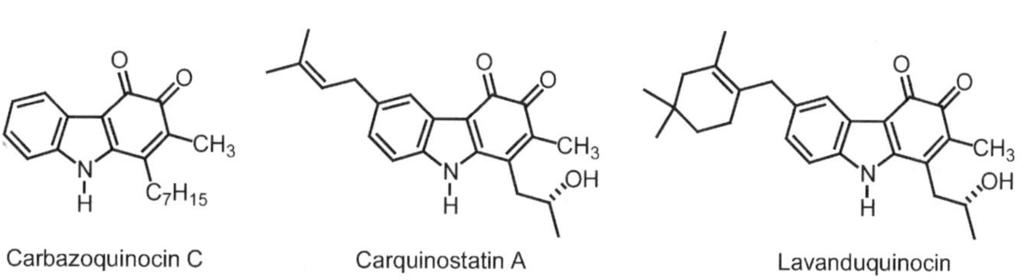

Scheme 4 Carbazole-3,4-quinone alkaloids

violaceus 2448-SVT2 and shows a strong inhibitory activity against lipid peroxidation [44]. Carquinostatin A, isolated from *Streptomyces exfoliatus* 2419-SVT2 [45], and lavanduquinocin, from *Streptomyces viridochromogenes* 2941-SVS3 [46], are potent neuronal cell protecting substances containing a terpenoid side chain in the 6-position.

The pyranocarbazole alkaloids were all obtained from terrestrial plants (Scheme 5). Girinimbine has a pyrano[3,2-*a*]carbazole framework and was isolated first from the stem bark of *Murraya koenigii* [47–49]. Subsequently, girinimbine was obtained from the roots of *Clausena heptaphylla* [50]. Murrayacine was also isolated from both plants [48–52]. Dihydroxygirinimbine [53] and pyrayafoline A [17, 54] were obtained from *Murraya euchrestifolia*. The pyran ring of dihydroxygirinimbine contains a *trans*-1,2-diol moiety. However, the absolute configuration is not known.

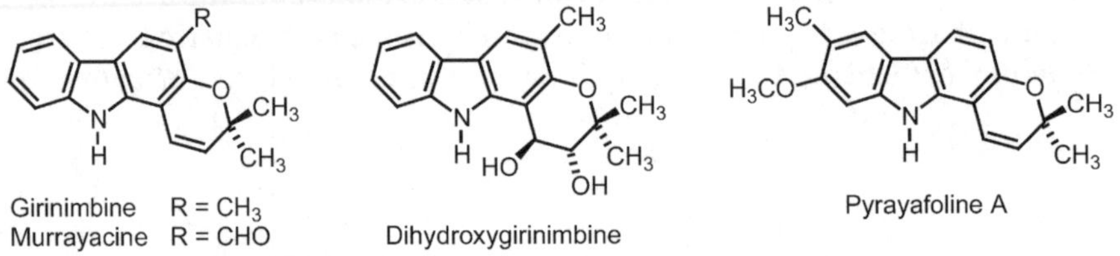

Scheme 5 Pyranocarbazole alkaloids

The heteroaryl-annulated carbazole alkaloids have attracted high interest because of their pharmacological activities [55] (Scheme 6). Furukawa isolated the furo[3,2-*a*]carbazole alkaloid furostifoline from *Murraya euchrestifolia* [56]. The pyrido[4,3-*b*]carbazole alkaloid ellipticine was originally obtained from the leaves of *Ochrosia elliptica* Labill [57]. Ellipticine shows significant antitumor activity [58]. This finding induced a strong interest in pyrido[4,3-*b*]carbazole alkaloids and their synthetic analogs [7, 59]. The indolo[2,3-*a*]carbazole alkaloids exhibit a broad range of potent biological activities (antifungal, antimicrobial, hypotensive, antitumor, and inhibition of protein kinase C and topoisomerase I), which prompted the development of several synthetic approaches [7, 60]. Nakanishi isolated staurosporinone (or K-252c), which represents the aglycon of staurosporine, from *Nocardiopsis* strain K-290 [61].

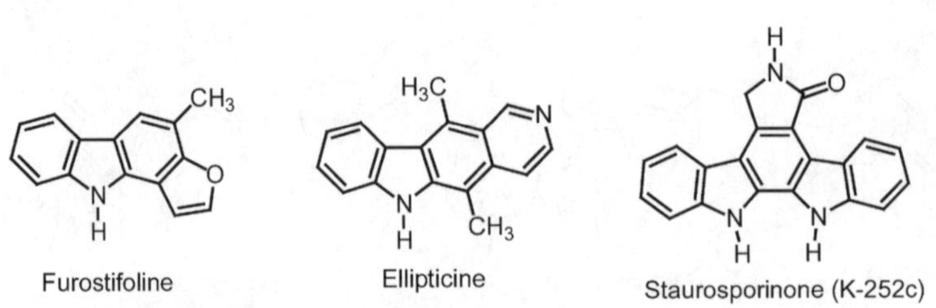

Scheme 6 Heteroarylcarbazole alkaloids

The dimeric carbazole alkaloids often occur along with the corresponding monomeric carbazoles in terrestrial plants [7, 62] (Scheme 7). Clausenamine-A was obtained by Wu from the stem bark of *Clausena excavata* [63]. Clausenamine-A and its synthetic analogs, like bis(*O*-demethylmurrayafoline-A), show cytotoxic activities against diverse human cancer cell lines [64] and exhibit moderate antimalarial activity [65, 66]. Furukawa isolated 1,1′-bis(2-hydroxy-3-methylcarbazole) and bismurrayaquinone-A, the first dimeric carbazolequinone alkaloid found in nature, from *Murraya koenigii* [67].

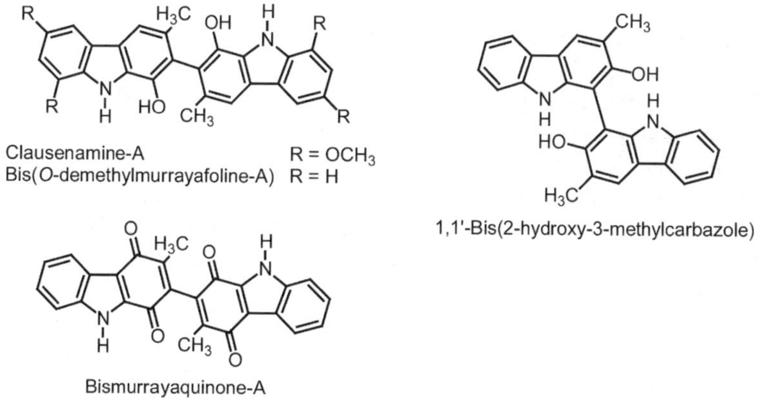

Scheme 7 Dimeric carbazole alkaloids

1.2
Convergent Synthesis of Carbazoles by Oxidative Coupling of Arylamines

The structural diversity and the useful biological activities of natural carbazole alkaloids induced the development of novel synthetic routes to the carbazole framework and their application in natural product synthesis [68–78]. Organometallic approaches to carbazoles received special attention, since they have led in many cases to highly convergent total syntheses of carbazole alkaloids [7, 75]. The present article describes the synthesis of carbazoles using a transition metal-mediated or -catalyzed oxidative coupling of arylamines with an unsaturated six-membered hydrocarbon as common general strategy (Scheme 8).

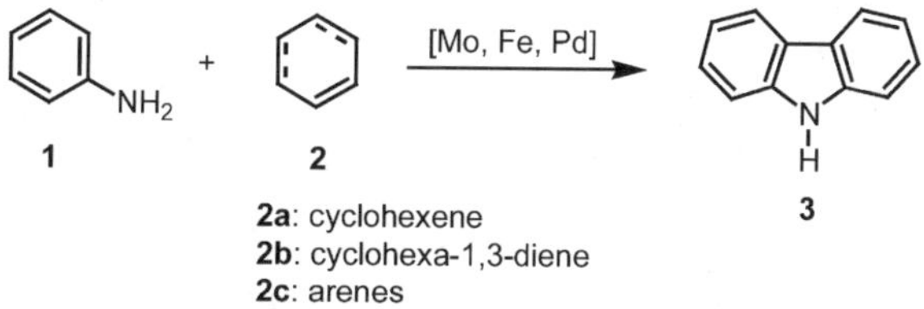

Scheme 8 Synthesis of carbazoles by oxidative coupling of arylamines

A molybdenum-mediated oxidative coupling of aniline **1** with cyclohexene **2a** provides carbazole **3**. Alternatively, the same overall transformation of aniline **1** to carbazole **3** is achieved by iron-mediated oxidative coupling with cyclohexa-1,3-diene **2b** or by palladium-catalyzed oxidative coupling with arenes **2c**. The use of appropriately substituted anilines and unsaturated six-membered hydrocarbons opens up the way to highly convergent organometallic syntheses of carbazole alkaloids.

2 Iron-Mediated Synthesis of Carbazoles

The iron-mediated construction of the carbazole framework proceeds via consecutive C–C and C–N bond formation as key steps [70,71]. The C–C bond formation is achieved by electrophilic substitution of the arylamine with a tricarbonyliron-coordinated cyclohexadienyl cation. The parent iron complex salt for electrophilic substitutions, tricarbonyl[η^5-cyclohexadienylium]iron tetrafluoroborate **6a**, is readily available by azadiene-catalyzed complexation and subsequent hydride abstraction (Scheme 9).

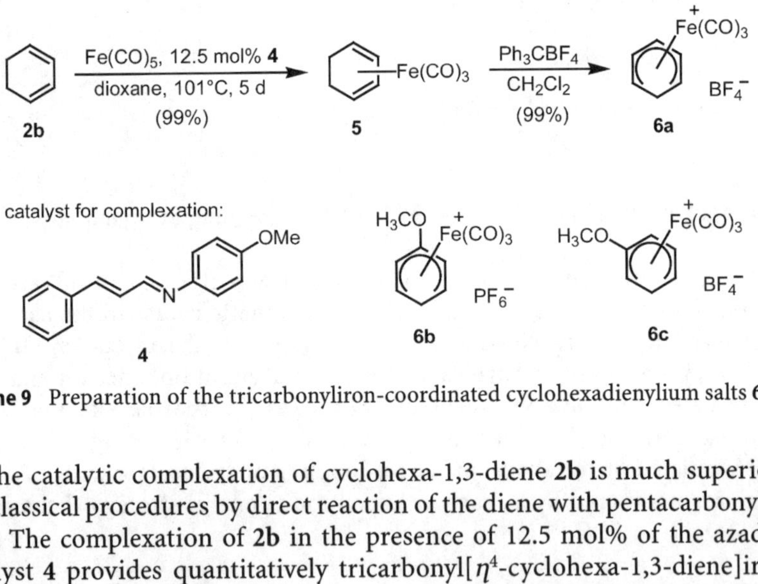

Scheme 9 Preparation of the tricarbonyliron-coordinated cyclohexadienylium salts **6**

The catalytic complexation of cyclohexa-1,3-diene **2b** is much superior to the classical procedures by direct reaction of the diene with pentacarbonyliron [79]. The complexation of **2b** in the presence of 12.5 mol% of the azadiene catalyst **4** provides quantitatively tricarbonyl[η^4-cyclohexa-1,3-diene]iron **5** [80]. Hydride abstraction of complex **5** using triphenylcarbenium tetrafluoroborate provides the iron complex salt **6a** [81]. Similar sequences afford the corresponding 3-methoxy-substituted and 2-methoxy-substituted complex salts **6b** and **6c** [82,83].

An oxidative cyclization leads to the C–N bond formation and furnishes the carbazole nucleus. The three modes of the iron-mediated carbazole synthesis differ in the procedures which are used for the oxidative cyclization [77,78].

2.1 Arylamine Cyclization

The carbazole construction using iron-mediated arylamine cyclization for the C–N bond formation was applied to the synthesis of 4-deoxycarbazomycin B [84]. The synthesis of this model compound demonstrates also the course of the two key steps which are involved in the iron-mediated carbazole synthesis (Scheme 10).

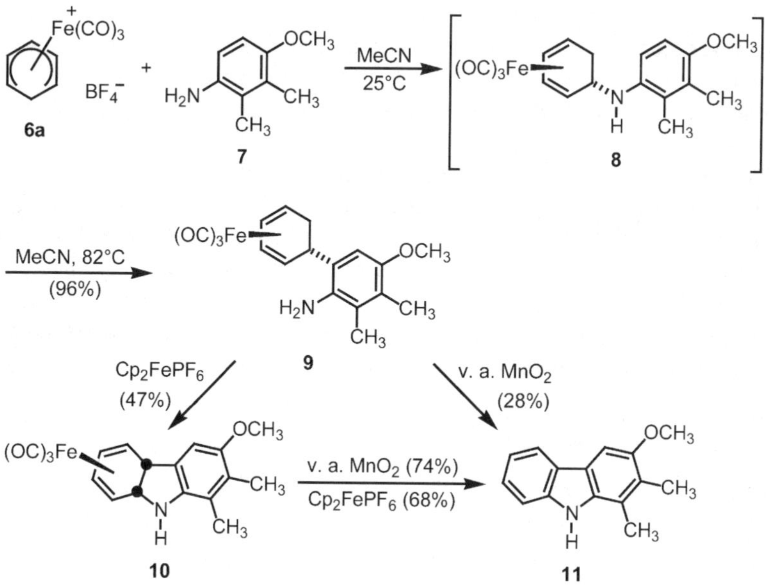

Scheme 10 Iron-mediated synthesis of 4-deoxycarbazomycin B **11**

The electrophilic substitution of 4-methoxy-2,3-dimethylaniline **7** by reaction with the complex salt **6a** in acetonitrile at room temperature initially leads to the *N*-alkylated arylamine **8** as the kinetic product. On heating the reaction mixture under reflux compound **8** rearranges to complex **9**, which is the result of regioselective electrophilic aromatic substitution at the position ortho to the amino group, and represents the thermodynamic product of this reaction [85]. The iron-mediated arylamine cyclization of complex **9** by oxidation with very active manganese dioxide [86] provides directly, with concomitant aromatization and demetalation, 4-deoxycarbazomycin B **11** [84]. A selective cyclodehydrogenation of complex **9** by treatment with the SET oxidizing agent ferricenium hexafluorophosphate affords the tricarbonyliron-complexed 4a,9a-dihydrocarbazole **10**, which represents the intermediate of the direct transformation of complex **9** to compound **11**. Thus, on oxidation with very active manganese dioxide, complex **10** is readily converted to **11**. The aromatization and demetalation of complex **10** to **11** can also be achieved by treatment with a second equivalent ferricenium hexafluorophosphate [84]. Investigations with deuterium-labeled analogs of complex **9** have shown that the regioselectivity of the iron-mediated oxidative cyclization can be directed by the choice of the oxidizing agent. Two-electron oxidants (manganese dioxide) initially lead to attack at C4 of the cyclohexadiene ligand (iron-coordinated sp^2 carbon atom), while one-electron oxidants (ferricenium hexafluorophosphate) result in attack at C6 of the cyclohexadiene ligand (sp^3 carbon atom) [87]. However, a subsequent proton-catalyzed isomerization of the kinetic product may occur. This observation is of importance for the iron-mediated synthesis of carbazole alkaloids, when using unsymmetrically substituted cyclohexadiene ligands. The methodology featuring the iron-mediated arylamine cyclization represents a threefold dehydrogenative coupling of an arylamine with cyclohexa-1,3-diene and has been applied to the total synthesis of a broad range of carbazole alkaloids.

The reaction of the complex salt **6a** with the arylamine **12** affords by regioselective electrophilic substitution the iron complex **13** [88] (Scheme 11). The oxidative cyclization of complex **13** with very active manganese dioxide provides directly mukonine **14**, which by ester cleavage was converted to mukoeic acid **15** [89]. Further applications of the iron-mediated construction of the carbazole framework to the synthesis of 1-oxygenated carbazole alkaloids include murrayanine, koenoline, and murrayafoline A [89].

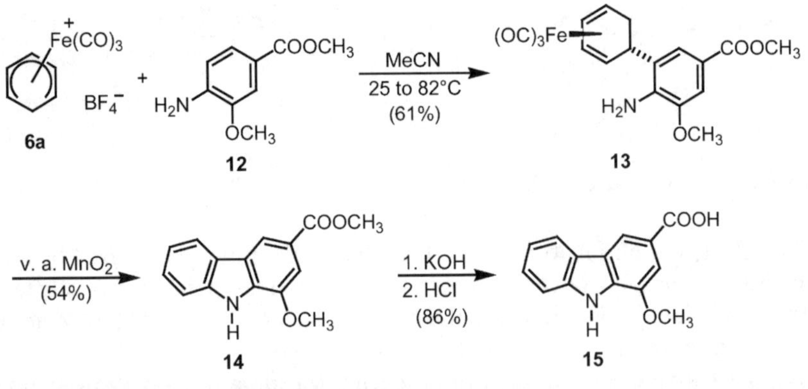

Scheme 11 Synthesis of mukonine **14** and mukoeic acid **15**

The iron-mediated synthesis of 2-oxygenated carbazole alkaloids is limited and provides only a moderate yield (11%) for the oxidative cyclization to 2-methoxy-3-methylcarbazole using iodine in pyridine as the reagent [90]. Ferricenium hexafluorophosphate is the superior reagent for the iron-mediated arylamine cyclization leading to 3-oxygenated carbazoles (Scheme 12). Electrophilic substitution of the arylamines **16** with the complex salt **6a** leads to the iron complexes **17**. Oxidative cyclization of the complexes **17** with an excess of ferricenium hexafluorophosphate in the presence of sodium carbonate affords hyellazole **19a** and *O*-methylcarazostatin **19b** along with the tricarbonyliron-coordinated 4b,8a-dihydrocarbazol-3-ones **18a** and **18b** as by-products. These by-products are additionally converted to the 3-methoxycarbazoles by demetalation with trimethylamine *N*-oxide and subsequent *O*-methylation. Thus, both compounds are readily available by the present route (hyellazole **19a:** three steps, 83% overall yield [91] and *O*-methylcarazostatin **19b:** three steps, 82% overall yield [92]).

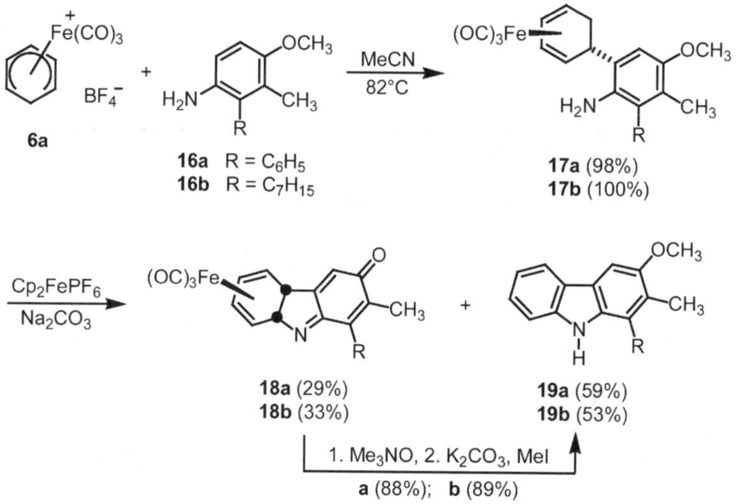

Scheme 12 Synthesis of hyellazole **19a** and *O*-methylcarazostatin **19b**

The total synthesis of the carbazomycins emphasizes the utility of the iron-mediated synthesis for the construction of highly substituted carbazole derivatives. The reaction of the complex salts **6a** and **6b** with the arylamine **20** leads to the iron complexes **21**, which prior to oxidative cyclization have to be protected by chemoselective *O*-acetylation to **22** (Scheme 13). Oxidation with very active manganese dioxide followed by ester cleavage provides carbazomycin B **23a** [93] and carbazomycin C **23b** [94]. The regioselectivity of the cyclization of complex **22b** to a 6-methoxycarbazole is rationalized by previous results from deuterium labeling studies [87] and the regiodirecting effect of the 2-methoxy substituent of the intermediate tricarbonyliron-coordinated cyclohexadienylium ion [79c, 79d]. Starting from the appropriate arylamine, the same sequence of reactions has been applied to the total synthesis of carbazomycin E (carbazomycinal) [95].

The carbazole-1,4-quinol alkaloids are also accessible by the iron-mediated arylamine cyclization (Scheme 14). Electrophilic substitution reaction of the arylamine **24** with the complex salts **6a** and **6b** affords the iron complexes **25**. Protection to the acetates **26** and oxidative cyclization with very active manganese dioxide leads to the carbazoles **27**, which are oxidized to the carbazole-

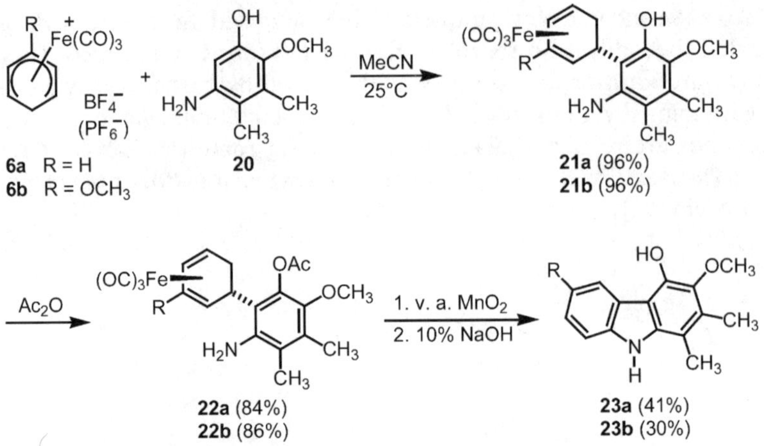

Scheme 13 Synthesis of carbazomycin B **23a** and carbazomycin C **23b**

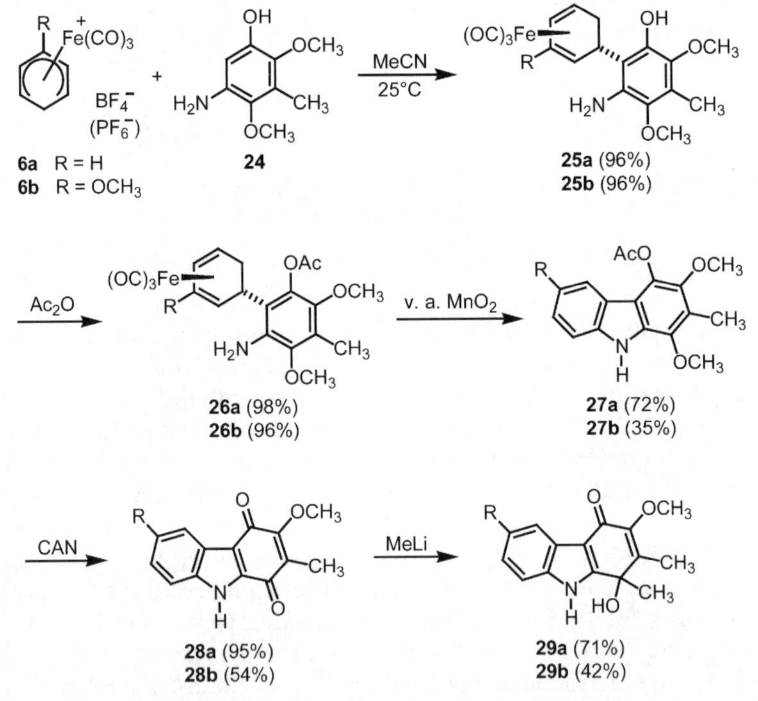

Scheme 14 Synthesis of carbazomycin G **29a** and carbazomycin H **29b**

1,4-quinones **28** using ceric ammonium nitrate. Conversion to carbazomycin G **29a** and carbazomycin H **29b** by regioselective addition of methyllithium at C1 completes the synthesis [96].

The total synthesis of the furo[3,2-*a*]carbazole alkaloid furostifoline is achieved in a highly convergent manner by successive formation of the carbazole nucleus and annulation of the furan ring (Scheme 15). Electrophilic substitution of the arylamine **30** using the complex salt **6a** provides complex **31**. In this case, iodine in pyridine was the superior reagent for the oxidative cyclization to the carbazole **32**. Finally, annulation of the furan ring by an Amberlyst 15-catalyzed cyclization affords furostifoline **33** [97].

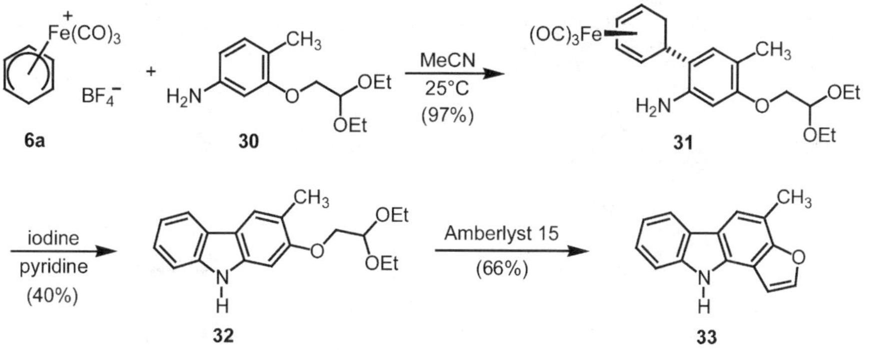

Scheme 15 Synthesis of furostifoline **33**

The double iron-mediated arylamine cyclization provides a highly convergent route to indolo[2,3-*b*]carbazole (Scheme 16). Double electrophilic substitution of *m*-phenylenediamine **34** by reaction with the complex salt **6a** affords the diiron complex **35**, which on oxidative cyclization using iodine in pyridine leads to indolo[2,3-*b*]carbazole **36** [98]. Thus, it has been demonstrated that the bidirectional annulation of two indole rings can be applied to the synthesis of indolocarbazoles.

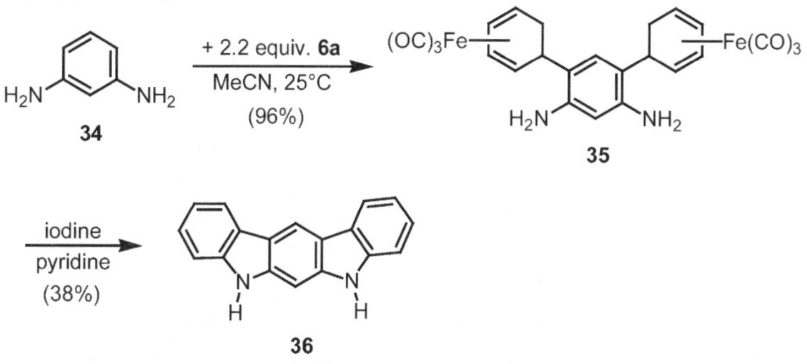

Scheme 16 Synthesis of indolo[2,3-*b*]carbazole **36**

2.2 Quinone Imine Cyclization

For the quinone imine cyclization of iron complexes to carbazoles the arylamine is chemoselectively oxidized to a quinone imine before the cyclodehydrogenation [99]. The basic strategy of this approach is demonstrated for the total synthesis of the 3-oxygenated tricyclic carbazole alkaloids 4-deoxycarbazomycin B, hyellazole, carazostatin, and *O*-methylcarazostatin (Scheme 17).

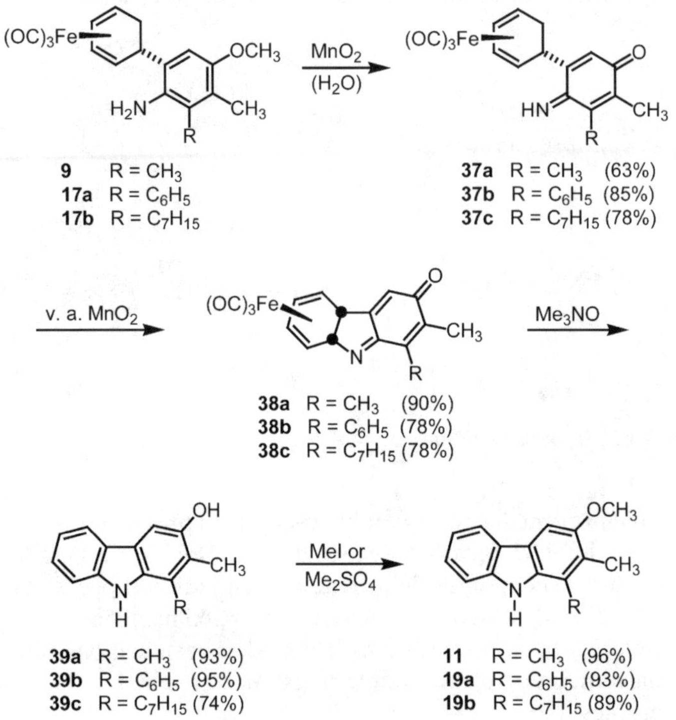

Scheme 17 Synthesis of carazostatin **39c** and the 3-methoxycarbazole alkaloids **11**, **19a**, and **19b**

Oxidation of the iron complexes **9** and **17** with commercial manganese dioxide (containing water) provides the noncyclized quinone imines **37**. A cyclodehydrogenation of the complexes **37** by treatment with very active manganese dioxide leads to the cyclized quinone imines **38**. These tricarbonyliron-coordinated 4b,8a-dihydrocarbazol-3-ones contain a dihydrobenzene ring as well as a quinone ring and overall have the oxidation state of an aromatic carbazole ring system. Therefore, demetalation using trimethylamine *N*-oxide occurs with concomitant tautomerization to the 3-hydroxycarbazoles **39**. Compound **39c** represents the natural free radical scavenger carazostatin. The *O*-methylation of the 3-hydroxycarbazoles **39** provides 4-deoxycarbazomycin B **11** [84], hyellazole **19a** [91], and *O*-methylcarazostatin **19b** [92]. Starting from the iron complex salt **6c**, the iron-mediated quinone imine cyclization leads to 3,7-dioxygenated carbazoles [99].

For the iron complexes resulting from 4,5-dioxygenated arylamines, the intermediate noncyclized quinone imines without isolation undergo a direct quinone imine cyclization [99]. This increased reactivity is exploited for the synthesis of carbazomycin A **44a** and carbazomycin D **44b** (Scheme 18). The reaction of the complex salts **6a** and **6b** with the arylamine **40a** affords the iron complexes **41**. Oxidation of the complexes **41** using commercial manganese dioxide (containing water) leads directly to the 4b,8a-dihydrocarbazol-3-one complexes **42** by generation of the quinone imine and subsequent in situ cyclization. The regioselectivity of the cyclization of complex **41b** is again explained by the previous deuterium labeling experiments [87] in context with the regioselective attack at a tricarbonyliron-coordinated 2-methoxycyclohexadienylium ion [79c, 79d]. Demetalation and tautomerization gives the 3-hydroxycarbazoles **43**, which on *O*-methylation provide carbazomycin A **44a** [93] and carbazomycin D **44b** [94].

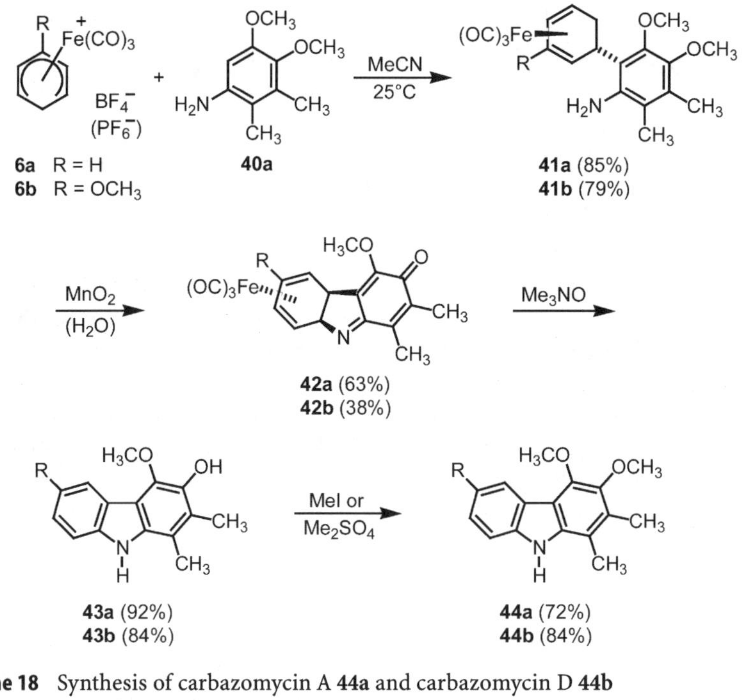

Scheme 18 Synthesis of carbazomycin A **44a** and carbazomycin D **44b**

2.3 Oxidative Cyclization by Air

More recently, an environmentally benign method using air as oxidant has been developed for the oxidative cyclization of arylamine-substituted tricarbonyliron–cyclohexadiene complexes to carbazoles (Scheme 19). Reaction of methyl 4-aminosalicylate **45** with the complex salt **6a** affords the iron complex **46**, which on oxidation in acidic medium by air provides the tricarbonyliron-complexed 4a,9a-dihydrocarbazole **47**. Aromatization with concomitant demetalation by treatment of the crude product with *p*-chloranil leads to mukonidine **48** [88]. The spectral data of this compound are in agreement with those reported by Wu [22].

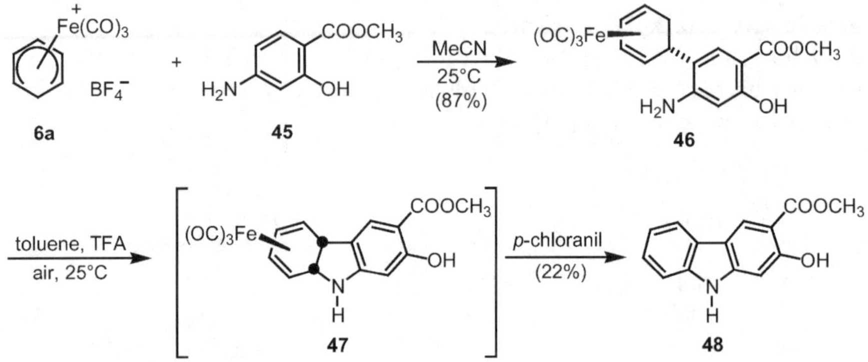

Scheme 19 Synthesis of mukonidine **48**

In a more direct approach, the electrophilic substitution of the arylamine by the complex salt **6a** is combined with the oxidative cyclization by air, thus furnishing the carbazole framework in a one-pot process via consecutive iron-mediated C–C and C–N bond formation (Scheme 20). Addition of two equivalents of the arylamines **40** to the complex salt **6a** in acetonitrile at room temperature leads directly to the tricarbonyliron-complexed 4a,9a-dihydrocarbazoles **49**. Demetalation and subsequent aromatization by catalytic dehydrogenation [100] provides carbazomycin A **44a** and the carbazole derivatives **50**. By this considerably improved synthesis, **44a** becomes available in three steps and 65% overall yield [101]. Ether cleavage of the carbazole **50b** using boron tribromide followed by oxidation with air to the *ortho*-quinone affords carbazoquinocin C **51** [102]. Compound **50c**, containing the (*R*)-2-acetoxypropyl side chain at the 1-position of the carbazole framework, represents a common precursor for the enantioselective total synthesis of carquinostatin A and lavanduquinocin [103–105] (Scheme 21). Electrophilic bromination of **50c** to the 6-bromocarbazole **52** and subsequent nickel-mediated coupling [106] with either prenyl bromide **53a** or *β*-cyclolavandulyl bromide **53b** provides the 6-substituted carbazoles **54**.

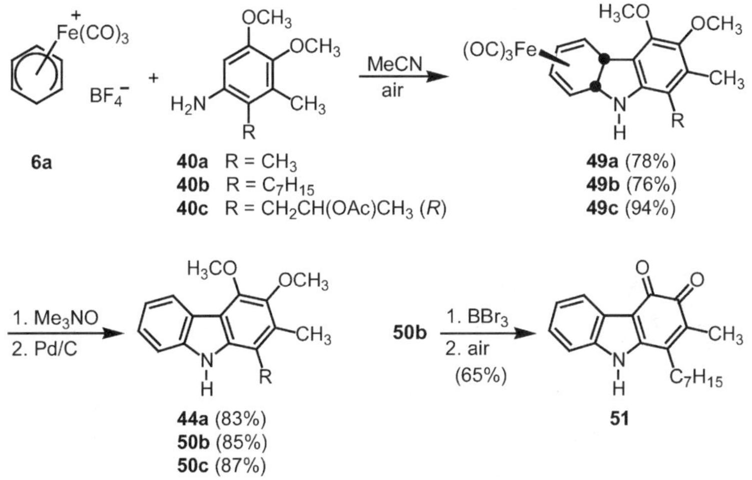

Scheme 20 Synthesis of carbazomycin A **44a**, carbazoquinocin C **51**, and the precursor **50c**

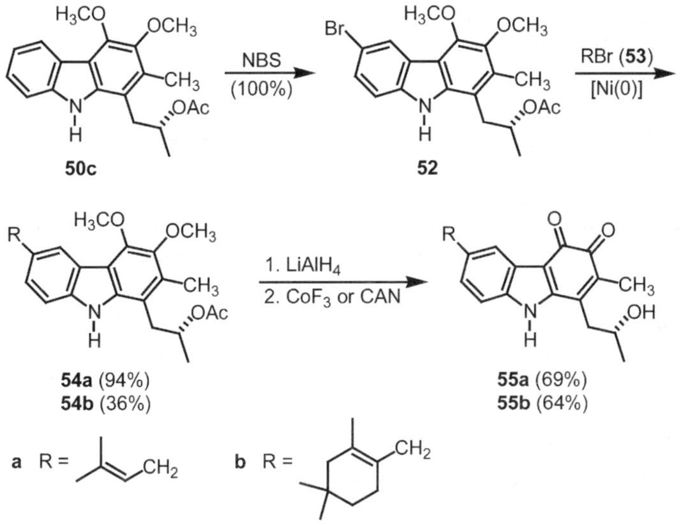

Scheme 21 Enantioselective synthesis of carquinostatin A **55a** and lavanduquinocin **55b**

Reductive removal of the acetyl group and oxidation to the *ortho*-quinone affords carquinostatin A **55a** [103] and lavanduquinocin **55b** [104].

The procedure is also applicable to acetoxy-substituted arylamines, thus broadening the scope of the carbazole synthesis (Scheme 22). Reaction of the complex salt **6a** with the 5-acetoxyarylamines **56** affords the tricarbonyliron-complexed 4a,9a-dihydrocarbazoles **57**, which on demetalation and subsequent catalytic dehydrogenation [100] give the 4-acetoxycarbazoles **58**. Removal of

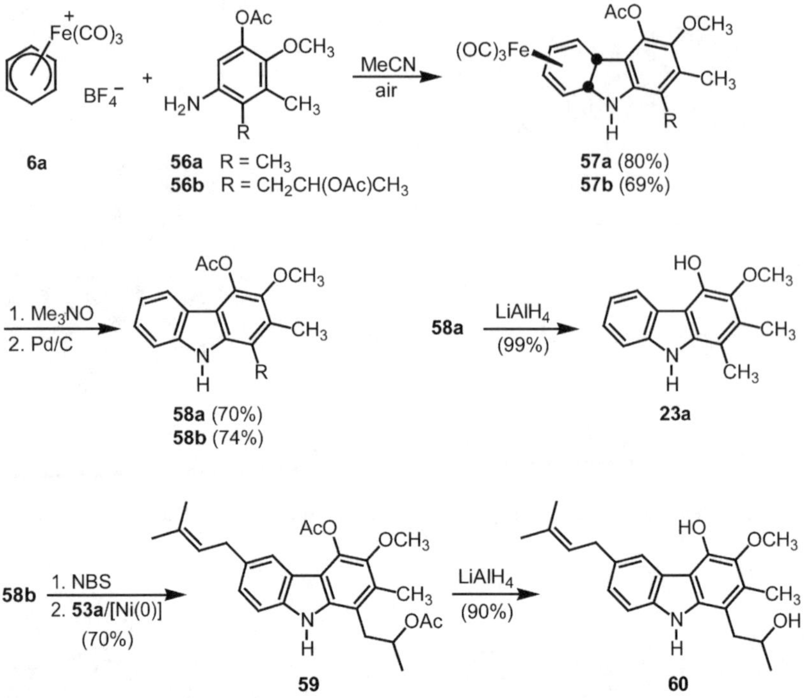

Scheme 22 Synthesis of carbazomycin B **23a** and (±)-neocarazostatin B **60**

the acetyl group by reduction of compound **58a** with lithium aluminum hydride completes an improved route to carbazomycin B **23a** (four steps, 55% overall yield) [101]. Electrophilic bromination of compound **58b** followed by nickel-mediated prenylation [106] provides the 6-prenylcarbazole **59**, which by ester cleavage is transformed into (±)-neocarazostatin B **60** [107].

3
Molybdenum-Mediated Synthesis of Carbazoles

Despite many applications of the iron-mediated carbazole synthesis, the access to 2-oxygenated tricyclic carbazole alkaloids using this method is limited due to the moderate yields for the oxidative cyclization [88, 90]. In this respect, the molybdenum-mediated oxidative coupling of an arylamine and cyclohexene **2a** represents a complementary method. The construction of the carbazole framework is achieved by consecutive molybdenum-mediated C–C and C–N bond formation. The cationic molybdenum complex, required for the electrophilic aromatic substitution, is easily prepared (Scheme 23).

Wohl–Ziegler bromination of cyclohexene **2a** and treatment of the resulting 3-bromocyclohexene with tri(acetonitrile)tricarbonylmolybdenum, followed

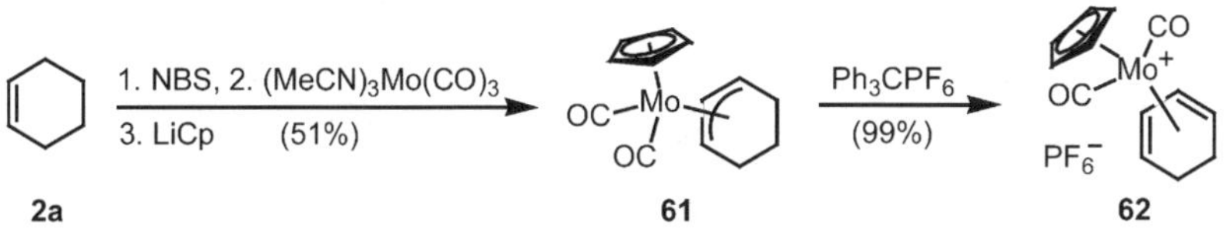

Scheme 23 Preparation of the molybdenum complex salt **62**

by lithium cyclopentadienide, provides dicarbonyl[η^3-cyclohexenyl][η^5-cyclopentadienyl]molybdenum **61** [108]. An alternative synthesis of complex **61** uses 3-cyclohexenyl diphenylphosphinate as starting material [109]. Hydride abstraction of complex **61** with triphenylcarbenium hexafluorophosphate affords the molybdenum complex salt **62** [108]. A broad range of nucleophiles can be added to this cationic molybdenum–cyclohexa-1,3-diene complex [108, 110]. A model study directed toward the synthesis of 4-deoxycarbazomycin B **11** first showed that the complex salt **62** can also be applied in the electrophilic aromatic substitution of arylamines [111] (Scheme 24).

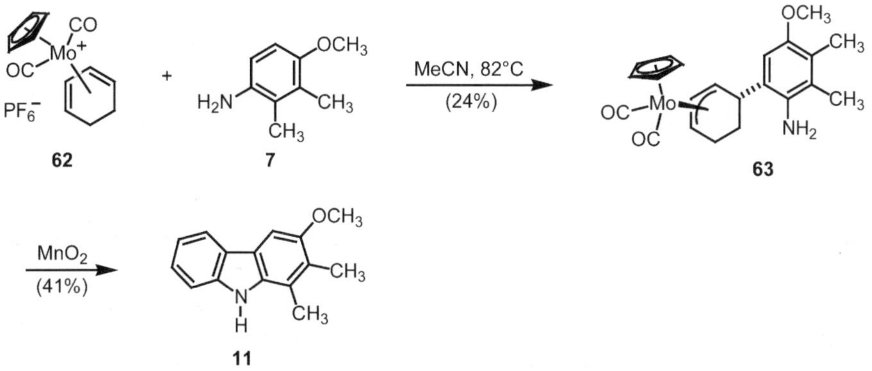

Scheme 24 Molybdenum-mediated synthesis of 4-deoxycarbazomycin B **11**

Reaction of 4-methoxy-2,3-dimethylaniline **7** with the complex salt **62** in acetonitrile under reflux affords the molybdenum complex **63**. The best reagent for the molybdenum-mediated arylamine cyclization with concomitant aromatization and demetalation is commercial activated manganese dioxide (activated by azeotropic removal of water) [112]. The oxidative cyclization of complex **63** using this activated commercial manganese dioxide leads directly to 4-deoxycarbazomycin B **11** [111]. The electrophilicity of the molybdenum complex salt **62** is lower than of the iron complex salt **6a**. Therefore, 2-oxygenated tricyclic carbazole alkaloids are better targets for this method, because they require 3-oxygenated arylamines as starting materials, which are more electron-rich at the *ortho*-amino position (Scheme 25).

Electrophilic substitution of 3-methoxy-4-methylaniline **64** by the complex salt **62** leads to the molybdenum complex **65**. Oxidative cyclization of complex

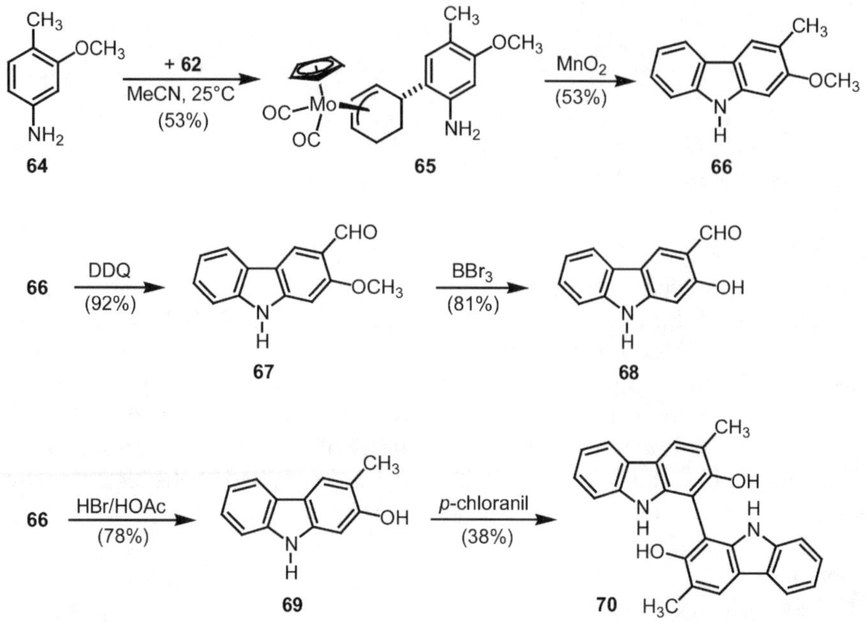

Scheme 25 Synthesis of the 2-oxygenated tricyclic carbazole alkaloids **66–69** and 1,1′-bis(2-hydroxy-3-methylcarbazole) **70**

65 with activated commercial manganese dioxide provides 2-methoxy-3-methylcarbazole **66**. This natural product [21] serves as a relay compound for the synthesis of further 2-oxygenated tricyclic carbazole alkaloids and a dimeric carbazole alkaloid [111]. Oxidation of the methyl group using 2,3-dichloro-5,6-dicyano-1,4-benzoquinone (DDQ) gives *O*-methylmukonal (glycosinine) **67**, which on cleavage of the ether affords mukonal **68**. Ether cleavage of compound **66** leads to 2-hydroxy-3-methylcarbazole **69**. Oxidative biaryl coupling of the carbazole **69**, using *p*-chloranil, provided 1,1′-bis(2-hydroxy-3-methylcarbazole) **70** [111].

The molybdenum-mediated arylamine cyclization was also applied to the total synthesis of pyrano[3,2-*a*]carbazole alkaloids (Scheme 26). Reaction of the 5-aminochromene **71** with the complex salt **62** affords the complex **72**, which on oxidative cyclization provides girinimbine **73**, a key compound for the transformation into further pyrano[3,2-*a*]carbazole alkaloids. Oxidation of **73** with DDQ leads to murrayacine **74**, while epoxidation of **73** using *meta*-chloroperbenzoic acid (MCPBA) followed by hydrolysis provides dihydroxygirinimbine **75** [113].

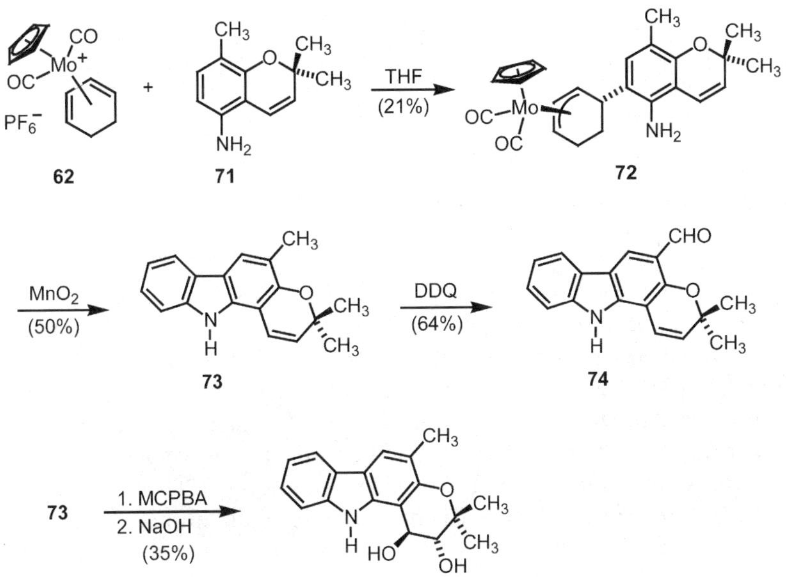

Scheme 26 Synthesis of girinimbine **73**, murrayacine **74**, and dihydroxygirinimbine **75**

4
Palladium-Catalyzed Synthesis of Carbazoles

The formation of carbon–carbon bonds by palladium-promoted reactions has been widely used in organic synthesis [114–116]. A major advantage is that most of these coupling reactions can be performed with catalytic amounts of palladium. Palladium(II)-catalyzed reactions, e.g., the Wacker process, are distinguished from palladium(0)-catalyzed reactions, e.g., the Heck reaction, since they require oxidative regeneration of the catalytically active palladium(II) species in a separate step [117]. Several groups have applied palladium-mediated and -catalyzed coupling reactions to the construction of the carbazole framework.

4.1
Cyclization of *N,N*-Diarylamines

The oxidative cyclization of *N,N*-diarylamines to carbazoles has been achieved by thermal or photolytic induction [7, 75]. However, the yields for this transformation are mostly moderate. Better results are obtained by the palladium(II)-mediated oxidative cyclization of *N,N*-diarylamines (Scheme 27). Oxidative cyclization by heating of the *N,N*-diarylamines **76** in the presence of a stoichiometric amount of palladium(II) acetate in acetic acid under reflux provides the corresponding 3-substituted carbazoles **77** in 70–80% yield [118]. The cou-

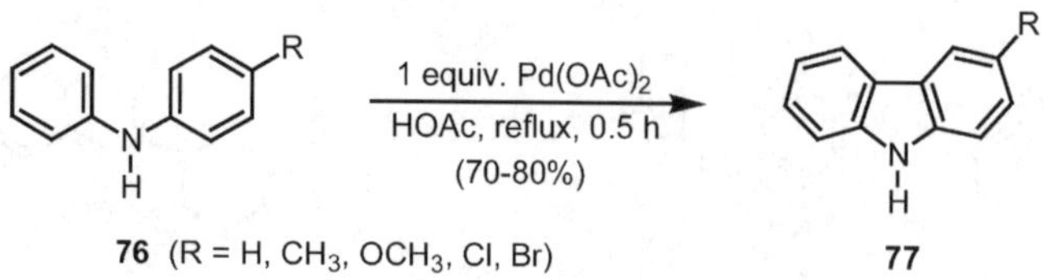

Scheme 27 Palladium(II)-mediated oxidative cyclization of *N,N*-diarylamines

pling is initiated by electrophilic attack of palladium(II) at the aromatic rings to an arylpalladium intermediate, which undergoes reductive elimination to generate the carbazole and palladium(0). By reoxidation of palladium(0) to palladium(II), this transformation becomes catalytic in the transition metal. The feasibility of achieving a catalytic cyclization for this type of reaction was demonstrated first by reoxidation of palladium(0) with copper(II) [119, 120], a catalytic system known from the Wacker oxidation [117]. Alternatively, *tert*-butyl hydroperoxide can be used as reoxidant for palladium(0) [121].

The cyclodehydrohalogenation of 2-halo-*N,N*-diarylamines is analogous to the classical Heck reaction [114–116] and represents a palladium(0)-catalyzed process (Scheme 28). Cyclization of the diarylamine **78** with a palladium(0) catalyst, generated in situ by reduction of palladium(II) with triethylamine, affords carbazole-1-carboxylic acid **79** in 73% yield [122].

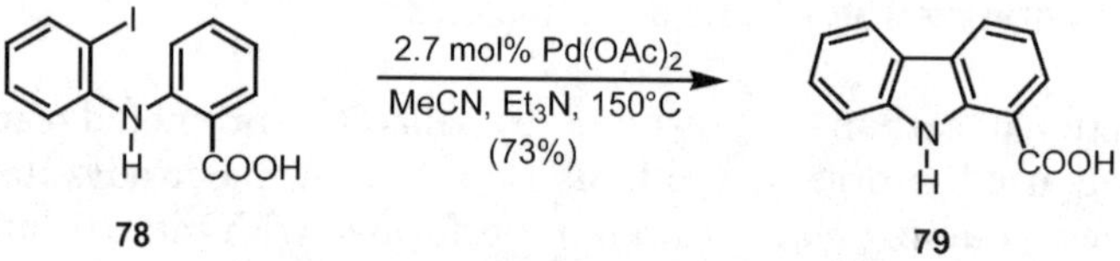

Scheme 28 Palladium(0)-catalyzed cyclization of 2-iodo-*N,N*-diarylamines

The required diarylamines, like **76** and **78**, can be prepared using the Goldberg coupling reaction followed by hydrolysis of the resulting diarylamide [123] or by the Buchwald–Hartwig amination [124]. Since the Buchwald–Hartwig amination represents a palladium(0)-catalyzed process, it can be combined with the cyclodehydrohalogenation of 2-halo-*N,N*-diarylamines. This leads to a simple construction of the carbazole framework by consecutive palladium(0)-catalyzed C–N and C–C bond formation [125]. A further recent palladium(0)-catalyzed carbazole synthesis uses the double Buchwald–Hartwig amination of 2,2′-dihalo-1,1′-biaryls with primary amines [126]. However, both procedures are still restricted to the synthesis of *N*-substituted carbazoles.

The palladium(II)-mediated oxidative cyclization of *N,N*-diarylamines is useful for convergent total syntheses of a range of structurally different carbazole alkaloids. Goldberg coupling of 2,3-dimethoxyacetanilide **80** and 2-bromo-5-methylanisole **81** and subsequent alkaline hydrolysis affords the diarylamine **82**

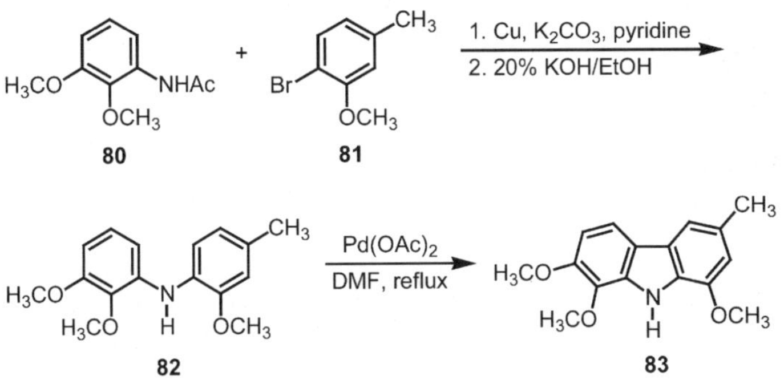

Scheme 29 Synthesis of murrayastine **83**

(Scheme 29). Cyclodehydrogenation of **82** with stoichiometric amounts of palladium(II) acetate in *N,N*-dimethylformamide (DMF) under reflux provides murrayastine **83** [17]. The same sequence of reactions can be applied to the synthesis of murrayaline A [17].

The Goldberg coupling between 5-acetylamino-2,2-dimethylchromene **84** and 5-bromo-2-methylanisole **85** followed by hydrolysis leads to the diarylamine **86**, which on palladium(II)-mediated oxidative cyclization affords pyrayafoline A **87** [17] (Scheme 30). Starting from 7-acetylamino-2,2-dimethylchromene, the method has been applied to the synthesis of *O*-methylpyrayafoline B [54].

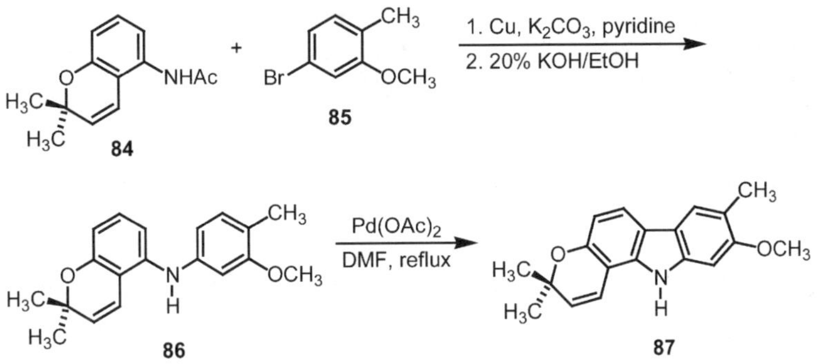

Scheme 30 Synthesis of pyrayafoline A **87**

Using an appropriate isoquinoline precursor, a highly convergent synthesis of ellipticine is achieved (Scheme 31). The Goldberg coupling of acetanilide **88** and 6-bromo-5,8-dimethylisoquinoline **89** and subsequent acidic hydrolysis gives the diarylamine **90**. Palladium(II)-mediated cyclodehydrogenation of **90** provides ellipticine **91** [127]. Although the yield of the cyclization step is only moderate, the overall yield of this two-step approach is still acceptable. The sequence of Goldberg coupling and palladium(II)-mediated cyclodehydro-

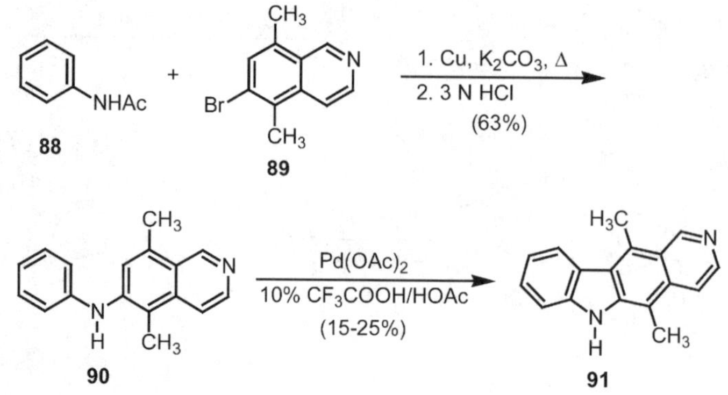

Scheme 31 Synthesis of ellipticine **91**

genation of the resulting *N*,*N*-diarylamine has been applied to the synthesis of tricyclic carbazoles as precursors of ellipticine and olivacine derivatives [123]. In this approach, the annulation of the pyridine ring is accomplished using a modified Pomeranz–Fritsch cyclization.

Buchwald–Hartwig amination of iodobenzene **92** with 2-benzyloxy-4-methylaniline **93** affords the diarylamine **94** in high yield (Scheme 32). In this case the Goldberg coupling gives poor yields. Oxidative cyclization of compound **94** using stoichiometric amounts of palladium(II) acetate in acetic acid under reflux leads to the carbazole **95**, which by reductive debenzylation provides

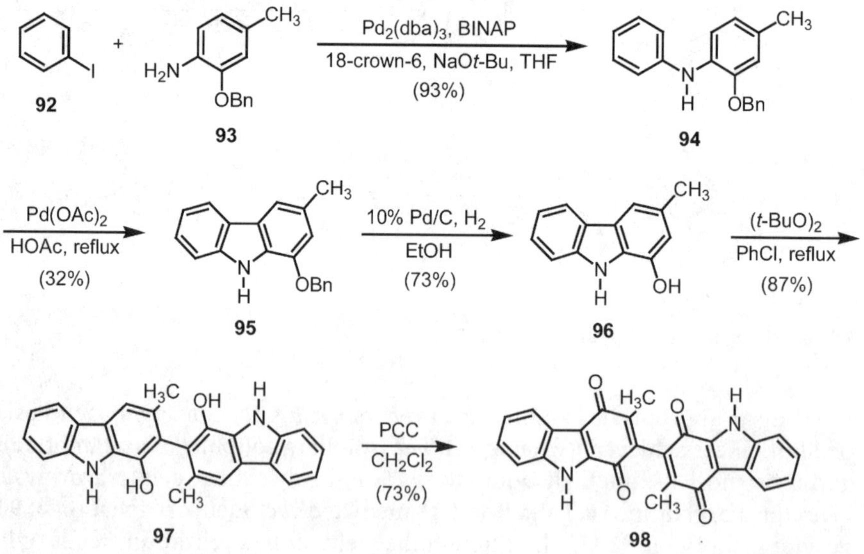

Scheme 32 Synthesis of *O*-demethylmurrayafoline A **96**, bis(*O*-demethylmurrayafoline-A) **97**, and bismurrayaquinone-A **98**

O-demethylmurrayafoline A **96** [66, 128]. Oxidative biaryl coupling of the 1-hydroxycarbazole **96** using di-*tert*-butyl peroxide affords bis(*O*-demethylmurrayafoline-A) **97** [65a, 66, 128]. Oxidation of the bisphenol **97** with pyridinium chlorochromate (PCC) leads to bismurrayaquinone-A **98** [129]. The low yield of the palladium(II)-mediated oxidative cyclization was explained by the electron-donating effect of the benzyloxy substituent. The projected palladium(II)-mediated cyclodehydrogenation to a precursor of clausenamine-A failed completely. Therefore, a different approach involving a reversal of the bond formations and palladium(0) catalysis was developed [64, 66]. Using a Suzuki cross coupling and subsequent cyclization of the resulting biaryl derivative by an intramolecular Buchwald–Hartwig amination provides the monomer required for the total synthesis of clausenamine-A.

The skeleton of carbazomadurin A was constructed utilizing three palladium-catalyzed reactions (Scheme 33). Buchwald–Hartwig amination of the aryl triflate **99** with the arylamine **100** gives the diarylamine **101**, which by oxidative cyclization with palladium(II) acetate affords the carbazole **102**. Following an exchange of the protecting groups, the palladium(0)-catalyzed Stille coupling of the 1-bromocarbazole **103** with the vinylstannane **104** leads to the carbazole **105**. Finally, reduction with diisobutylaluminum hydride (DIBAL) followed by cleavage of the silyl ethers using tetrabutylammonium fluoride (TBAF) provides carbazomadurin A **106** [130].

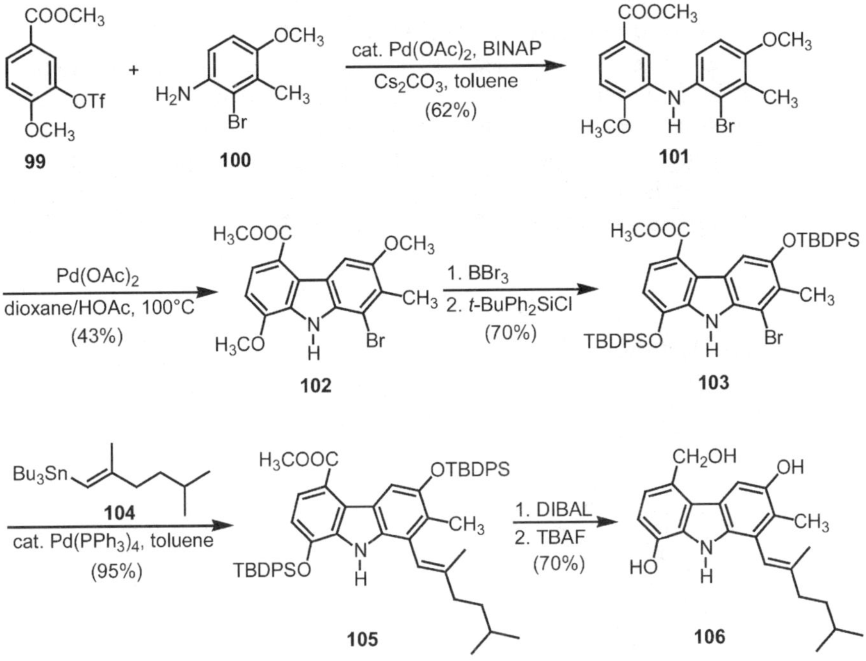

Scheme 33 Synthesis of carbazomadurin A **106**

4.2
Cyclization of Arylamino-1,4-benzoquinones

The palladium(II)-mediated oxidative cyclization is also applied to the synthesis of carbazole-1,4-quinone alkaloids. The required arylamino-1,4-benzoquinones are readily prepared by arylamine addition to the 1,4-benzoquinone and in situ reoxidation of the resulting hydroquinone [131].

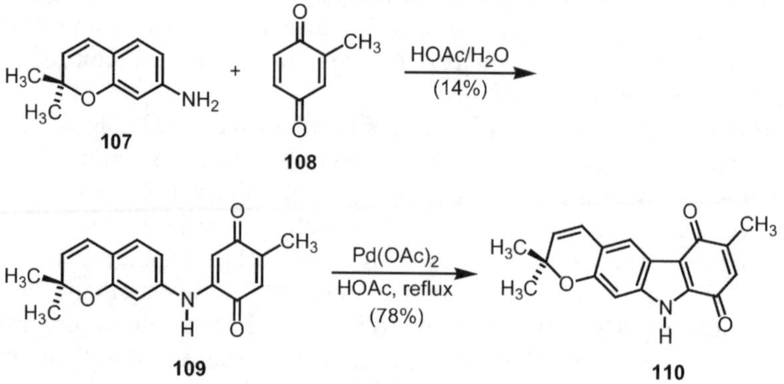

Scheme 34 Synthesis of pyrayaquinone A **110**

Addition of 7-amino-2,2-dimethylchromene **107** to 2-methyl-1,4-benzoquinone **108** in acetic acid/water leads to the 2-arylamino-5-methyl-1,4-benzoquinone **109** in moderate yield (Scheme 34). Oxidative cyclization of compound **109** using a stoichiometric amount of palladium(II) acetate in acetic acid under reflux provides pyrayaquinone A **110** [42, 132].

Acidic hydrolysis of 5-acetylamino-2,2-dimethylchromene **84**, the precursor for the synthesis of pyrayafoline A **87** (cf. Scheme 30), gives 5-amino-2,2-di-

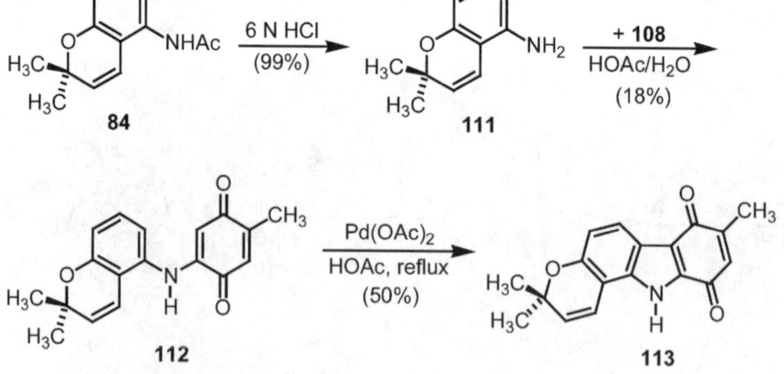

Scheme 35 Synthesis of pyrayaquinone B **113**

methylchromene **111** (Scheme 35). Addition of compound **111** to 2-methyl-1,4-benzoquinone **108** provides the 2-arylamino-5-methyl-1,4-benzoquinone **112**, which by cyclodehydrogenation with palladium(II) acetate affords pyrayaquinone B **113** [42, 132].

This approach using stoichiometric amounts of palladium(II) acetate has also been applied to the total synthesis of murrayaquinone A [132] and clausenaquinone A [41]. However, the more recent procedure using catalytic amounts of palladium(II) is superior [119] (Scheme 36). Addition of the arylamines **114** to 2-methyl-1,4-benzoquinone **108** by using Musso's conditions [133] affords the 2-arylamino-5-methyl-1,4-benzoquinones **115** in improved yields. The oxidative cyclization becomes catalytic in palladium by reoxidation of palladium(0) to palladium(II) with copper(II) [119], a procedure also used for the Wacker oxidation [117]. Application of this palladium(II)-catalyzed process to the oxidative cyclization of the benzoquinones **115** provides the carbazole-1,4-quinone alkaloids murrayaquinone A **116a**, koeniginequinone A **116b**, and koeniginequinone B **116c** [134]. A palladium(II)-catalyzed cyclodehydrogenation of 2-anilino-5-methyl-1,4-benzoquinone **115a** to murrayaquinone A **116a** is also achieved using *tert*-butyl hydroperoxide as an oxidant for palladium(0) [121].

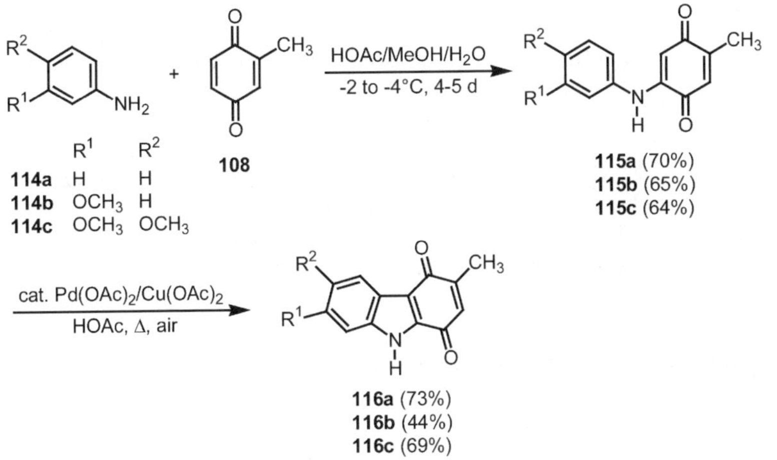

Scheme 36 Palladium(II)-catalyzed synthesis of murrayaquinone A **116a**, koeniginequinone A **116b**, and koeniginequinone B **116c**

Addition of the arylamines **117** to 2-methoxy-3-methyl-1,4-benzoquinone **118** affords regioselectively the 5-arylamino-2-methoxy-3-methyl-1,4-benzoquinones **119** (Scheme 37). Palladium(II)-catalyzed oxidative cyclization leads to the carbazole-1,4-quinones **28** [135, 136], previously obtained by the iron-mediated approach (cf. Scheme 14). Regioselective addition of methyllithium to the quinones **28** provides carbazomycin G **29a** and carbazomycin H **29b** [96, 135]. Reduction of **29a** with lithium aluminum hydride followed by elimination of water on workup generates carbazomycin B **23a** [135]. Addition of heptylmag-

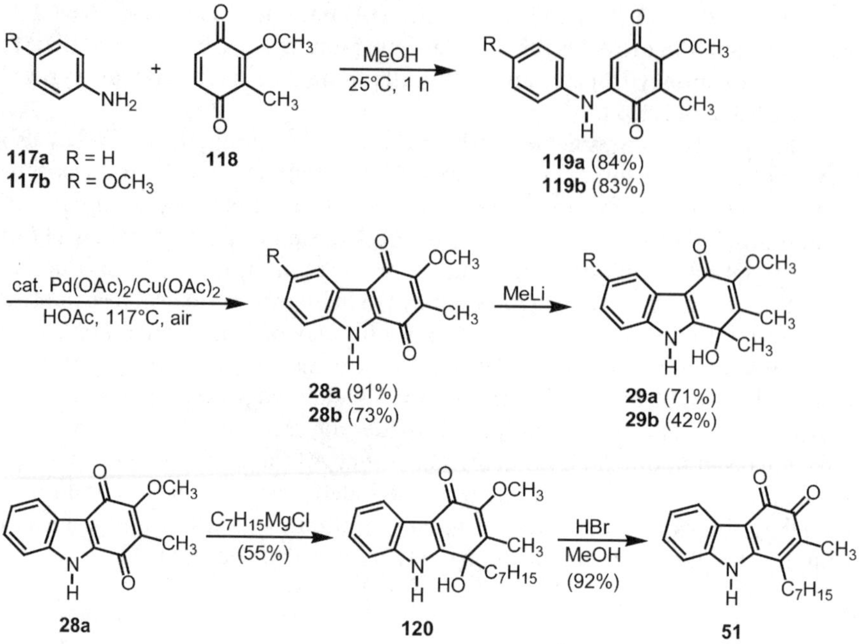

Scheme 37 Palladium(II)-catalyzed synthesis of carbazomycin G **29a**, carbazomycin H **29b**, and carbazoquinocin C **51**

nesium chloride at C1 of 3-methoxy-2-methylcarbazole-1,4-quinone **28a** affords the carbazole-1,4-quinol **120**, which by treatment with hydrogen bromide is converted to carbazoquinocin C **51** [136]. Thus, **28a** represents a versatile precursor for carbazole alkaloid synthesis.

4.3 Cyclization of Arylamino-1,2-benzoquinones

An even more direct approach to carbazole-3,4-quinone alkaloids is provided by the palladium(II)-mediated oxidative coupling of *ortho*-quinones with arylamines, which gives access to this class of natural products in a three-step route [137].

Ether cleavage of 4-heptyl-3-methylveratrole **121** using boron tribromide affords 4-heptyl-3-methylcatechol **122** (Scheme 38). Oxidation of the catechol **122** with *o*-chloranil to 4-heptyl-3-methyl-1,2-benzoquinone **123** and subsequent immediate addition of aniline leads to 5-anilino-4-heptyl-3-methyl-1,2-benzoquinone **124**. Unlike the very labile disubstituted *ortho*-quinone **123**, compound **124** is stable and can be isolated. Palladium(II)-mediated oxidative cyclization of the anilino-1,2-benzoquinone **124** provides carbazoquinocin C **51**.

The palladium(II)-mediated *ortho*-quinone cyclization is also applied to the synthesis of racemic carquinostatin A (±)-**55a** (Scheme 39). Ether cleavage of

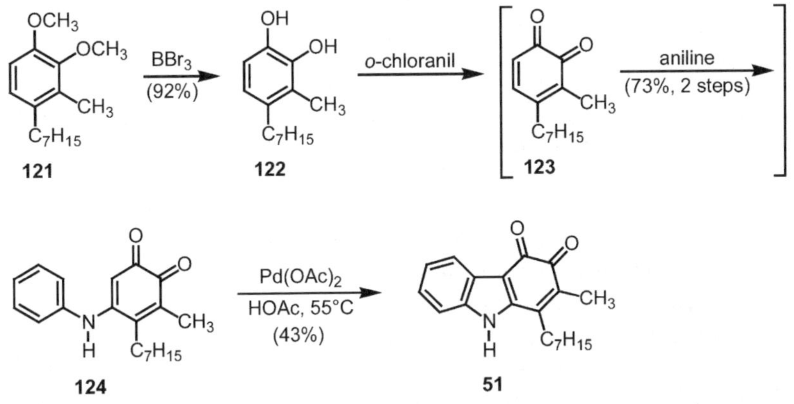

Scheme 38 *Ortho*-quinone cyclization to carbazoquinocin C **51**

the veratrole **125** to the catechol **126**, followed by oxidation to the *ortho*-quinone **127** and addition of 4-prenylaniline provides the arylamino-*ortho*-benzoquinone **128**. Palladium(II)-mediated cyclodehydrogenation of compound **128** and removal of the acetyl group by reduction affords (±)-carquinostatin A (±)-**55a.**

The *ortho*-quinone cyclization to the carbazole-3,4-quinones carbazoquinocin C **51** and (±)-carquinostatin A (±)-**55a** is more convergent than the previous approaches to these natural products, which required transformation to the *ortho*-quinone and/or introduction of substituents (prenyl or heptyl) following the carbazole ring formation (cf. Schemes 20, 21, and 37).

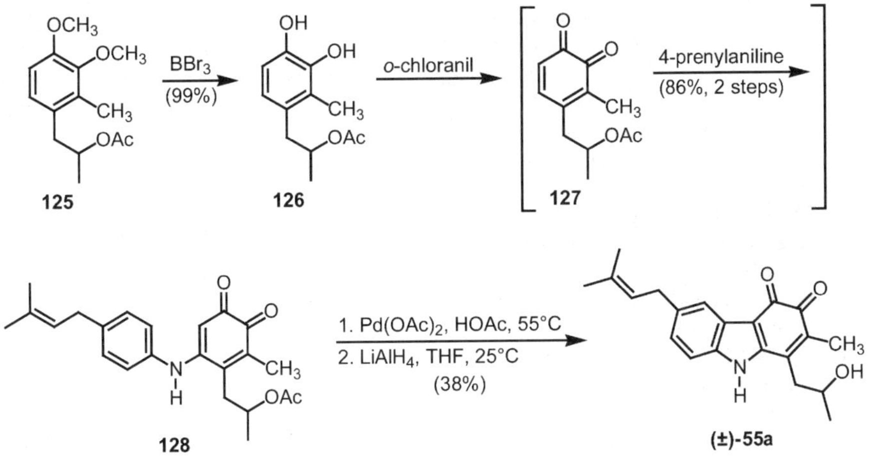

Scheme 39 *Ortho*-quinone cyclization to (±)-carquinostatin A (±)-**55a**

5 Conclusion

The oxidative coupling of arylamines and unsaturated six-membered hydrocarbons, mediated or catalyzed by transition metals, provides convergent routes to biologically active carbazole alkaloids and is widely applied to natural product synthesis. The different methods involving this strategy are the oxidative coupling of arylamines with (a) cyclohexa-1,3-dienes (iron-mediated), (b) cyclohexene (molybdenum-mediated), and (c) arenes or benzoquinones (palladium-catalyzed). Although, in principle, many of the carbazole alkaloids could be prepared by all three methods, the applicability and the utility is dependent on the substitution pattern and on the functional groups. The iron-mediated carbazole synthesis is highly useful for 1-oxygenated, 3-oxygenated, and 3,4-dioxygenated tricyclic carbazoles, as well as carbazolequinones and furocarbazoles. The molybdenum-mediated carbazole synthesis is preferentially applied to 2-oxygenated tricyclic carbazoles and pyranocarbazoles. The palladium-catalyzed carbazole synthesis is of advantage for highly functionalized carbazoles and carbazolequinones. In conclusion, the three synthetic approaches are complementary and are applied depending on the substitution of the carbazole alkaloid.

References

1. Kapil RS (1971) In: Manske RHF (ed) The alkaloids, vol 13. Academic, New York, p 273
2. Chakraborty DP (1977) In: Herz W, Grisebach H, Kirby GW (eds) Progress in the chemistry of organic natural products, vol 34. Springer, Wien, p 299
3. Husson HP (1985) In: Brossi A (ed) The alkaloids, vol 26. Academic, New York, p 1
4. Bhattacharyya P, Chakraborty DP (1987) In: Herz W, Grisebach H, Kirby GW, Tamm C (eds) Progress in the chemistry of organic natural products, vol 52. Springer, Wien, p 159
5. Chakraborty DP, Roy S (1991) In: Herz W, Grisebach H, Kirby GW, Steglich W, Tamm C (eds) Progress in the chemistry of organic natural products, vol 57. Springer, Wien, p 71
6. Chakraborty DP (1993) In: Cordell GA (ed) The alkaloids, vol 44. Academic, New York, p 257
7. Knölker H-J, Reddy KR (2002) Chem Rev 102:4303
8. Chakraborty DP, Roy S (2003) In: Herz W, Grisebach H, Kirby GW, Steglich W, Tamm C (eds) Progress in the chemistry of organic natural products, vol 85. Springer, Wien, p 125
9. Chakraborty DP, Barman BK, Bose PK (1965) Tetrahedron 21:681
10. Bhattacharyya P, Chakraborty DP (1973) Phytochemistry 12:1831
11. Das KC, Chakraborty DP, Bose PK (1965) Experientia 21:340
12. Ramsewak RS, Nair MG, Strasburg GM, DeWitt DL, Nitiss JL (1999) J Agric Food Chem 47:444
13. (a) Chowdhury BK, Chakraborty DP (1969) Chem Ind 549; (b) Chowdhury BK, Chakraborty DP (1971) Phytochemistry 10:1967
14. Bhattacharyya P, Maiti AK, Basu K, Chowdhury BK (1994) Phytochemistry 35:1085
15. Fiebig M, Pezzuto JM, Soejarto DD, Kinghorn AD (1985) Phytochemistry 24:3041
16. (a) Wu T-S, Ohta T, Furukawa H, Kuoh C-S (1983) Heterocycles 20:1267; (b) Furukawa H, Wu T-S, Ohta T, Kuoh C-S (1985) Chem Pharm Bull 33:4132

17. Furukawa H, Ito C, Yogo M, Wu T-S (1986) Chem Pharm Bull 34:2672
18. Bhattacharyya P, Jash SS, Chowdhury BK (1986) Chem Ind 246
19. Bhattacharyya P, Chakraborty A (1984) Phytochemistry 23:471
20. Chakraborty DP, Roy S, Guha R (1978) J Indian Chem Soc 55:1114
21. Bhattacharyya P, Chowdhury BK (1985) Indian J Chem 24B:452
22. Wu T-S, Huang S-C, Lai J-S, Teng C-M, Ko F-N, Kuoh C-S (1993) Phytochemistry 32:449
23. Ruangrungsi N, Ariyaprayoon J, Lange GL, Organ MG (1990) J Nat Prod 53:946
24. Jash SS, Biswas GK, Bhattacharyya SK, Bhattacharyya P, Chakraborty A, Chowdhury BK (1992) Phytochemistry 31:2503
25. Ito C, Nakagawa M, Wu T-S, Furukawa H (1991) Chem Pharm Bull 39:2525
26. Yamasaki K, Kaneda M, Watanabe K, Ueki Y, Ishimaru K, Nakamura S, Nomi R, Yoshida N, Nakajima T (1983) J Antibiot 36:552
27. (a) Sakano K, Nakamura S (1980) J Antibiot 33:961; (b) Kaneda M, Sakano K, Nakamura S, Kushi Y, Iitaka Y (1981) Heterocycles 15:993
28. Saha C, Chakraborty A, Chowdhury BK (1996) Indian J Chem 35B:667
29. Cardellina JH, Kirkup MP, Moore RE, Mynderse JS, Seff K, Simmons CJ (1979) Tetrahedron Lett 4915
30. Kato S, Kawai H, Kawasaki T, Toda Y, Urata T, Hayakawa Y (1989) J Antibiot 42:1879
31. Jackson PM, Moody CJ, Mortimer RJ (1991) J Chem Soc Perkin Trans I 2941
32. Iwatsuki M, Niki E, Kato S, Nishikori K (1992) Chem Lett 1735
33. Kotoda N, Shin-ya K, Furihata K, Hayakawa Y, Seto H (1997) J Antibiot 50:770
34. (a) Sakano K, Ishimaru K, Nakamura S (1980) J Antibiot 33:683; (b) Naid T, Kitahara T, Kaneda M, Nakamura S (1987) J Antibiot 40:157
35. Kondo S, Katayama M, Marumo S (1986) J Antibiot 39:727
36. Hook DJ, Yacobucci JJ, O'Connor S, Lee M, Kerns E, Krishnan B, Matson J, Hesler G (1990) J Antibiot 43:1347
37. Kato S, Shindo K, Kataoka Y, Yamagishi Y, Mochizuki J (1991) J Antibiot 44:903
38. (a) Furukawa H (1994) J Indian Chem Soc 71:303; (b) Bouaziz Z, Nebois P, Poumaroux A, Fillion H (2000) Heterocycles 52:977
39. Takeya K, Itoigawa M, Furukawa H (1989) Eur J Pharmacol 169:137
40. Saha C, Chowdhury BK (1998) Phytochemistry 48:363
41. Wu T-S, Huang S-C, Wu P-L, Lee K-H (1994) Bioorg Med Chem Lett 4:2395
42. Furukawa H, Yogo M, Ito C, Wu T-S, Kuoh C-S (1985) Chem Pharm Bull 33:1320
43. Kaneda M, Naid T, Kitahara T, Nakamura S, Hirata T, Suga T (1988) J Antibiot 41:602
44. Tanaka M, Shin-ya K, Furihata K, Seto H (1995) J Antibiot 48:326
45. Shin-ya K, Tanaka M, Furihata K, Hayakawa Y, Seto H (1993) Tetrahedron Lett 34:4943
46. Shin-ya K, Shimizu S, Kunigami T, Furihata K, Furihata K, Seto H (1995) J Antibiot 48:574
47. (a) Chakraborty DP, Barman BK, Bose PK (1964) Sci Cult 30:445; (b) Dutta NL, Quasim C (1969) Indian J Chem 7:307
48. Chakraborty DP, Das KC, Chowdhury BK (1971) J Org Chem 36:725
49. Joshi BS, Kamat VN, Gawad DH (1970) Tetrahedron 26:1475
50. Joshi BS, Kamat VN, Gawad DH, Govindachari TR (1972) Phytochemistry 11:2065
51. Chakraborty DP, Das KC (1968) Chem Commun 967
52. Ray S, Chakraborty DP (1976) Phytochemistry 15:356
53. Furukawa H, Wu T-S, Kuoh C-S (1985) Heterocycles 23:1391
54. Ito C, Nakagawa M, Wu T-S, Furukawa H (1991) Chem Pharm Bull 39:1688
55. Kirsch GH (2001) Curr Org Chem 5:507
56. Ito C, Furukawa H (1990) Chem Pharm Bull 38:1548
57. (a) Goodwin S, Smith AF, Horning EC (1959) J Am Chem Soc 81:1903; (b) Woodward RB, Iacobucci GA, Hochstein FA (1959) J Am Chem Soc 81:4434

58. (a) Dalton LK, Demerac S, Elmes BC, Loder JW, Swan JM, Teitei T (1967) Aust J Chem 20:2715; (b) Svoboda GH, Poore GA, Monfort ML (1968) J Pharm Sci 57:1720; (c) Pindur U, Haber M, Sattler K (1992) Pharm Uns Zeit 21:21
59. (a) Gribble GW (1990) In: Brossi A (ed) The alkaloids, vol 39. Academic, New York, p 239; (b) Potier P (1992) Chem Soc Rev 21:113; (c) Álvarez M, Joule JA (2001) In: Cordell GA (ed) The alkaloids, vol 57. Academic, New York, p 235
60. (a) Gribble GW, Berthel SJ (1993) In: Atta-ur-Rahman (ed) Studies in natural product chemistry, vol 12. Elsevier, Amsterdam, p 365; (b) Prudhomme M (1997) Curr Pharm Des 3:265; (c) Pindur U, Kim Y-S, Mehrabani F (1999) Curr Med Chem 6:29
61. (a) Nakanishi S, Matsuda Y, Iwahashi K, Kase H (1986) J Antibiot 39:1066; (b) Yasuzuwa T, Iida T, Yoshida M, Hirayama N, Takahashi M, Shirahata K, Sano H (1986) J Antibiot 39:1072
62. (a) Furukawa H (1993) Trends Heterocycl Chem 3:185; (b) Tasler S, Bringmann G (2002) Chem Record 2:114
63. Wu T-S, Huang S-C, Wu P-L (1996) Tetrahedron Lett 37:7819
64. Zhang A, Lin G (2000) Bioorg Med Chem Lett 10:1021
65. (a) Bringmann G, Ledermann A, Francois G (1995) Heterocycles 40:293; (b) Bringmann G, Ledermann A, Holenz J, Kao M-T, Busse U, Wu HG, Francois G (1998) Planta Med 64:54
66. Lin G, Zhang A (2000) Tetrahedron 56:7163
67. Ito C, Thoyama Y, Omura M, Kajiura I, Furukawa H (1993) Chem Pharm Bull 41:2096
68. Pindur U (1990) Chimia 44:406
69. Bergman J, Pelcman B (1990) Pure Appl Chem 62:1967
70. Knölker H-J (1991) In: Dötz KH, Hoffmann RW (eds) Organic synthesis via organometallics. Vieweg, Braunschweig, p 119
71. Knölker H-J (1992) Synlett 371
72. Kawasaki T, Sakamoto M (1994) J Indian Chem Soc 71:443
73. Moody CJ (1994) Synlett 681
74. Pindur U (1995) In: Moody CJ (ed) Advances in nitrogen heterocycles, vol 1. JAI, Greenwich (CT), p 121
75. Knölker H-J (1995) In: Moody CJ (ed) Advances in nitrogen heterocycles, vol 1. JAI, Greenwich (CT), p 173
76. Hibino S, Sugino E (1995) In: Moody CJ (ed) Advances in nitrogen heterocycles, vol 1. JAI, Greenwich (CT), p 205
77. Knölker H-J (1998) In: Beller M, Bolm C (eds) Transition metals for organic synthesis, vol 1. Wiley-VCH, Weinheim, p 534
78. Knölker H-J (1999) Chem Soc Rev 28:151
79. (a) Hallam BF, Pauson PL (1958) J Chem Soc 642; (b) Birch AJ, Cross PE, Lewis J, White DA, Wild SB (1968) J Chem Soc A 332; (c) Birch AJ, Chamberlain KB, Haas MA, Thompson DJ (1973) J Chem Soc Perkin Trans I 1882; (d) Pearson AJ (1985) Metallo-organic chemistry. Wiley, Chichester, chaps 7 and 8
80. (a) Knölker H-J, Baum G, Foitzik N, Goesmann H, Gonser P, Jones PG, Röttele H (1998) Eur J Inorg Chem 993; (b) Knölker H-J, Baum E, Gonser P, Rohde G, Röttele H (1998) Organometallics 17:3916
81. Fischer EO, Fischer RD (1960) Angew Chem 72:919
82. Birch AJ, Kelly LF, Thompson DJ (1981) J Chem Soc Perkin Trans I 1006
83. (a) Knölker H-J, Ahrens B, Gonser P, Heininger M, Jones PG (2000) Tetrahedron 56:2259; (b) Knölker H-J (2000) Chem Rev 100:2941
84. Knölker H-J, Bauermeister M, Pannek J-B, Bläser D, Boese R (1993) Tetrahedron 49:841
85. (a) Knölker H-J, Bauermeister M, Pannek J-B (1992) Chem Ber 125:2783; (b) Birch AJ, Liepa AJ, Stephenson GR (1982) J Chem Soc Perkin Trans I 713

86. Fatiadi AJ (1976) Synthesis 65
87. Knölker H-J, Budei F, Pannek J-B, Schlechtingen G (1996) Synlett 587
88. Knölker H-J, Wolpert M (2003) Tetrahedron 59:5317
89. Knölker H-J, Bauermeister M (1993) Tetrahedron 49:11221
90. Knölker H-J, Bauermeister M (1994) J Indian Chem Soc 71:345
91. Knölker H-J, Baum E, Hopfmann T (1999) Tetrahedron 55:10391
92. Knölker H-J, Hopfmann T (2002) Tetrahedron 58:8937
93. Knölker H-J, Bauermeister M (1993) Helv Chim Acta 76:2500
94. Knölker H-J, Schlechtingen G (1997) J Chem Soc Perkin Trans I 349
95. Knölker H-J, Bauermeister M (1991) Heterocycles 32:2443
96. Knölker H-J, Fröhner W, Reddy KR (2003) Eur J Org Chem 740
97. Knölker H-J, Fröhner W (2000) Synthesis 2131
98. Knölker H-J, Reddy KR (2000) Tetrahedron 56:4733
99. Knölker H-J, Bauermeister M, Pannek J-B, Wolpert M (1995) Synthesis 397
100. Knölker H-J, Baum G, Pannek J-B (1996) Tetrahedron 52:7345
101. Knölker H-J, Fröhner W (1999) Tetrahedron Lett 40:6915
102. Knölker H-J, Fröhner W (1997) Tetrahedron Lett 38:1535
103. Knölker H-J, Baum E, Reddy KR (2000) Tetrahedron Lett 41:1171
104. Knölker H-J, Baum E, Reddy KR (2000) Chirality 12:526
105. Knölker H-J, Braier A, Bröcher DJ, Cämmerer S, Fröhner W, Gonser P, Hermann H, Herzberg D, Reddy KR, Rohde G (2001) Pure Appl Chem 73:1075
106. Billington DC (1985) Chem Soc Rev 14:93
107. Knölker H-J, Fröhner W, Wagner A (1998) Tetrahedron Lett 39:2947
108. Faller JW, Murray HH, White DL, Chao KH (1983) Organometallics 2:400
109. McCallum JS, Sterbenz JT, Liebeskind LS (1993) Organometallics 12:927
110. (a) Pearson AJ, Khan MdNI (1984) J Am Chem Soc 106:1872; (b) Pearson AJ, Khan MdNI (1984) Tetrahedron Lett 25:3507; (c) Pearson AJ, Khan MdNI, Clardy JC, Cun-heng H (1985) J Am Chem Soc 107:2748
111. Knölker H-J, Goesmann H, Hofmann C (1996) Synlett 737
112. Pearson AJ, Ong CW (1982) J Org Chem 47:3780
113. Knölker H-J, Hofmann C (1996) Tetrahedron Lett 37:7947
114. Heck RF (1985) Palladium reagents in organic synthesis. Academic, New York
115. Tsuji J (1995) Palladium reagents and catalysts – innovations in organic synthesis. Wiley, Chichester
116. Li JJ, Gribble GW (2000) Palladium in heterocyclic chemistry – a guide for the synthetic chemist. Pergamon, Oxford
117. (a) Tsuji J (1995) Palladium reagents and catalysts – innovations in organic synthesis. Wiley, Chichester, chap 3; (b) Omae I (1998) Applications of organometallic compounds. Wiley, Chichester, chap 20.6; (c) Iida H, Yuasa Y, Kibayashi C (1980) J Org Chem 45:2938
118. Åkermark B, Eberson L, Jonsson E, Pettersson E (1975) J Org Chem 40:1365
119. Knölker H-J, O'Sullivan N (1994) Tetrahedron 50:10893
120. Li JJ, Gribble GW (2000) Palladium in heterocyclic chemistry – a guide for the synthetic chemist. Pergamon, Oxford, chaps 1.1 and 3.2
121. (a) Åkermark B, Oslob JD, Heuschert U (1995) Tetrahedron Lett 36:1325; (b) Hagelin H, Oslob JD, Åkermark B (1999) Chem Eur J 5:2413
122. Ames DE, Opalko A (1984) Tetrahedron 40:1919
123. (a) Jackson AH, Jenkins PR, Shannon PVR (1977) J Chem Soc Perkin Trans I 1698; (b) Hall RJ, Marchant J, Oliveira-Campos AMF, Queiroz M-JRP, Shannon PVR (1992) J Chem Soc Perkin Trans I 3439; (c) Oliveira-Campos AMF, Queiroz M-JRP, Raposo

MMM, Shannon PVR (1995) Tetrahedron Lett 36:133; (d) Chunchatprasert L, Dharmasena P, Oliveira-Campos AMF, Queiroz M-JRP, Raposo MMM, Shannon PVR (1996) J Chem Res S 84, M 0630; (e) Raposo MMM, Oliveira-Campos AMF, Shannon PVR (1997) J Chem Res S 354, M 2270
124. (a) Hartwig JF (1997) Synlett 329; (b) Wolfe JP, Wagaw S, Marcoux JF, Buchwald SL (1998) Acc Chem Res 31:805; (c) Hartwig JF (1998) Angew Chem 110:2154, Angew Chem Int Ed 37:2046; (d) Yang BH, Buchwald SL (1999) J Organomet Chem 576:125; (e) Wolfe JP, Tomori H, Sadighi JP, Yin J, Buchwald SL (2000) J Org Chem 65:1158
125. (a) Iwaki T, Yasuhara A, Sakamoto T (1999) J Chem Soc Perkin Trans I 1505; (b) Bedford RB, Cazin CSJ (2002) Chem Commun 2310
126. Nozaki K, Takahashi K, Nakano K, Hiyama T, Tang H-Z, Fujiki M, Yamaguchi S, Tamao K (2003) Angew Chem 115:2097, Angew Chem Int Ed 42:2051
127. Miller RB, Moock T (1980) Tetrahedron Lett 21:3319
128. Lin G, Zhang A (1999) Tetrahedron Lett 40:341
129. Bringmann G, Ledermann A, Stahl M, Gulden K-P (1995) Tetrahedron 51:9353
130. Knölker H-J, Knöll J (2003) Chem Commun 1170
131. (a) Luly JR, Rapoport H (1981) J Org Chem 46:2745; (b) Bittner S, Krief P (1990) Synthesis 350
132. Yogo M, Ito C, Furukawa H (1991) Chem Pharm Bull 39:328
133. Musso H, Döpp D (1966) Chem Ber 99:1470
134. Knölker H-J, Reddy KR (2003) Heterocycles 60:1049
135. Knölker H-J, Fröhner W (1998) J Chem Soc Perkin Trans I 173
136. Knölker H-J, Fröhner W, Reddy KR (2002) Synthesis 557
137. Knölker H-J, Reddy KR (1999) Synlett 596

Top Curr Chem (2005) 244: 149–213
DOI 10.1007/b96891

Recent Advances in Charge-Accelerated Aza-Claisen Rearrangements

Udo Nubbemeyer (✉)

Institut für Organische Chemie und Abteilung für Lehramtskandidaten der Chemie, Fachbereich Chemie und Pharmazie, Johannes Gutenberg-Universität Mainz, Duesbergweg 10–14, 55099 Mainz, Germany
nubbemey@mail.uni-mainz.de

1 **Introduction** 150

2 **Aliphatic Simple Aza-Claisen Rearrangements** 151

3 **Simple Aromatic Aza-Claisen Rearrangements** 159

4 **Amide Enolate Rearrangements** 166

5 **Zwitterionic Aza-Claisen Rearrangements** 173
5.1 Alkyne Carbonester Claisen Rearrangements 175
5.2 Ketene Claisen Rearrangements 178
5.2.1 Stereochemical Results: 1,3-Chirality Transfer and Internal Asymmetric Induction 182
5.2.2 Stereochemical Results: External Asymmetric Induction/Remote Stereocontrol 193
5.2.3 Stereochemical Results: External Asymmetric Induction/Auxiliary Control . . 196
5.3 Allene Carbonester Claisen Rearrangement 201

6 **Alkyne Aza-Claisen Rearrangements** 203

7 **Iminoketene Claisen Rearrangements** 206

References 211

Abstract Aza-Claisen rearrangements (3-aza-Cope rearrangements) have gained an increasing interest in synthetic organic chemistry. Originally, the exceptionally high reaction temperatures of this hetero variant of the well-known 3,3-sigmatropic reaction limited their applicability to selected molecules. Since about 1970, charge acceleration enabled a significant reduction of the reaction temperature to be achieved, and cation- and anion-promoted rearrangements found their way into the syntheses of more complex molecules. The first total syntheses of natural products were reported. The development of zwitterionic aza-Claisen rearrangements allowed the reactions to be run at room temperature or below, and the charge neutralization served as the highly efficient driving force. After overcoming several teething troubles, the method was established as a reliable conversion displaying various stereochemical advantages. The first successful total syntheses of natural products incorporating the aza-Claisen rearrangement as a key step emphasized the synthetic potential. To date the aza-Claisen rearrangements are far from being exhausted. Still, an enantioselectively catalyzed variant has to be developed. This review summarizes one decade of investigation efforts in this area.

Keywords Aza-Claisen rearrangement · 3-Aza-Cope rearrangement · Chirality transfer · Asymmetric induction · Charge acceleration

Abbreviations

Acc	Acceptor
BINOL	Binaphthol
BOX	Bis(oxazolinyl)
Cy	Cyclohexyl
DDQ	2,3-Dichloro-5,6-dicyano-1,4-benzoquinone
DMAD	Dimethyl acetylene dicarboxylate
LiHMDS	Lithium hexamethyldisilazide
MCPBA	3-Chloroperbenzoic acid
MPA	Mycophenolic acid
PG	Protecting group
Pht	Phthaloyl
PMB	4-Methoxybenzyl
Proton sponge	1,8-Bis-(dimethylamino) naphthalene
TBS	TBDMS, *tert*-butyldimethylsilyl
TFA	Trifluoroacetic acid
TOSMIC	*p*-Toluenesulfonylmethyl isocyanide

1 Introduction

3,3-Sigmatropic rearrangements are defined as uncatalyzed processes to migrate a sigma bond of two connected allyl systems from position 1 to position 3. That means both allyl systems suffer from an allyl inversion. Though described for the first time in 1940, the Cope rearrangement can be considered as the basic type of such a process, since C–C bonds only are reorganized during the course of the reaction [1]. More than two decades earlier, in 1912, L. Claisen first described the rearrangement of aromatic allyl vinyl ethers to generate *o*-allyl phenols [2]. This so-called Claisen rearrangement is characterized by the replacement of the C3 carbon of the rearrangement system against a heteroatom X. The basic Claisen rearrangement bears X=O; consequently, such a process can be termed as a 3-oxa Cope rearrangement. Analyzing the literature, rearrangement systems displaying other heteroatoms X in position 3 can be found as hetero Claisen and 3-hetero Cope rearrangements. Focussing on systems with X=N, names such as aza- and amino-Claisen as well as 3-aza-Cope rearrangement occur in the literature. Furthermore, the term aza/amino Claisen rearrangement is widely used for nitrogen introduction processes rearranging 1-aza-3-oxy-Cope systems (imidates) to generate carbamates. Finally, the Fischer indole synthesis represents a special type of aza-Claisen rearrangement incorporating two N atoms in the 3 and 4 positions of the rearrangement system.

Intending to set a firm basis concerning the notion of the sigmatropic rearrangements, the following review will use the term Claisen rearrangement

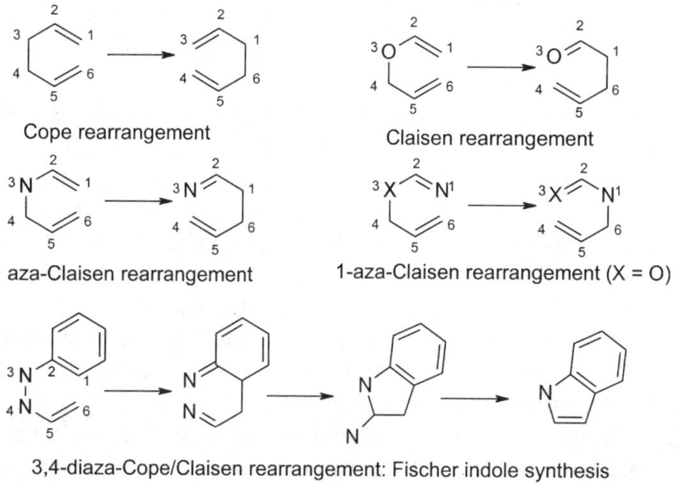

Fig. 1 Nomenclature of aza-Claisen rearrangements

for 3,3-sigmatropic core systems incorporating a heteroatom X in position 3, i.e., an aza-Claisen-type rearrangement is characterized by X=nitrogen (Fig. 1).

For a long period, aza-Claisen rearrangements were regarded as a sophisticated variant of the widely used oxygen analog. Because of the much more drastic reaction conditions, few applications have been published. Most investigation had been restricted to fundamental research on aza-Claisen rearrangements. About 25 years ago, the perception tended to change, because the nitrogen atom in the central position of the sigmatropic core systems was discerned as an ideal anchor for catalysts such as protons, Lewis acids, and for chiral auxiliaries in enantioselective rearrangements. Charge acceleration allowed a significant reduction of the reaction temperature, recommending the process now as a suitable key step in complex molecule syntheses. Finally, the development of very mild zwitterionic variants enabled the aza-Claisen rearrangement to be classified as a powerful method in synthetic organic chemistry. This review reports on recent advances in aza-Claisen rearrangements [3].

2 Aliphatic Simple Aza-Claisen Rearrangements

In analogy to the oxygen analogs, the simple aliphatic aza-Claisen rearrangements represent the basis of such reactions. Generally, the replacement of the oxygen offers two advantages. The vinyl double bond of the sigmatropic framework can be built up with a high *E* selectivity, since a bulky C1 substituent and the chain-branched nitrogen will adopt a maximal distance around the enamine moiety. Furthermore, only two valences of the nitrogen are occupied by allyl and vinyl substituents of the rearrangement system; the third one potentially can bear an optically active subunit intending to run the rearrangement under the influence of an external chirality-inducing side chain (auxiliary control).

The major problem of the aza-Claisen rearrangement is the extremely high temperature, excluding the presence of a variety of functional groups, upon running the reaction. Uncatalyzed simple allyl vinyl amines undergo the 3,3-sigmatropic conversion at about 250 °C, while the somewhat more activated aromatic analogs require 200–210 °C [4]. Hence, charge acceleration was found to be a promising tool to achieve a significant decrease of the reaction temperature. Only 80–120 °C was required upon running the sigmatropic rearrangement in the presence of a proton and a Lewis acid. Alternatively, a comparable temperature-decreasing effect was observed after conversion of the central nitrogen into a peralkylated ammonium salt.

Intending to use the aliphatic simple aza-Claisen rearrangement to generate new C–C bonds, several prerequisites had to be considered. The first problem to be solved was the smooth and selective generation of the allyl vinyl amine backbone. The second challenge was the 3,3-sigmatropic reaction: the rearrangement should pass a single, highly ordered transition state to give rise to diastereoselective formation of the product. Since both reactant and product might suffer from imine–enamine equilibration, some difficulties concerning the unique sense of the rearrangement and the stable configuration of a stereogenic α center (with respect to the new imine) have to be taken into account.

A set of systematic investigations has been published by Stille et al. [5]. Starting from allylamine **1**, an optimized three-step sequence of initial imine **3** formation with aldehyde **2**, *N*-acylation to **5** using acid chloride **4**, and subsequent $LiAlH_4$ reduction delivered the desired rearrangement systems **6** in 60–96% yield overall. Even though the enamide **5** formation was found to be unselective with respect to the vinyl double bond, the enamine **6** was isolated with substantially higher *E* selectivity pointing out some epimerization during the course of the reduction. The aza-Claisen rearrangement proceeded upon heating the enamines **6** to >100 °C (dioxane, toluene) in the presence of 0.3 to 0.8 eq. of HCl. Always, the intermediately formed imines **7** were reduced by means of $LiAlH_4$ to give the corresponding amines **8** as stable products. The success of the rearrangement strongly depended on the substitution pattern of the aldehyde **2** involved: branched starting material **6** (R^2, $R^3 \neq H$) underwent a smooth conversion to give the new amines **8**. In contrast, nonbranched derivatives (R^2 or R^3=H, small rings) suffered from competing reactions such as oligomerization and reductive amination [5a]. The replacement of the originally used HCl by different Lewis acids led to an extension of the original limitations. In the presence of one equivalent of Me_3Al a series of *N*-allyl enamines could be rearranged with high yield (Table 1, entries 1–6; Scheme 1) [5b].

The rearrangement of unsymmetrical allylamines **9** was investigated to exclude any competing 1,3-rearrangement during the course of the reaction. Allyl vinyl amines **10** were generated via condensation starting from allylamine **9** and isobutyraldehyde **2**. The substrates **10** were subjected to the acid-accel-

Table 1 [5]

Entry	Educt	R/R′	R^1	R^2	R^3	Acid/yield (%)/dr			Ref.
1	**6**	*i*Pr/H	H	Me	Me	HCl/81	$TiCl_4$/71	Me_3Al/95	a,b
2	**6**	*i*Pr/H	H	Et	H	HCl/0	$TiCl_4$/0	Me_3Al/84	a,b
3	**6**[b]	*i*Pr/H	H	Ph	Me	HCl/77	$TiCl_4$/88	Me_3Al/92	a,b
4	**6**	*i*Pr/H	$(CH_2)_4$		H	HCl/99	$TiCl_4$/92	Me_3Al/96	a,b
5	**6**	*i*Pr/H	$(CH_2)_3$		H	HCl/10	$TiCl_4$/3	Me_3Al/83	a,b
6	**6**	Cy/H	H	Me	Me	HCl/85	$TiCl_4$/73	Me_3Al/99	b
7	**10a**	*n*Pr/H	H	Me	Me	HCl/69	$TiCl_4$/79	Me_3Al/80	c
8	**10a**	Ph/H	H	Me	Me	HCl/81	$TiCl_4$/80	–	c
9	**10b**	*n*Pr/H	H	Me	Me	HCl/76	$TiCl_4$/78	Me_3Al/87	c
10	**10b**	*n*Pr/H	H	Me	Me	–	–	Me_3Al/56	c
11[a]	**10c**	*n*Bu/H	$(CH_2)_4$		H	HCl/69 54:46	$TiCl_4$/72 55:45	Me_3Al/94 67:33	c,d
12	**10c**	*n*Bu/H	H	Et	H	–	–	Me_3Al/88 62:38	d
13	**10c**[b]	*n*Bu/H	H	Ph	Me	HCl/54 95:5	$TiCl_4$/48 80:20	Me_3Al/86 68:32	d
14[a]	**10c**	Me/Me	$(CH_2)_4$		H	HCl/98 [89:3]:8	$TiCl_4$/84 [65:11]:24	Me_3Al/98 [73:7]:20	d
15	**10c**	Me/Me	H	Et	H	–	–	Me_3Al/78 >98:2	d
16	**10c**[b]	Me/Me	H	Ph	Me	–	–	Me_3Al/95 >95:5	d

[a] Diastereoselective final DIBALH reduction.
[b] *E*:*Z*=86:14–90:10.

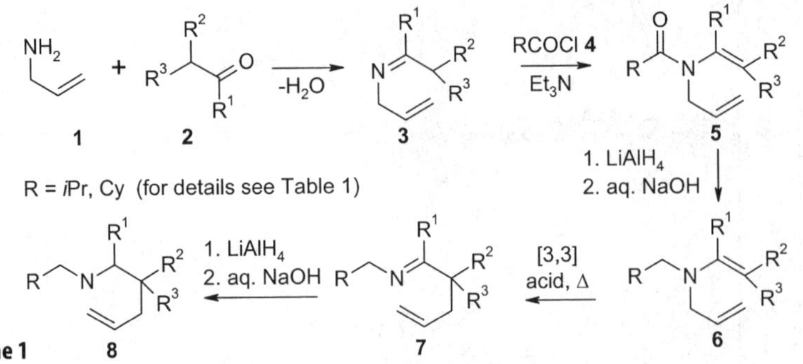

Scheme 1

erated rearrangement conditions. Smooth conversions led to the intermediates **11**, which were finally reduced with $LiAlH_4$ to give the stable amines **12**. Careful analyses (GC, ^{1}H NMR) of the products **12** indicated that all rearrangements proceeded stereospecifically (**a**→**a**, **b**→**b**). No crossover products originating from competing 1,3-sigmatropic and stepwise reactions could be detected (**a**→**b**, **b**→**a**) (Table 1, entries 7–10). The rearrangement of **10a** gave **11a** with exclusive *E* double bond geometry (Scheme 2) [5c].

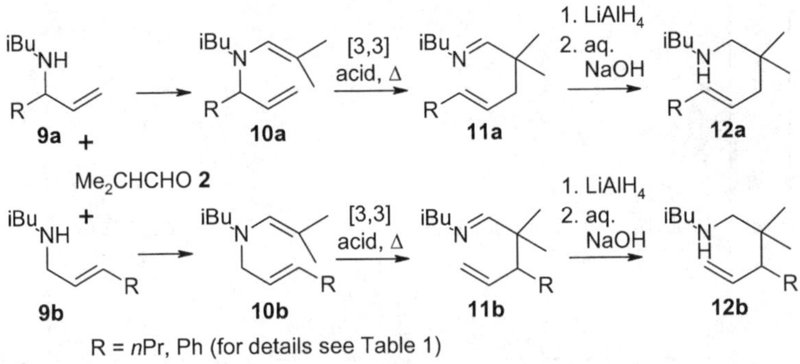

Scheme 2

The ability to generate defined configured allyl and enamine moieties in the 3,3-sigmatropic rearrangement framework raised the question of achieving internal and external asymmetric induction upon running aza-Claisen reactions [5d]. First investigations focused on conversions of type **10c** reactants (R=*n*Bu/R′=H). Allyl alcohols **13** were converted into the corresponding allylamines via Overman trichloroacetimidate rearrangement and a consecutive saponification resulting in allylamine **14** [6]. A three-step sequence (as displayed in Scheme 1) led to the allyl vinyl systems **10c** with high *E* selectivity concerning the enamine subunit and 56 to 89% yield.

Though the rearrangements of the reactants **10c** delivered single regioisomers, the products **12c** (R=*n*Bu/R′=H) were isolated as a mixture of diastereomers after the rearrangement reduction sequence. Likewise, the 3,3-sigmatropic process must have passed chair- and boat-like transition states or, alternatively, the products **11c** suffered from some epimerization via imine–enamine equilibrium (Table 1, entries 11–13). Intending to destabilize the boat-like transition states, the aza-Claisen systems **10c** (R=Me/R′=Me) were heated in the presence of an appropriate acid, and the rearranged crude imines **11c** were immediately reduced by $LiAlH_4$ and DIBALH. Analyzing the products **12c** (R=Me/R′=Me), substantially higher diastereoselectivities could be achieved, indicating the nearly exclusive passing of chair-like transition states during the course of the rearrangement. It was noteworthy that the conversion of the Ph/Me-substituted reactant **10c** (R^1=H, R^2/R^3=Ph/Me) displayed a comparatively high selectivity on generating **12c** even though **10c** was used as an *E*/*Z* vinyl amine mixture. Obviously, the presence of the acid allowed a fast *E*/*Z* interconversion prior to the sigmatropic process, enabling the isomer to be rearranged predominantly with minimized repulsive interactions. The second minor isomer found after running this special sequence was the corresponding *Z* olefin built up via a minor chair transition state (Scheme 3).

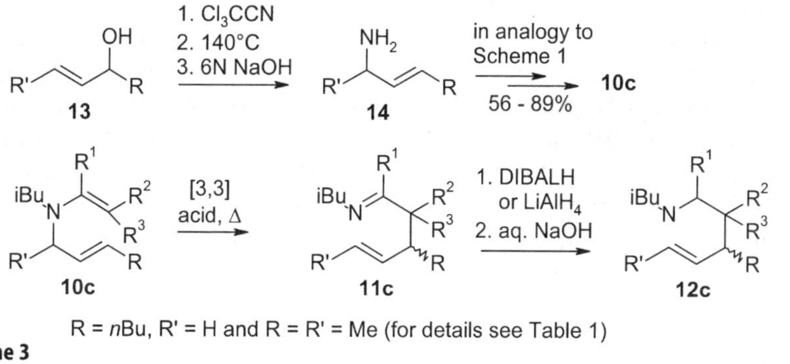

Scheme 3

First investigations using a chiral substituent attached to the free valence of the nitrogen for efficient asymmetric induction gave disappointing results [5d]. The rearrangement of *N*-allyl enamines **15** in the presence of proton and Lewis acids produced γ,δ-unsaturated imines as mixtures of diastereomers with acceptable yield and regioselectivity. The de determined after the final $LiAlH_4$ reduction to amine **16** varied between 8 and 20%. The passing of a single transition state conformation failed. Further efforts investigating sterically more demanding auxiliaries to enable a more effective 1,5-asymmetric induction (1,4-induction including an intermediately built chiral ammonium center) were strongly recommendable (Scheme 4).

Finally, ring-expansion reactions starting from optically active 1,2-divinyl pyrrolidines **19** promised a smooth generation of nine-membered azonine derivatives **20** bearing defined configured centers and double bonds [5d]. Firstly, *N*-BOC prolinal **17** was converted into the divinyl pyrrolidine **18** via Wittig olefination (mixture of *E* and *Z* olefins), protective group removal, and a final con-

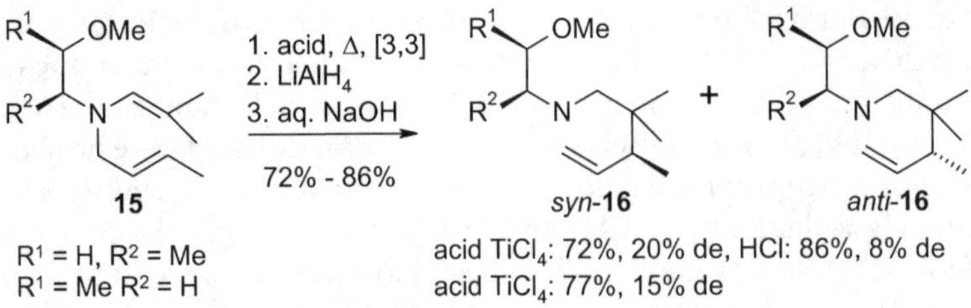

Scheme 4

densation with phenyl acetaldehyde with 69% (R=Ph) and 79% (R=Me) yield. The rearrangement/$LiAlH_4$ sequence of **18** (R=Me) delivered the azonine **20** (via **19**) as a mixture of diastereomers; concerning the stereogenic centers, the olefin was exclusively *Z*, indicating the passing of a boat-like transition state. Since the final diastereomer pattern in **20** represented the same ratio as present in the starting material **18**, the result could be interpreted as a sigmatropic rearrangement characterized by a complete 1,3-chirality transfer. However, a thermodynamic reason cannot be excluded, since the reactant **19** and the intermediately formed γ,δ-unsaturated imines might have suffered from some equilibration as observed in several reactions discussed above. In contrast, subjecting the divinyl pyrrolidine **18** (R=Ph) to the rearrangement conditions, no azonine derivative **19** could be isolated. The *E*/*Z* mixture of diastereomers **18** was completely converted into the *E* material **18**. The result was rationalized by a reversible aza-Claisen rearrangement: the initial aza-Claisen reaction led to the nine-membered ring *cis*-**19** which, after appropriate conformational relax-

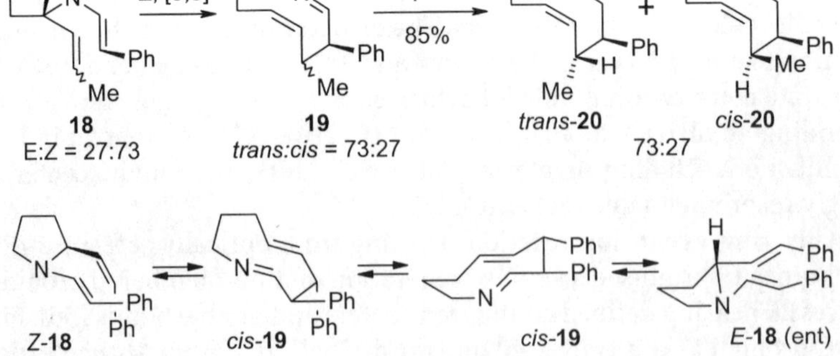

Scheme 5

ation, suffered from a final 1-aza-Cope rearrangement to regenerate the starting material *E*-**18**, but with inverse absolute configuration (not proven). This special result indicates that the unique sense of the aza-Claisen rearrangement should be taken into account when planning to use such a process as a key step in the total syntheses of complicated compounds (Scheme 5).

The regioselective enamine formation starting from unsymmetric ketimines **21** has been investigated by Welch [7]. Some simple ketones were converted into the corresponding *N*-allyl imines **21**. Treatment with methyl trifluorosulfonate at low temperatures caused *N*-methylation; the iminium salts **22** formed were then deprotonated by means of a proton sponge. Kinetically motivated deprotonation gave rise to the formation of the least substituted enamines **23**, which underwent aza-Claisen rearrangement upon warming up to room temperature. Acid hydrolysis gave the product ketones **24** with 13 to 55% yield (GC, NMR analyzed). The passing of a 3,3-sigmatropic process has been proven by deuterium labeling experiments (replacement of marked protons by deuterium) (Scheme 6).

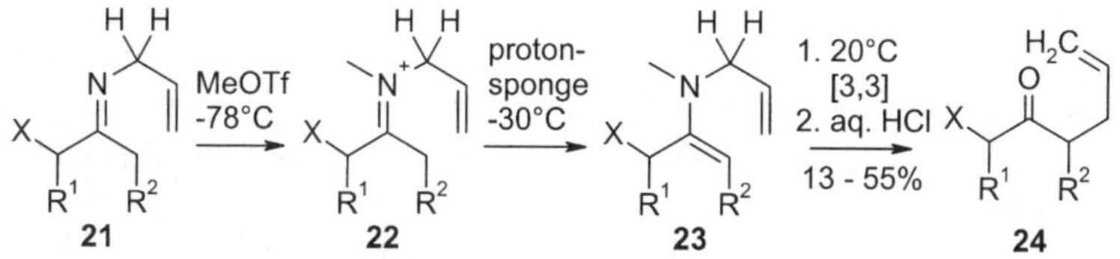

R^1 = Me, R^2 = H; R^1-R^2 = $(CH_2)_n$ with n = 2, 3, 4, X = H, Me, F, Cl

Scheme 6

The synthesis of the tricyclic core of the cytotoxic marine alkaloid madangamine required an efficient method to generate the central quaternary carbon function. Weinreb employed an aza-Claisen rearrangement in the presence of a palladium catalyst [8]. After treatment of ketone **25** subsequently with TOSMIC and DIBALH, the carbaldehyde **26** formed was reacted with diallylamine in the presence of $Pd(OCOCF_3)_2/PPh_3$. Initially, the enamine **27** was formed, which underwent diastereoselective aza-Claisen rearrangement. The γ,δ-unsaturated imine **28** was cleaved with aqueous HCl and the corresponding aldehyde **29** was isolated in 68% yield. Several further steps allowed completion of the synthesis of the core fragment **30** of the natural product (Scheme 7).

Tetrasubstituted *N*-allyl enammonium salts underwent 3,3-sigmatropic rearrangements at ambient temperature or upon heating to about 80 °C. Maryanoff used such a reaction to generate tricyclic tetrahydroisoquinoline derivatives **35** bearing a quaternary center with defined configuration [9]. Initially, indolizidine **31** was allylated. The open-book shape of **31** forced the allyl bromide to attack the *exo* face, building up a chiral ammonium center with defined configuration in **32**. TFA-mediated H_2O elimination delivered the vinyl ammonium subunit in **33**. The quinolizidinine system **33** (*n*=2) rearranged

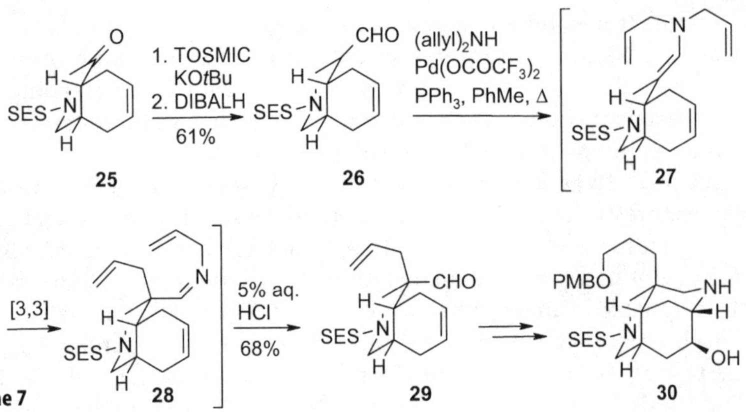

Scheme 7

slowly at 23 °C to give a single diastereomer **34** (n=2); $NaBH_4$ reduction gave the amine **35** (n=2) with 90% yield. The bridgehead H and allyl group exclusively displayed the *cis* arrangement, indicating a complete 1,3-transfer of the chiral information from the ammonium center toward the new quaternary carbon position. In contrast, the indolizidinine systems **33** (n=1) required higher reaction temperatures to induce the rearrangement. After about 2 h at 100 °C, iminium salt **34** (n=1) was formed as a 95:5 mixture of allyl/bridgehead H with *cis* and *trans* relative configuration. The final $NaBH_4$ reduction produced the indolizidinines **35** (n=1) in a 91:9 ratio of diastereomers and 73% yield. The loss of chiral information has been ascribed to a 5 to 10% portion passing a dissociative, nonconcerted reaction path at the elevated temperatures. This hypothesis was supported by deuterium labeling experiments (>90% allyl inversion) (Scheme 8).

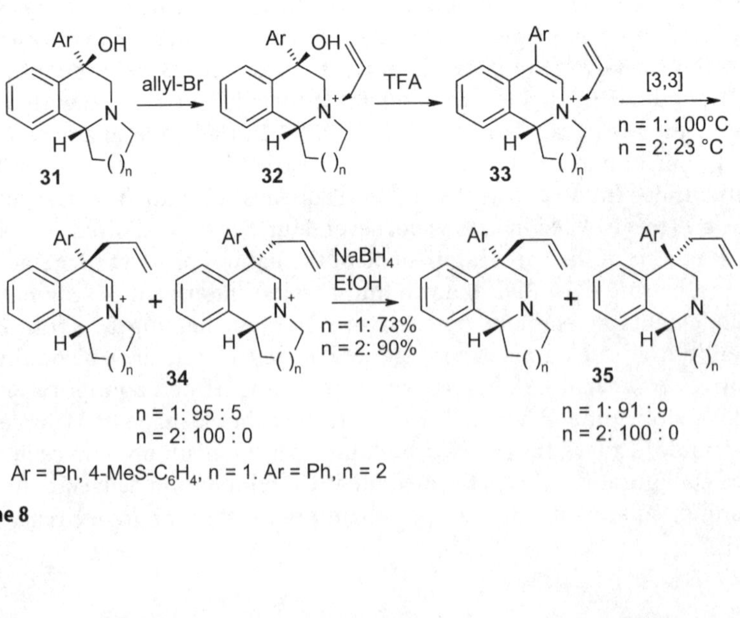

Scheme 8

R. Winter reported on aza-Claisen rearrangements in transition metal complexes [10]. *Cis* ruthenium dichloride complexes **36** added butadiyne in the presence of nonnucleophilic $NaPF_6$ to generate a cationic butatrienylidene intermediate, which was trapped by a regioselective *γ*-addition of dialkyl allylamines to produce a vinyl allyl ammonium salt **37**. This material underwent an immediate aza-Claisen rearrangement. The formation of a resonance-stabilized iminium salt **38** was thought to serve as the driving force upon running this process. While most simple allylamines gave smooth reactions at ambient temperature, the rearrangement of sterically more demanding compounds failed. The rearrangement of a propargylamine proceeded at elevated temperature, and the product *β*,*γ*-allenylamine was obtained with 90% yield (Scheme 9). Extensive computational studies gave a somewhat reduced activation barrier of the allyl vinyl ammonium system compared to the uncharged analog [11].

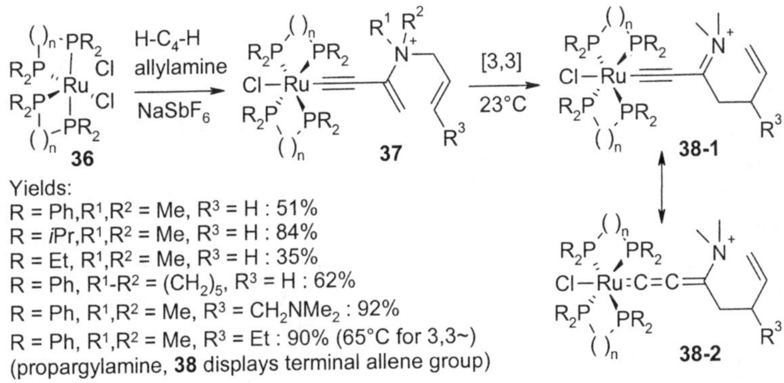

Scheme 9

3
Simple Aromatic Aza-Claisen Rearrangements

In analogy to the simple aliphatic aza-Claisen rearrangements, the aromatic systems required 200 to 350 °C reaction temperature to undergo the 3,3-sigmatropic conversion. Such drastic conditions often caused competing transformations such as the loss of the allylic moiety. Charge acceleration promised to run the process at lower temperatures making the reaction more attractive for extended synthetic use. Stille conducted the first systematic investigations concerning ammonium Claisen rearrangements [12]. Broad variation of suitable acids, solvent, concentration, and reaction time showed that $AlCl_3$, $ZnCl_2$, and $BF_3{\cdot}Et_2O$ in refluxing toluene (110 °C) and refluxing mesitylene (140 °C) led to the best results. The treatment of *N*-allylanilines **39** with the acid caused the formation of an intermediate ammonium salt **40**, which underwent the 3,3-sigmatropic rearrangement to generate the 2-allylaniline **41**. Bicyclic derivatives **42** were found in some attempts as a minor side product. 4-Alkoxy substituents decelerated the rate, and the corresponding 3-alkoxy regioisomers enhanced the rearrangement. In this latter case, the regiochemistry could not be influenced, and mixtures of 1,2,3- and 1,2,5-trisubstituted products **41** were obtained. Detailed information is given in Table 2 (entries 1–9) (Scheme 10).

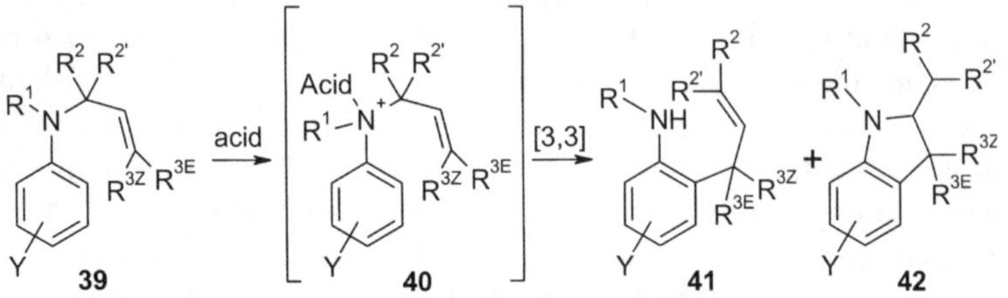

Scheme 10 For detailed information see Table 2

Ward reported on a mild proton-catalyzed variant of aromatic aza-Claisen rearrangements. A series of *N*-(1,1-disubstituted-allyl) anilines **39** were rearranged in the presence of 10 mol% of *p*-TsOH to give **41**. The reaction proceeded with complete allyl inversion. The formation of side products such as carbinols originating from a H_2O addition to the nascent double bond could be suppressed by running the conversion in aqueous acetonitrile. In agreement with the findings reported by Stille (*vide supra*), electron-withdrawing substituents Y in the *p*-position accelerated and electron-rich functional groups Y retarded the rearrangement (Table 2, entries 10–17, Scheme 10) [13].

Majumdar published several aza-Claisen rearrangements of 2-cyclohexenyl-1-anilines **39** ($R^{2'}$–R^{3Z}=$(CH_2)_3$, Table 2, entries 22–28) [14]. The reaction was carried out upon heating the reactant in EtOH/HCl. The corresponding 2-cyclohexenylanilines **41** were obtained with 50 to 90% yield. The cyclization to give indole derivatives **42** could be achieved in a separate step: treatment of the rearrangement products **41** with $Hg(OAc)_2$ in a suitable alcohol in the presence of acetic acid induced formation of the tetrahydrocarbazole **42**. The tricyclic products **42** were synthesized with 70–85% yield. Finally, carbazoles could be obtained after DDQ dehydrogenation.

Lai used a BF_3-catalyzed rearrangement of *N*-allylaniline **39** and *N*-allylindoline **43** as a key step in benzimidazole analogs **46** (X=N) and indole analogs **46** (X=C) of mycophenolic acid (MPA) [15]. MPA was known as an immunosuppressant and to have some antipsoriasis activity. With the intention of synthesizing metabolically more stable compounds, the replacement of the central isobenzofuranone subunit by an indole core was planned. Firstly, a suitable functionalized 7-alkyl indole **45** had to be generated. Upon heating of *N*-allylindoline **43** (R=H) in sulfolane to about 200–210 °C in the presence of $BF_3{\cdot}Et_2O$, the aza-Claisen rearrangement via **44** delivered 7-allylindoline **45** (R=H) in 47% yield. Generally, such reaction conditions enabled rearrangement of a set of aniline **39** (Scheme 10, Table 2, entries 18–21) and indoline deriva-

Table 2 [12–15]

Entry	Y	R^1	$R^2/R^{2'}$	R^{3E}/R^{3Z}	Acid/yield (%)			Ref.
1	H	Me	H	H/H	$AlCl_3$/68	$ZnCl_2$/45	$BF_3 \cdot Et_2O$/58	12
2	H	Bn	H	H/H	$AlCl_3$/15	$ZnCl_2$/15	$BF_3 \cdot Et_2O$/13	12
3	4-MeO	Me	H	H/H	–	$ZnCl_2$/58	$BF_3 \cdot Et_2O$/55	12
4	4-MeO	Bn	H	H/H	–	$ZnCl_2$/53	$BF_3 \cdot Et_2O$/35	12
5[a]	3-MeO	Me	H	H/H	–	$ZnCl_2$/70	$BF_3 \cdot Et_2O$/99	12
6[a]	3-MeO	Bn	H	H/H	–	$ZnCl_2$/57	$BF_3 \cdot Et_2O$/38	12
7[a]	3-MeO	*i*Bu	H	H/H	–	$ZnCl_2$/98	$BF_3 \cdot Et_2O$/80	12
8[a]	3-MeO	Me	H	*n*Pr/H	–	$ZnCl_2$/70	$BF_3 \cdot Et_2O$/75	12
9	4-MeO	Me	H	*n*Pr/H	–	$ZnCl_2$/50	$BF_3 \cdot Et_2O$/79	12
10	H	H	Me/Me	H/H	*p*TsOH/70	–	–	13
11	H	H	Me/Et	H/H	*p*TsOH/53	–	–	13
12	H	H	Et/Et	H/H	*p*TsOH/90	–	–	13
13	H	H	$(CH_2)_5$	H/H	*p*TsOH/98	–	–	13

Conditions: 1–2 eq. $AlCl_3$, mesitylene 140°C, 1–2 eq. $ZnCl_2$, mesitylene 140°C, 1–2 eq. $BF_3 \cdot Et_2O$, toluene 110°C, 10% *p*TsOH, $MeCN/H_2O$ 10:1, 65°C, HCl, EtOH, reflux.

[a] Product: mixture of regioisomers, 1,2,3-and 1,2,5-trisubstituted anilines, ratios varying between 36:64 to and 17:83.

[b] Conditions: 1 eq. $BF_3 \cdot Et_2O$, sulfolane, 160–195°C, 1 h.

Table 2 (continued)

Entry	Y	R^1	$R^2/R^{2'}$	R^{3E}/R^{3Z}	Acid/yield (%)			Ref.
14	4-EtO_2C	H	$(CH_2)_5$	H/H	*p*TsOH/78	–	–	13
15	4-MeO	H	$(CH_2)_5$	H/H	*p*TsOH/88	–	–	13
16	H	H	Bn/H	H/H	*p*TsOH/95	–	–	13
17	4-EtO_2C	H	Me/Me	H/H	*p*TsOH/91	–	–	13
18[b]	2-NH_2	H	H/H	H/H	–	–	$BF_3 \cdot Et_2O$/53	15a
19[b]	2-NO_2	H	H/H	H/H	–	–	$BF_3 \cdot Et_2O$/57	15a
20[b]	2-MeO_2C	H	H/H	H/H	–	–	$BF_3 \cdot Et_2O$/52	15a
21[b]	3-MeO	H	H/H	H/H	–	–	$BF_3 \cdot Et_2O$/43	15a
22	H	Me	$(CH_2)_3$	$(R^{2'}–R^{3Z})$	HCl/90	–	–	14
23	4-Me	Me	$(CH_2)_3$	$(R^{2'}–R^{3Z})$	HCl/88	–	–	14
24	4-Br	Me	$(CH_2)_3$	$(R^{2'}–R^{3Z})$	HCl/50	–	–	14
25	2-Me	Me	$(CH_2)_3$	$(R^{2'}–R^{3Z})$	HCl/50	–	–	14
26	H	Et	$(CH_2)_3$	$(R^{2'}–R^{3Z})$	HCl/90	–	–	14
27	4-Me	Et	$(CH_2)_3$	$(R^{2'}–R^{3Z})$	HCl/90	–	–	14
28	4-Me	*i*Pr	$(CH_2)_3$	$(R^{2'}–R^{3Z})$	HCl/66	–	–	14

tives **43** (Scheme 11) in 31–52% yield. Several further steps allowed completion of the syntheses of three target molecules **46** (X=C) [15b,c] displaying different substitution patterns; one of them (R′=$CONH_2$) showed significant antitumor activity. One benzimidazole compound **46** (X=N) [15a] was synthesized via the same sequence starting from a nitrogen derivative of 2,3-dehydro-**45** (X=N). Recently, Ganesan employed such aza-Claisen rearrangement as a key step in the total synthesis of (+)-okaranine J, which displays potent insecticidal activity (Scheme 11) [15d].

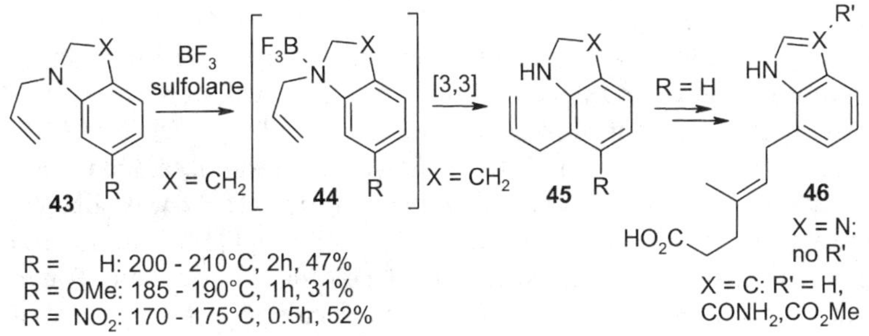

Scheme 11

Microwave-assisted aza-Claisen rearrangements of *N*-allylanilines **47** proceeded in very short reaction times in the presence of Zn^{2+} montmorillonite as a catalyst [16]. In contrast to most of the (Lewis) acid-accelerated reactions a 3,3-sigmatropic rearrangement/cationic cyclization tandem process was passed to generate 2-methylindole derivatives **49** with high yield. The nonmicrovave-assisted reactions led to mixtures of simple rearrangement products **48** and rearrangement/cyclization products **49** (80 °C, 2 h) (Scheme 12) [16b].

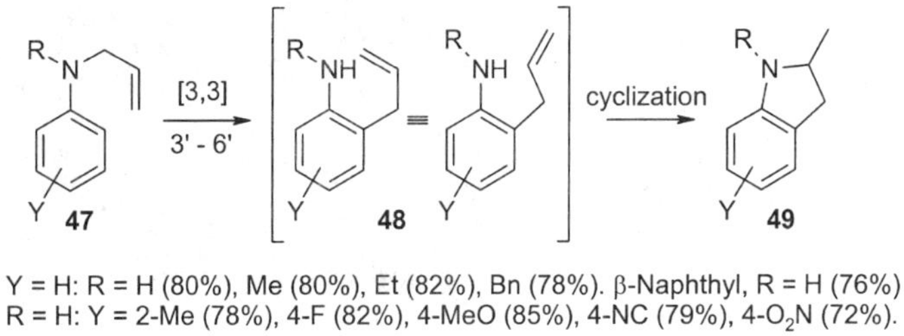

Scheme 12

Teleocidines are known as tumor-promoting compounds characterized by an indolactam V core structure. Such indolactams adopt two stable conformations at room temperature. With the intention to investigate the biologically active conformer, Irie and Wender synthesized conformationally restricted analogs of indolactam V **50** (R=Me) [17]. Starting from desmethylindolactam

V **50** (R=H), the reaction with allyl bromide gave the corresponding *N*-allyl compound **51**. Heating of the material in the presence of substoichiometric amounts of $ZnCl_2$ gave the 5-allylated product **52** with 26.4% yield. Best results concerning a 3,3-sigmatropic rearrangement/cationic cyclization tandem reaction were achieved upon heating **51** to 140 °C (xylene) in the presence of $AlCl_3$: a mixture of **52**, β-Me-**53**, α-Me-**53**, and **54** was obtained in 29.3% yield and a 43:5.4:9.2:1 ratio. Furthermore, cationic cyclization to give **53**/**54** could be achieved by subjecting 5-allyl indolactam **52** to the $AlCl_3$-promoted reaction conditions (Scheme 13). The examination of tumor promotion resulted in a significant activity upon testing of β-Me-**53** and **54** [17a]. A substantial yield enhancement could be achieved by conducting the generation of the conformationally restricted indolactam analog **54** via a two-step procedure [17b]. After the rearrangement of *N*-allyl reactant **51** in the presence of 0.45 mol% $AlCl_3$ (sealed tube, 140 °C), the 5-allyl indolactam **52** was obtained in 36% yield. Then, the OH group of **52** was acetylated and the OAc-**52** was subjected to a Pd(II)-catalyzed amination to give OAc-**54** in 67% yield [18]. Tumor-promoting activity was enhanced after introduction of a terpenoid side chain. The reaction of β-Me-**53** with prenyl bromide led to the formation of a mixture of three regioisomers in about 35% overall yield. The generation of the 7-prenyl

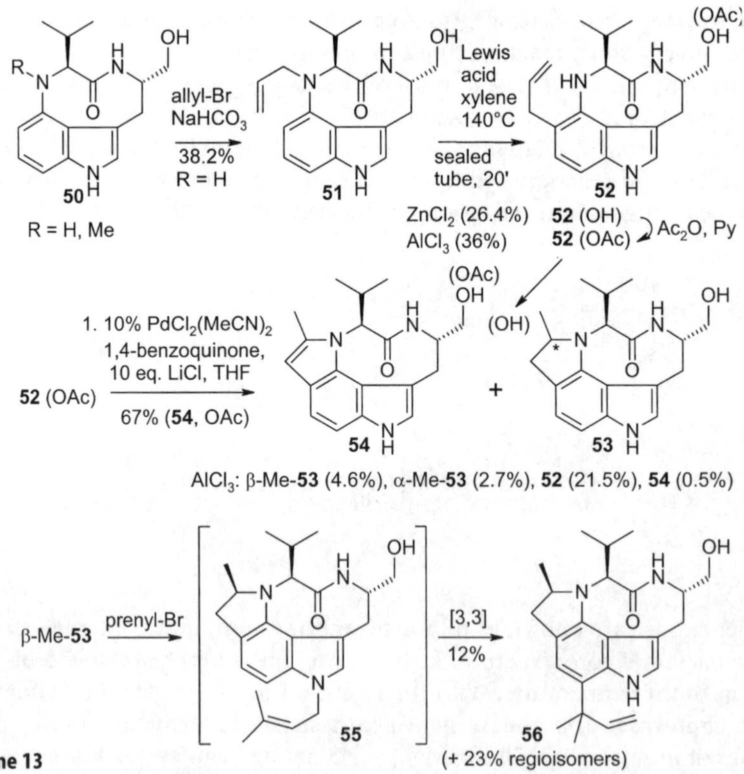

Scheme 13

compound **56** (one third of the adduct) was explained by the initial indole *N*-allylation (→**55**) and a subsequent aza-Claisen rearrangement. The direct C7 allylation would have built up the allyl-inverted analog only (Scheme 13).

Desmethylindolactam G (**50**, R=H, replace *i*Pr by H) was subjected to the same sequence (**50**→**54**). The first allylation step succeeded with 52% yield, but the aza-Claisen rearrangement/cationic cyclization gave only 9.5% of the product mixture **52–54** (replace *i*Pr by H).

Photocyclization of allyl tetrahydroquinoline **60** (*n*=1) and allyl indoline **60** (*n*=0) delivered tricyclic compounds **62** with high regioselectivity. The reactant C-allyl systems **60** were produced by means of aza-Claisen rearrangements [19]. Initially, *N*-allylation succeeded upon treatment of the amine **57** with allyl bromide **58** to give the substrates **59**. Subsequent heating to 140–150 °C in the presence of $ZnCl_2$ caused the 3,3-sigmatropic rearrangements to **60**. The indole derivative **60** (*n*=0) was obtained in 42% yield along with 9% of the *p*-product **61** (*n*=0) generated via a final Cope rearrangement (for quinoline: no yield given, Scheme 14). It should be pointed out that the reactions incorporating the cinnamyl system (R=Ph) always produced compounds with a 3-phenyl side chain, indicating the passing of dissociative reaction paths prior to a 3,3-sigmatropic rearrangement.

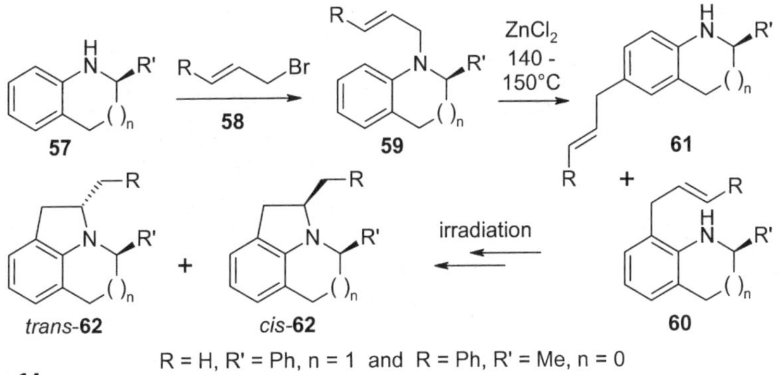

Scheme 14

Pyrimidine annulated heterocycles fused at positions 5 and 6 to uracil were synthesized via a three-step sequence starting from uracil **63** [20]. Firstly, the reaction with 3-bromocyclohexene gave the *N*-allyl-vinyl core system **64** in 80% yield. Upon heating **64** in EtOH in the presence of HCl, aza-Claisen rearrangement gave rise to the C-cyclohexenyl uracil **65** in 38% yield. Final bromination (→**66**) and dehydrogenation steps (→**67**) allowed synthesis of the desired tricyclic fused uracil systems (Scheme 15).

Simple aromatic aza-Claisen rearrangement without charge acceleration by addition of an acid required other activating factors to enable the reaction to be run at acceptable temperatures. The reduction of ring strain was found to serve as a useful promoter to induce aromatic aza-Claisen rearrangement [21].

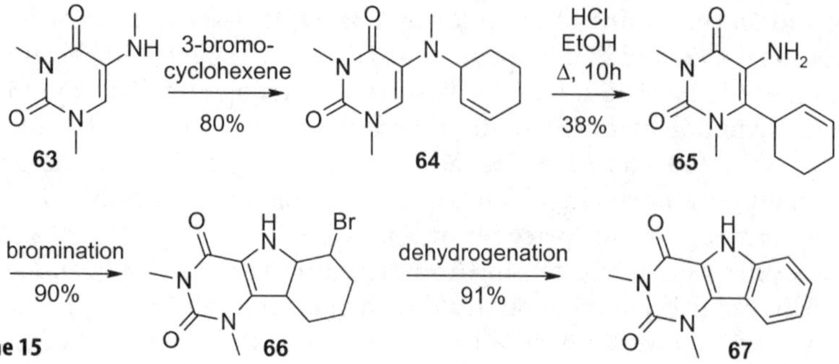

Scheme 15

4-Phenoxy-1,2-benzoquinone **68** was treated with 2-vinylazetidine **69** (n=2) and 2-vinylaziridine **69** (n=1). The azetidine adduct **70** (n=2) could be isolated in 61% yield. Heating in toluene induced aza-Claisen rearrangement to give the benzoquinone annulated azocine **71** (n=2) with 100% yield. In contrast, the corresponding aziridine adduct **70** (n=1) could not be trapped, and the azepine **71** (n=1) was isolated in 55% yield indicating an efficient aza-Claisen rearrangement run at ambient temperature (Scheme 16).

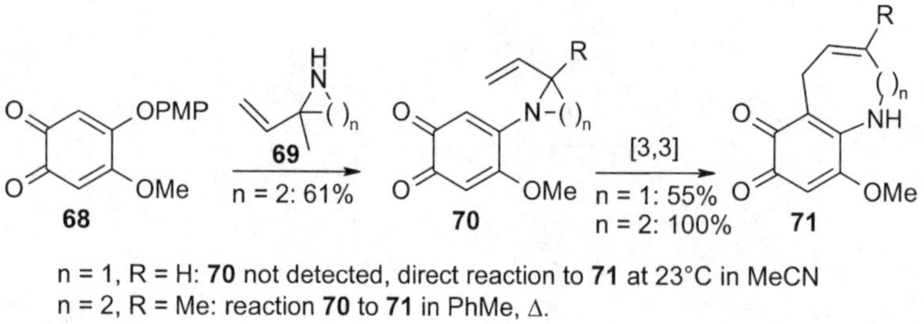

Scheme 16

4 Amide Enolate Rearrangements

The acceleration of simple aza-Claisen rearrangements by means of the addition of Lewis and proton acids offered the advantage of achieving a significant decrease of the originally very high reaction temperatures (>200 °C), enabling such a process to be employed in synthetically useful sequences. One major problem of the cation-mediated rearrangement was the potential passing of a dissociative reaction path via allyl cation and neutral enamine subunits. In several cases the regioselectivity of the sigmatropic reaction turned out to be a problem, since mixtures of 3,3- and 1,3-rearrangement products were found after running the conversion involving nonsymmetrical *N*-allyl moieties in the reactant core. The *N*-allyl amide enolate rearrangement promised to be much more flexible. In analogy to the oxygen analogs (Ireland ester enolate and silyl ketene acetal rearrangement), relatively mild reaction conditions should be applicable. In contrast to the oxygen systems, the fragmentation to generate a ketene and an amide anion was unlikely to occur. Furthermore, the bulky nitrogen forced the nascent amide acetal to adopt the *Z* configuration, offering the advantage of achieving high internal asymmetric induction (high simple diastereoselectivity) upon forming the product γ,δ-unsaturated amides.

Basic systematic investigation of amide enolate rearrangements concerning reaction conditions and diastereoselectivity have been published by Tsunoda et al. [22]. *E* and *Z* *N*-crotyl propionic acid amides **72** were initially deprotonated with LDA at –78 °C in THF to form the corresponding *Z*-enolates **73** [22a,b]. After exchanging THF against a high-boiling nonpolar solvent, rearrangement was induced upon heating. The corresponding *anti* γ,δ-unsaturated amides **74** were generated with high yield. A high diastereoselectivity was achieved starting from the reactant olefin *E*-**72** via a chair-like transition state *E-c*. In contrast, the corresponding *Z* crotyl reactant *Z*-**72** gave the product **74** with high yield too, but with a low diastereoselectivity of *syn*-**74**:*anti*-**74**=63:37, indicating the passing of competing chair- and boat-like transition states *Z-c* and *Z-b* (Scheme 17).

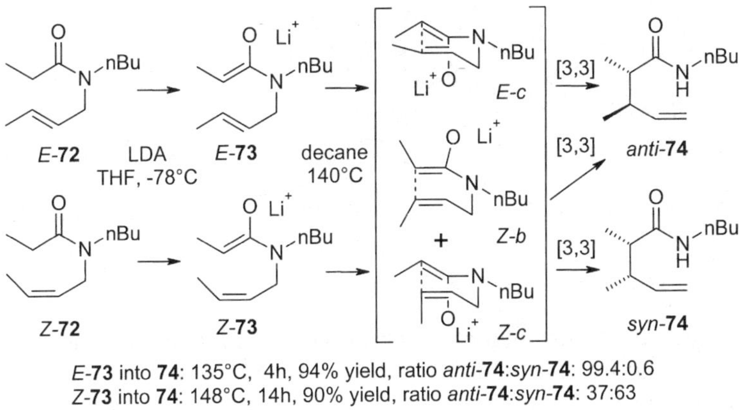

Scheme 17

The high yield and the high internal asymmetric induction obtained via the rearrangement of *E*-**72** to *anti*-**74** raised a question concerning the efficiency of an additional external asymmetric induction [22a,b]. The best prerequisite provided amide **75** placing the chiral center of the phenethyl side chain (attached to the nitrogen) adjacent to the 3,3-sigmatropic rearrangement core system. The chiral auxiliary-directed rearrangement of the *N*-crotylamide *E*-**75** led to the corresponding γ,δ-unsaturated amides *anti*-**76** with 85% yield; no *syn*-**76** amides were found, indicating the exclusive passing of chair-like transition states *E-c*. The careful analysis of the product *anti*-**76** gave a composition of *anti*-$\mathbf{76}_a$:*anti*-$\mathbf{76}_b$ of 8.5:91.5. The outcome could be rationalized by the predominant passing of a chair-like transition state *E*-c_b causing the formation of *anti*-76_b with 83% de (Scheme 18). The removal of the auxiliary was succeeded by initial acylation of the secondary amide **76** with acid chloride **77**. Then, neighboring group-assisted saponification of **78** gave the optically active γ,δ-unsaturated acid **79** with 63–82% yield overall. The chiral auxiliary phenethylamine could be recovered in two further steps in 71% yield overall starting from **80** (Scheme 18). Investigating a series of chiral auxiliaries attached to amide *E*-**75**, most *S*-configured auxiliaries predominantly led to product *anti*-76_b (yield: 66–90%, de: 70–84%). As an exception, (2*S*)-3,3-dimethylbutyl-2-amine resulted in *anti*-76_a as the major product (38% yield, 80% de) [22c]. Recently, a related example has been published by Davies [22i].

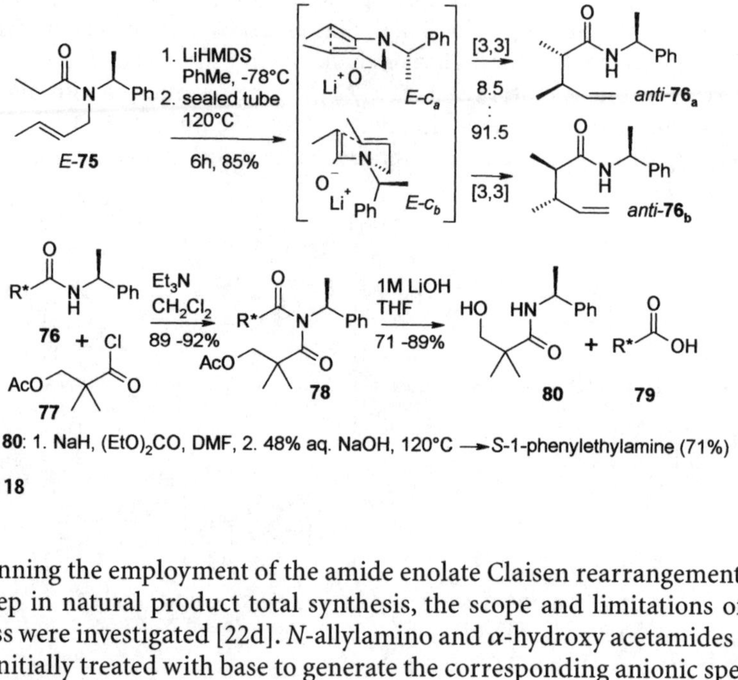

Scheme 18

Planning the employment of the amide enolate Claisen rearrangement as a key step in natural product total synthesis, the scope and limitations of the process were investigated [22d]. *N*-allylamino and α-hydroxy acetamides *E*-**81** were initially treated with base to generate the corresponding anionic species. If possible, lithium chelate formation (enolate O and α heteroatom) should fix the *Z* enolate geometry. The glycolic acid amides *E*-**81** (R^1=OH, OTBS) underwent smooth rearrangement upon heating to give the γ,δ-unsaturated amides **82** with 59–95% yield. The chiral auxiliary-directed rearrangements led to de values of 34 to 73%. The *N*-crotyl glycine derivatives displayed different reactivity *E*-**81** (R^1=NH_2, NHBOC). Despite varying the conditions, the NHBOC did not give any rearrangement product. In contrast, *E*-**81** (R^1=NH_2) underwent rearrangement at ambient temperature to build up **82** with 81–89% yield and 78% de after running the auxiliary-directed conversion (Table 3, Scheme 19).

First applications in total syntheses started from amide **83**. The rearrangement under standard conditions led diastereoselectively to amide **84** in 77% yield; no traces of other diastereomers were found. Amide **84** was converted

Table 3 [22]

Entry	R^1	R^2	R^3	Time (h)	*T* (°C)	Yield (%)	Ratio **82**$_a$:**82**$_b$	Ref.
1	OH	H	*n*Pr	15	80	74	50:50	d
2	OTBS	H	*n*Pr	15	100	59	50:50	d
3	NH_2	H	*n*Pr	4	23	81	50:50	d
4	NHBOC	H	*n*Pr	20	23–140	0	–	d
5	OH	Ph	Me	15	80	95	86:13	d
6	OTBS	Ph	Me	6	120	62	67:33	d
7	NH_2	Me	Ph	15	23	89	11:89	d

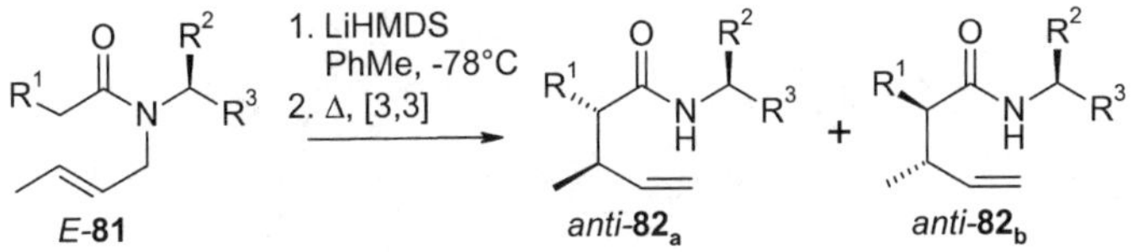

R^1 = OH, OTBS, NH_2, NHBOC. R^2 = H, R^3 = *n*Pr or R^2 = Me, R^3 = Ph (Table 3)

Scheme 19

into (–)-isoiridomyrmecin **85** via diastereoselective hydroboration, oxidative workup, and a final lactonization in the presence of *p*-TsOH (41.4% yield over three steps) [22f]. Alternatively, $LiAlH_4$ reduction of **84** resulted in amine **86** (80%), and a final five-step sequence enabled (+)-α-skytanthine **87** to be synthesized with 60% yield (Scheme 20) [22g].

As an extension of the above-mentioned method, a successful rearrangement starting from silyl enol ethers as an electron-rich *N*-allyl moiety has been published. Allylamide **88** was synthesized in two steps starting from (+)-phenethylamine and acrolein via a condensation–acylation sequence. The auxiliary-

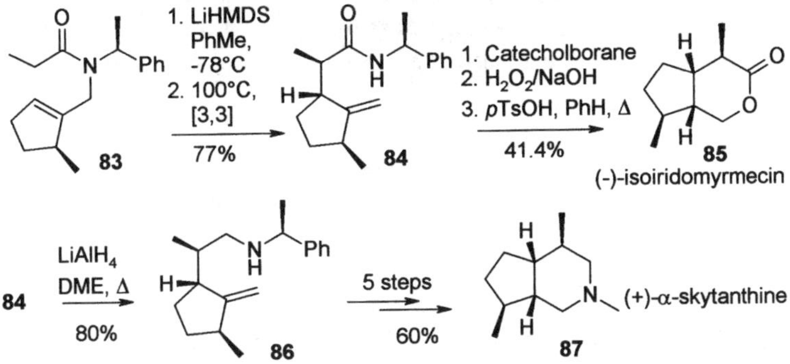

Scheme 20

directed aza-Claisen rearrangement of **88** produced a mixture of two diastereomers; the major compound **89** was isolated by means of column chromatography. Iodocyclization and subsequent reductive removal of the halide resulted in γ-butyrolactone **90**, which served as the western half of the bislactone framework of antimycine A_{3b} **91**. Starting from **90** the total synthesis of **91** (component of antimycine A mixture: oxidoreductase inhibitor) was completed in several further steps (Scheme 21).

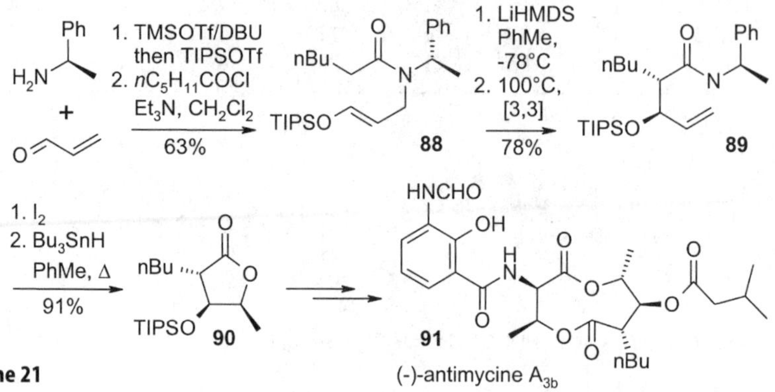

Scheme 21

The synthesis of the medium-sized lactam **94** started from 2-vinylpiperidine **92** [23]. The amide enolate aza-Claisen rearrangement led to the corresponding ten-membered ring lactam **94**. Reacting terminally unsubstituted olefins as in **92**, a complete 1,4-chirality transfer was observed, pointing out the highly efficient internal asymmetric induction. The stereochemical outcome of the process was rationalized by the passing of a chair-like transition state **93** minimizing repulsive interactions. As described above, the amide enolate in **93** should have been *Z* configured. One optically active azecinone **94** (R=Me, Et side chain) served as key intermediate in an asymmetric total synthesis of fluvirucin A_1 (Scheme 22).

Somfai enhanced the driving force of some amide enolate aza-Claisen rearrangements by choosing vinylaziridines as reactants [24]. The additional loss of ring strain offered the advantage of running most of the reactions at room temperature to synthesize unsaturated chiral azepinones. Various substitution

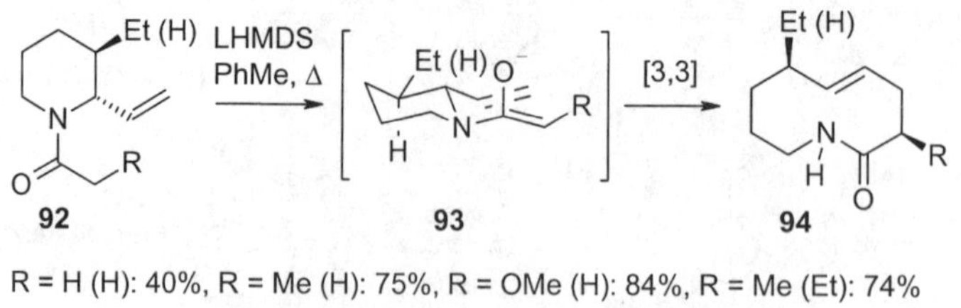

Scheme 22

patterns were tested to investigate the scope and limitations of this process (Table 4). The stereochemical properties of the 3,3-sigmatropic rearrangement enabled the transfer of stereogenic information of easily formed C–N bonds completely to new C–C bonds by means of the highly ordered cyclic *endo* transition states **96**. The configuration of the allyl double bond was retained throughout the reaction. Furthermore, the defined enolate geometry of the in situ formed ketene aminal double bond caused a high internal asymmetric induction leading to one predominant relative configuration of the newly generated stereogenic centers.

Rearranging divinylaziridines **95**, the passing of a boat-like transition state **96** explained the stereochemical outcome of the reactions to give the azepinones **97** in 60 to 85% yield. The rearrangement of an α-branched amide **95** (replace CH_2R^2 by $CH(R^2)CH_3$, Table 4, entry 14) only required a higher reaction temperature of about 65 °C to induce the conversion. A single diastereomer **97** was generated bearing β-methyl and α-amino functions. Surprisingly, the α-NHBOC group was found to be replaced by a urea subunit, indicating a defined substitution during the course of the reaction (neighboring group-assisted cleavage of the BOC protective group). The reactant divinylaziridines **95** were synthesized via ex-chiral pool sequences starting from optically active α-amino acids (Table 4, Scheme 23).

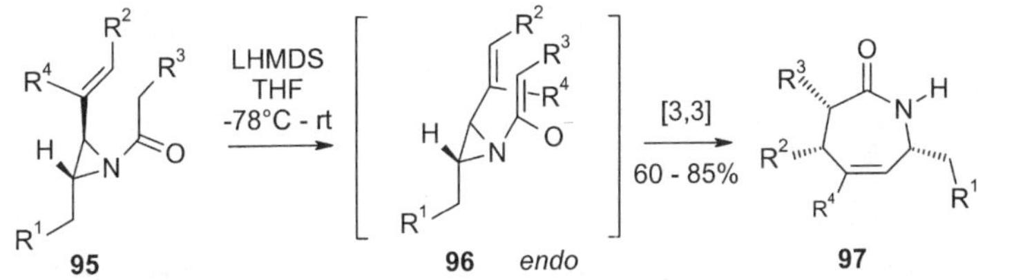

R^1 = H, OBn, Bn, R^2 = H, Me, OBn, R^3 = H, Me, OBn, NHBoc R^4 = H, CH_2OBn (Table 4)

Scheme 23

Neier et al. planned the development of a Diels–Alder cycloaddition/aza-Claisen rearrangement tandem process intending to construct up to four new stereogenic centers in a defined manner [25]. Initially, some test systems were checked but only a disappointing efficiency of such a sequence was found. Hence, cycloaddition and rearrangement were tested as separate reactions employing suitable model systems. Imides **98** (generated from the corresponding amides by *N*-acylation, 88–93%) were converted into the *Z*-*N*,*O*-ketene acetals **99** with 64–93% yield. The *N*-allyl amide acetals obtained were subjected to a set of different aza-Claisen rearrangement conditions to investigate the usefulness in regard to the planned tandem process, synthesizing imide **100** (Scheme 24).

Thermal aza-Claisen rearrangements were induced upon heating the reactant **99** in decalin to 135–190 °C. Though some product **100** could be obtained,

Table 4 [22]

Entry	R^1	Config./R^2	R^3	R^4	Yield (%)	Ref.
1	OBn	H	H	H	83	a
2	Bn	H	N	H	83	a
3	Bn	H	Me	H	85	a
4	Bn	H	OBn	H	81	a
5	Bn	H	NHBOC	H	76	a
6	H	α/CH_2OBn	H	H	73	a
7	OBn	α/Me	H	H	71	c
8[a]	OBn	β/Me	H	H	73	a
9	H	α/CH_2OBn	Me	H	60	c
10	OBn	α/Me	Me	H	64	c
11[a]	OBn	β/Me	Me	H	61	c
12	H	α/CH_2OBn	NHBOC	H	63	c
13	OPMB	H	H	CH_2OBn	85	c
14	Bn	NHBOC[b]	H	H	58	c

[a] Rearrangement of *Z/E*-olefin mixture (*Z/E*=1:13).
[b] Replace CH_2R^2 by $CH(R^2)CH_3$.

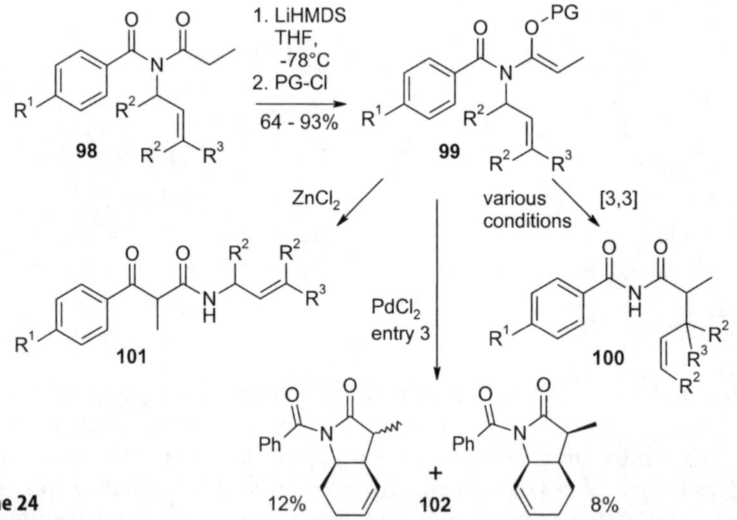

Scheme 24

best yields achieved were about 40%. The substituted allyl moieties in particular suffered from poor yield and the formation of diastereomer mixtures (Table 5, entries 2, 3). Lewis acid mediation was investigated by treating the reactants **99** with 0.7 mol% of $ZnCl_2$ in PhMe with heating to 85 °C. No product **100** was formed; β-ketoamides **101** were isolated pointing out the predomination of a competing reaction path. Further testing of various Lewis acids did not result in any rearrangement product **100**. The Overman rearrangement

Table 5 [25]

Entry	R^1	R^2	R^3	PG	Yield (%)[b]		
					>135°C	$ZnCl_2$, 85°C	$PdCl_2$
1	H	H	H	TBS	36	85	48
2	H	H	Me	TBS	13[c]	84	–
3	H	$(CH_2)_3$	H	TBS	11[c]	–	20[d]
4	Me	H	H	TBS	39	61	–
5	MeO	H	H	TBS	41	–	–
6	H	H	H	$P(O)(OEt)_2$	7	–	–
7	H	H	H	TES	40	–	–
8[a]	H	H	H	TBS	28	–	–

[a] Replace *N*-benzoyl by *N*-benzyl.
[b] Product **101**.
[c] Mixture of diastereomers.
[d] Cyclization products **102**.

conditions employed the soft electrophile $Pd(PhCN)_2Cl_2$ to accelerate the bond reorganization [26]. The reaction mechanism was rationalized likewise by a Pd(II)-catalyzed sigmatropic rearrangement or by a cyclization fragmentation sequence. However, rearrangement-type products should be formed. In the present investigations, the reaction of *N*,*O*-acetal **99** (entry 1, Table 5) gave some product **100** and 50 mol% Pd(II) catalyst was necessary to obtain a maximum yield of 48%. In contrast, the cyclohexenylamine derivative **99** (entry 3, Table 5) produced no amide **100**. Instead, bicyclic imides **102** were found with low yield, indicating the passing of a Pd(II)-mediated cyclization/β-hydride elimination tandem process (Table 5, Scheme 24).

5
Zwitterionic Aza-Claisen Rearrangements

Charge-accelerated ammonium and amide enolate aza-Claisen rearrangements allowed the originally very high reaction temperatures of 200–350 °C to be reduced to about 80–140 °C. Considering the significant nucleophilicity of some tertiary amines **103**, the addition to neutral electrophiles must cause an initial charge separation. Constructing such a zwitterion using an allylamine **103** as the nucleophile and a triple bond **104** or an allenic species **108** as an electrophile, the combination gives rise to an aza-Claisen rearrangement framework **105/109**. A consecutive 3,3-sigmatropic bond reorganization to **107/110** should profit from charge neutralization, holding out the prospect of a further decrease of the reaction temperature – a significant extension of the limitations in tolerating more complicated substitution patterns and functional groups within the reaction as well as higher stereoselectivities seems to be achievable. Additional base and (Lewis) acid catalysis can be employed to support the reaction. One problem should be pointed out: the addition step of nucleophile **103** and electrophile **104/108** potentially is reversible. Consequently, the fragmentation of **105/109** to regenerate the reactants occurs not as a negligible process (entropy, charge neutralization). Intending to use the zwitterionic intermediate **105/109** for Claisen rearrangement, a sufficient lifetime of the charge-separated species must be taken into account to achieve the highly ordered transition state as a prerequisite of the 3,3-sigmatropic process. Analyzing the systems suitable for the intermolecular zwitterionic aza-Claisen rearrangement (addition/rearrangement tandem process), three combinations were found to be of interest (Fig. 2).

The first one can be described as the Michael addition of the *N*-allylamine **103** to an acceptor-substituted triple bond of **104** to form an intermediate *N*-allyl enammonium enolate **105** (alkyne carbonester Claisen rearrangement). The anion stabilizing group is placed in position 1 of the rearrangement core system. 3,3-Sigmatropic rearrangement delivers an iminium enolate **106**, which undergoes immediate charge neutralization to form an acceptor-substituted enamine **107**.

The second mechanism starts with addition of the *N*-allylamine **103** to the cumulated acceptor system of a ketene **108** (Acc=O) to form an intermediate *N*-allyl ammonium enolate **109** (ketene Claisen rearrangement) [27]. The anion stabilizing group is predominantly placed in position 2 of the rearrangement core system; resonance stabilization can place the anion in position 1, too. Then, 3,3-sigmatropic rearrangement gives rise to an iminium analog carboxylate (**110**, $Acc^-=O^-$), which represents the zwitterionic mesomer of the charge neutral amide **110** (Acc=O).

The third mechanism starts with addition of the *N*-allylamine **103** to the cumulated acceptor system of an allene carbonester **108** (Acc=$CHCO_2Me$) to form an intermediate *N*-allyl ammonium amide enolate **109** (allene carbonester Claisen rearrangement). The anion stabilizing group is exclusively placed

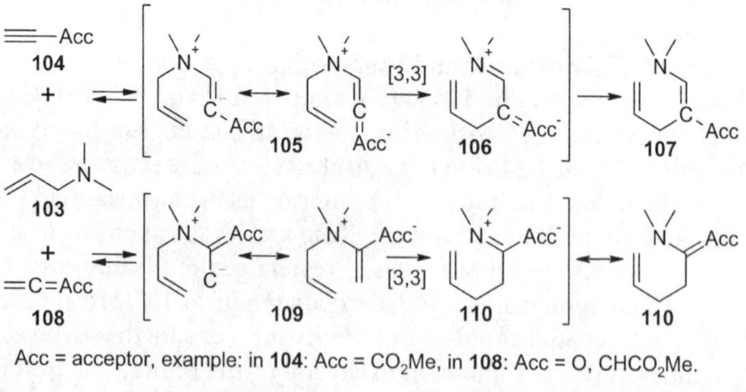

Acc = acceptor, example: in **104**: Acc = CO_2Me, in **108**: Acc = O, $CHCO_2Me$.

Fig. 2 Systems suitable for the intermolecular zwitterionic aza-Claisen rearrangement

in position 2 of the rearrangement core system. Then, 3,3-sigmatropic rearrangement generates an iminium enolate (**110**, Acc^{-}=$CH^{-}CO_2Me$), which represents the zwitterionic mesomer of the charge neutral vinylogous amide **110** (Acc=$CHCO_2Me$).

All reaction mechanisms presented here should be understood as hypotheses to rationalize the outcome of the processes. Alternative explanations such as stepwise and dissociative mechanisms cannot be excluded. However, as long as the constitution and configuration of the product can be described as that of a 3,3-sigmatropic rearrangement, the present hypotheses seem acceptable.

5.1
Alkyne Carbonester Claisen Rearrangements

Mariano et al. developed alkyne carbonester Claisen rearrangements as key steps in alkaloid syntheses. A convincing application of such a process was published in 1990 describing the total synthesis of the Rauwolfia alkaloid deserpidine **117** [28]. Isoquinuclidene **111** was synthesized in several steps starting from a dihydropyridine. The heating of a mixture of amine **111** and propiolic ester **112** in MeCN at about 80 °C induced the addition rearrangement sequence to give the isoquinoline derivative **114**. Careful optimization of the reaction conditions improved a crucial role of the diethyl ketal function to obtain a smooth reaction. The addition of the alkynoic acid proceeded *anti* with respect to the ketal building up zwitterion **113** bearing the suitable 3,3-sigmatropic rearrangement framework. Without any indole protecting group (PG=H) only 39% of the desired product was obtained. Rearranging the sulfonamide of **111** (PG=$PhSO_2$), acceptable 64% yields of ketone **115** were achieved after consecutive cleavage of the ketal. Heating in aqueous acetic acid induced Wenkert cyclization to generate tetracyclus **116** as a mixture of three diastereomers. Several further steps allowed completion of the total synthesis of deserpidine **117** (Scheme 25).

Vedejs and Gingras investigated intermolecular aza-Claisen rearrangements of acetylene dimethyldicarboxylates **119** and methyl propiolate **128** and various acyclic and cyclic *N*-allylamines **118, 124,** and **127** [29]. Proton and Lewis acids were found to accelerate the Michael addition step of the amine to the triple bond. Hence, the reaction temperature could be lowered to 23–0 °C in most of the experiments; optimized conversions were run at –40 to –60 °C. The reaction of allylamine **118** with DMAD **119** initially formed the Michael adduct **120**. The acid was thought to promote the addition and to stabilize **120**. A subsequent rearrangement enabled the isolation of enamine **121** in up to 99% yield. Acidic cleavage allowed removal of the enamine and buildup of the corresponding ketoesters. In the presence of a sterically demanding allyl system and, likewise, a nucleophilic counter-ion of the acid (e.g., benzoate from benzoic acid), the yield of rearrangement product **121** was decreased. A competing reaction activated the nucleophile to degrade the intermediate **120** by means of an S_N process. Up to 50% yield of enamine **123** and allyl compound **122** could be isolated (Scheme 26).

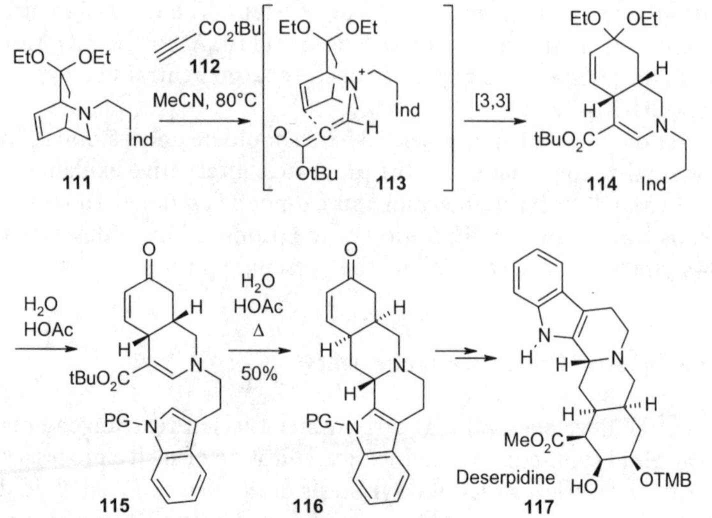

Ind = N-PG-3-Indoyl, yield **111** in **115**: PG = H: 39%, PG = $PhSO_2$: 64%

Scheme 25

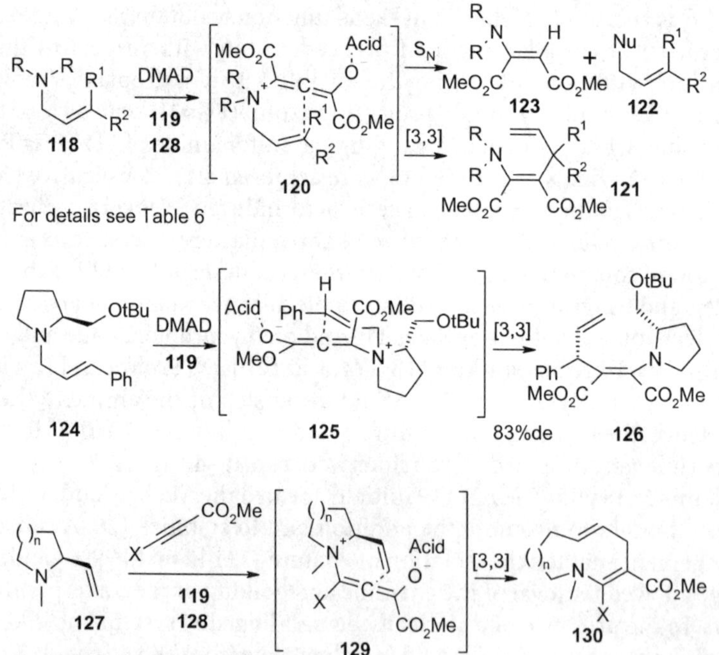

X = H: **128**, X = CO_2Me: **119**. n = 1,2

Scheme 26

Table 6 [29]

Entry	R	R^1	R^2	X	*T* (°C)	mol%/acid	Yield (%)
1	*n*Pr	H	H	CO_2Me	23	10/TsOH	99
2	*n*Pr	H	H	H	23	5/TsOH	95
3	*n*Pr	H	Me	CO_2Me	23	10/TsOH	95
4	$(CH_2)_4$	H	Me	CO_2Me	23	10/TsOH	89
5	$(CH_2)_4$	H	Ph	CO_2Me	23	5/TsOH	91
6	*n*Pr	Me	Me	CO_2Me	23	10/TsOH	64
7	$(CH_2)_5$	H	Ph	CO_2Me	23	10/TsOH	88
8	*n*Pr	Me	R[c]	CO_2Me	23	10/TsOH	73
9	*n*Pr	H	Me	CO_2Me	–40	27/$TiCl_2(OiPr)_2$	66
10	*n*Pr	H	Me	CO_2Me	–40	27/(BINOL)$TiCl_2$	93
11	$(CH_2)_4$	H	Ph	CO_2Me	–40	10/$TiCl_2(OiPr)_2$	83
12	$(CH_2)_4$	H	Ph	CO_2Me	–40	27/(BINOL)$TiCl_2$	92
13[a]	Et	H	Ph	CO_2Me	–60	10/$TiCl_2(OiPr)_2$	83
14[b]	**124**			CO_2Me	–60	10/$TiCl_2(OiPr)_2$	62
15	**127**	*n*=1		CO_2Me	–15	5.5/TsOH	57
16	**127**	*n*=2		CO_2Me	20	10/TsOH	71
17	**127**	*n*=2		H	65	–	71

[a] +25% AgOTf.
[b] +20% AgOTf.
[c] R=$CH_2CH_2CHCMe_2$.

Lewis acid catalysis offered the advantage of using chiral Lewis acid ligands with the intention of achieving some catalyst-directed asymmetric induction upon generating the enamine **121**. First experiments using (±)-BINOL catalyst (27 mol%) gave a significant rate acceleration, but no experiment with enantiopure material had been described. An auxiliary-directed asymmetric rearrangement was investigated starting from prolinol derivative **124**. The addition should have given the intermediate **125** bearing a chiral ammonium center. The consecutive 3,3-sigmatropic rearrangement gave the enamine **126** with 62% yield and 83% de, as determined after auxiliary cleavage/decarboxylation and final amidation with chiral 1-phenylethylamine. Here, the use of (+)-BINOL led to a slight decrease of the ee to about 80% (mismatched combination?). Ring expansions were studied upon rearranging 2-vinylpyrrolidine **127** (*n*=1) and piperidine **127** (*n*=2) with propiolic esters **119** and **128**, respectively. Best results were obtained using TsOH as accelerating acid. In most cases, NMR-scale experiments were conducted allowing generation of the medium-sized ring systems **130** via adduct **129** with 50–70% yield (Table 6, Scheme 26).

5.2 Ketene Claisen Rearrangements

The *N*-allyl ammonium enolate Claisen rearrangement requires the addition of a tertiary allylamine to the carbonyl center of a ketene to generate the sigmatropic framework. The first crucial point of this so-termed ketene Claisen rearrangement is the formation of the ketene with a lifetime sufficient for the consecutive attack of the nucleophile and the avoidance of competing 2+2-cycloadditions. Roberts tested stable diphenylketene **131** [30]. *N*-benzyl azanorbornene **132** (R=Ph) was treated with diphenylketene **131** upon heating to reflux in acetonitrile for 6 days. The passing of the addition (→**133**) rearrangement sequence gave the desired bicyclic material **134** with 59% yield. Ultrasonication allowed reduction of the reaction time to 12 h (61% yield). At 0 to 23 °C, the sterically less hindered *N*-methyl derivative **132** (R=H) suffered from the intermolecular Claisen rearrangement to give **134** in 53% yield. The use of stable ketenes was mandatory, otherwise only 2+2 cycloadducts of ketene and olefin and degradation products were observed (Scheme 27).

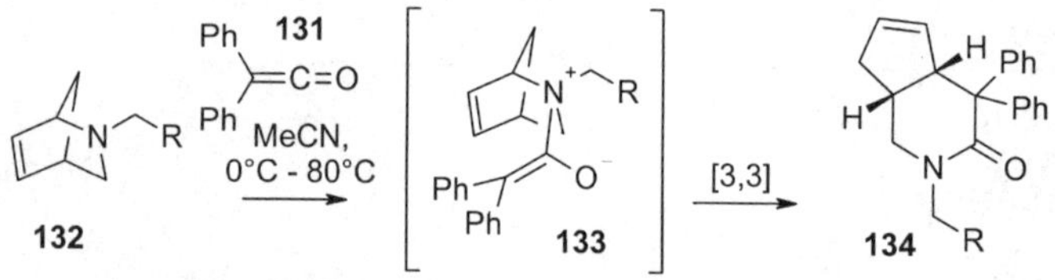

Scheme 27 R = Ph: 80°C, 6d, 59%. R = Ph, ultrasonication, 12h, 61%.
R = H: 0°C - 23°C, 13h, 53%

A significant acceleration of the ketene reactivity could be achieved using electron-deficient species such as dichloroketene. Pombo-Villar described a rearrangement of optically pure *N*-phenethyl azanorbornene **135** to synthesize the α,α-dichloro-δ-valerolactam **138** [31]. At 0 °C in CH_2Cl_2, dichloroketene – generated in situ from dichloroacetyl chloride **136** and Hünig's base – was reacted with amine **135**. The addition step led to the hypothetical zwitterion **137**, which underwent immediate 3,3-sigmatropic rearrangement. The bicyclic enantiopure lactam **138** was obtained with 61% yield. Higher reaction temperatures led to tarry side products, and the yield was significantly decreased. The α,α-dichloro function of **138** could be reduced by means of $Zn/NH_4Cl/MeOH$ to give the dechlorinated material **139**. Lactam **139** served as a starting material in a (–)-normethylskytanthine **140** synthesis (Scheme 28).

Edstrom used terminally unsubstituted 2-vinylpyrrolidine and piperidine, respectively, and dichloroketene to achieve the ring expansion to nine- and ten-membered lactams [32]. Starting from *N*-benzyl-2-vinylpyrrolidine (*n*=0) and piperidine (*n*=1) **141**, respectively, the ketene Claisen rearrangement using in situ generated dichloroketene led to the corresponding azoninone and azeci-

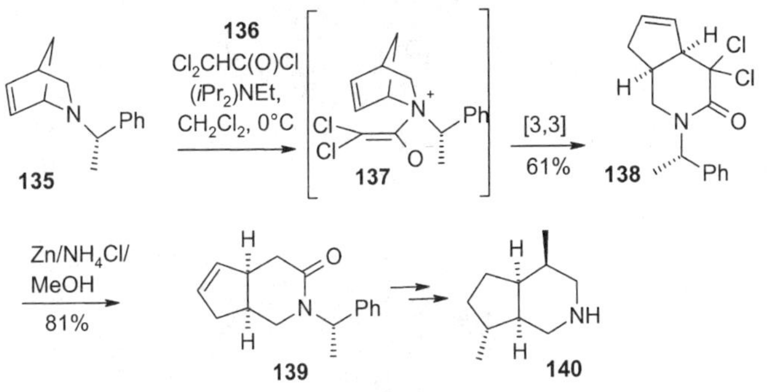

Scheme 28

none **144** (R^1=Bn) in 64 and 96% yield. Replacing the N protective group by the more electron-rich PMB substituent (R^1=PMB), the yields of **144** were observed to decrease to 52 and 54%, respectively. Here, Edstrom used trichloroacetyl chloride **142** and activated zinc to generate dichloroketene via a reductive dechlorination at 0 to 62 °C. Simultaneously, Lewis acidic $ZnCl_2$ was formed which might have activated the ketene and stabilized the zwitterionic intermediate **143** to support the rearrangement. Though the double bond included in the medium-sized ring was found to be exclusively *E* configured, both rearrangements suffered from a complete loss of chiral information because of the use of terminally symmetric substituted olefins in **141** (=CH_2) and ketenes **108** (=CCl_2, Acc=O) as reactants. The NMR spectra of the azecinone **144** (n=1) were characterized by the coexistence of two conformers. In contrast, the nine-membered ring **144** (n=0) was revealed as a single species. Both medium-sized lactams were used in transannular ring contractions to yield the corresponding quinolizidinones **146**/**149** (n=1) and indolizidinones **146** (n=0), respectively. The *E* double bonds suffered from an external attack of an electrophile (I^+, $PhSe^+$, Me_3Si^+) and the resultant onium ion underwent a regio- and stereoselective addition of the N center of the lactam to give an acylammonium salt **145**. Then, the benzyl group was removed (→**146**) by a von Braun-type degradation to form the corresponding benzyl halide. Surprisingly, the relative configuration of bridgehead hydrogen and the adjacent substituent were found to be *trans* on synthesizing quinolizidinones **149**. After dechlorination with Zn/Ag/HOAc to give the lactams **147**, the transannular ring contraction of the azoninone (n=0) took the expected path, generating indolizidinone **146** (R^2=H). In contrast, the analog reactions involving the azecinone **147** (n=1) gave bicyclus **149** by passing the hypothetical acylammonium ion **148** as a quasi-*syn* adduct of E and N at the double bond. Finally, the quinolizidinone **149** (n=1, E=I) were employed as a key intermediate in a total synthesis of D,L-epilupinine (Scheme 29, Table 7).

The mild reaction conditions and the obviously high potential driving force of the ketene Claisen rearrangement recommended the use of the process with more complex systems [33]. The first series of this type of reaction suffered from severe limitations (see Schemes 27–29, Fig. 3) [30–32]. On the one hand,

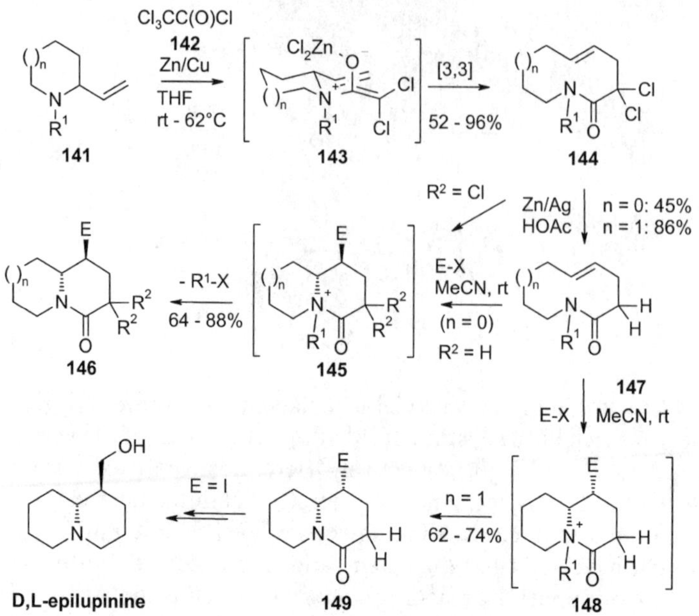

Scheme 29 n = 0, 1 R¹ = Bn, PMB, R² = H, Cl, E-X = PhSeCl, I_2, TMSI (Table 7)

Table 7 [28]

Entry	R^1	*n*	Yield (%) **144 (147)**	Yield (%) **146 [146], (149)**[a]		
				E=I	E=PhSe	E=TMS
1	Bn	0	64 (45)	88 [87]	79 [64]	72
2	PMB	0	52	–	–	–
3	Bn	1	96 (86)	85 (62)	84 (74)	–
4	PMB	1	54	–	–	–

[a] Yield: **146**: R^2=Cl, [**146**]: R^2=H, (**149**): R^2=H.

predominantly electron-deficient ketenes **108** added to the allylamines **103**, and useful yields of the amides were exclusively achieved by reacting dichloroketene [31, 32]. On the other hand, the rearrangement was restricted to either monosubstituted olefins in the amino fragment (**146**) or the driving force had to be increased by a loss of ring strain (**132, 135**) during the process. The reaction path was rationalized as pointed out in Fig. 3. Initially, the ketene **108** was generated from a suitable precursor **150** (i.e., acid halide). Then, a reversible addition of the ketene **108** to allylamine **103** gave rise to the intermediate zwitterion **109**. Here, dichloroketene was found reactive enough to push the equilibrium toward adduct **109** and diphenylketene **131** was stable enough to

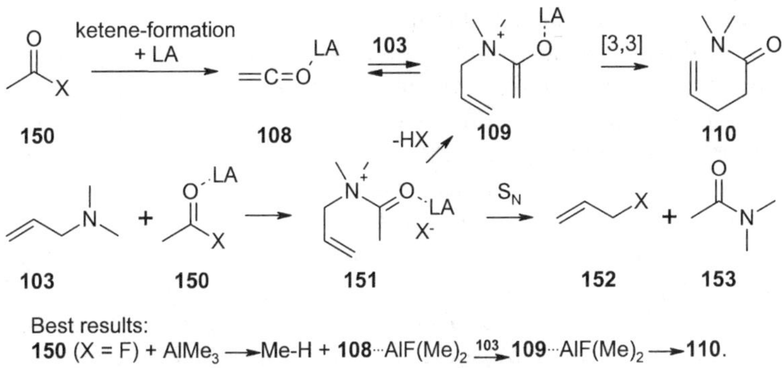

Fig. 3 Ketene Claisen rearrangements

survive several addition/elimination cycles without suffering from competing reactions (diketene formation, etc.). Finally, zwitterion **109** underwent the sigmatropic rearrangement to give amide **110** (Fig. 3). The careful analysis of a range of conversions indicated that two further competing processes have to be mentioned [33a,d]:

1. The tertiary amines **103** and the acid chlorides **150** (X=Cl) initially formed acylammonium salts **151**, which underwent a von Braun-type degradation by an attack of the nucleophilic chloride ion ($X^-=Cl^-$) at the allyl system to give allyl chlorides **152** and carboxylic acid amide functions **153** [34].
2. The reaction of acyl chlorides **150** led to the corresponding ketenes **108** while the allylamines **103** were deactivated as ammonium salts **103**-HCl (Schotten–Baumann conditions).

Three changes concerning the processing led to a pioneering surmounting of the limitations in converting allylamines **103** into the corresponding amides **110** [33a,d]:

1. Addition of stoichiometric amounts of a Lewis acid (LA), especially trimethyl aluminum to the reaction mixture. A range of α-substituted carboxylic acid halides **150** (X=Cl, F) as precursors of the ketenes **108** could be used, overcoming the restriction concerning the ketene component, but up to now, the rearrangement failed using α,α-difunctionalized carboxylic acid halides. The Lewis acid might have increased the acidity of the α-protons by interacting with the carbonyl group in **108**, **150**, and **151**, facilitating the formation of the intermediate zwitterions **109**, and/or the Lewis acid had stabilized the zwitterionic intermediate **109** suppressing the elimination of ketene **108**. Furthermore, allylamines **103** bearing 1,2-disubstituted double bonds could be successfully rearranged overcoming a restriction concerning the carbon framework [33a,b,d].
2. The replacing of the acyl chlorides **150** (X=Cl) by the corresponding acyl fluorides **150** (X=F) as the substituents of the ketenes **108**. The von Braun-

type degradation as the major competing reaction observed was efficiently suppressed. The fluoride counter-ion was known to be less nucleophilic but more basic. In the presence of trimethyl aluminum, the potential formation of a stable Al–F bond (F–$AlMe_2$, methane evolved) should have eliminated the fluoride as a latent nucleophile. The acyl fluorides **150** were found to be less reactive compared to the corresponding acid chlorides, causing some difficulties in the rearrangement with *n*-alkyl carboxylic acid derivatives. Such transformations needed longer reaction times, and the yield of the corresponding rearrangement products was moderate [33d,e].

3. The use of a second base to trap all proton acids generated during the course of the rearrangement. In most cases, a two-phase system of solid potassium carbonate as a suspension in dichloromethane or chloroform gave the best results, even though excessive HX formation during the course of the reaction could be avoided by employing the combination acid fluoride/Me_3Al (formation of methane and dimethylaluminum fluoride).

5.2.1 Stereochemical Results: 1,3-Chirality Transfer and Internal Asymmetric Induction

Employing the optimized reaction conditions upon running various reactions, the stereochemical advantages of the Claisen rearrangements were combined with an efficient synthesis of the azoninones **157** and **158** bearing defined *E*-configured double bonds in the medium-sized rings (Scheme 30) [33]. As is known for all Claisen rearrangements, a complete 1,3-chirality transfer was observed on treating *E*-allylamines **154** (R^1, R^4=H) with acetyl chloride **155** (R^5=H) [33a]. Both enantiomers of the core framework were constructed starting from the same L-(–)-proline derivative choosing either an *E* (R^4=H) or a *Z* (R^3=H) allylamine **154** [33f]. Furthermore, a high internal asymmetric induction could be observed involving α-substituted acyl halides **155** ($R^5\neq$H) in the synthesis of the lactams. In most cases the diastereomeric excess was >5:1 in favor of the 3,4-*trans* lactam **156** (entries 4–14, Table 8). The phenylacetyl halide rearrangement (R^5=Ph, entry 7, Table 8) only gave a nearly equal mixture of *cis* and *trans* azoninones **157** and **158** (R^5=Ph). The stereochemical outcome of the rearrangement of **154** (R^1=H) was explained by the passing of a chair-like transition state **156** (cα) with minimized repulsive interactions and a defined *Z* enolate geometry (as is known for all amide enolates) [33b,d]. However the passing of the chair-like transition state **156** (cβ) could not be excluded: both **156** (cα) and (cβ) resulted in the same diastereomer *pS*-**158**!

Surprisingly, the rearrangement of the 4-*t*-butyldimethylsilyloxy-2-vinylpyrrolidines **154** (R^1=OTBS, R^3, R^4=H) took another course. The stereochemical outcome had to be rationalized by the passing of a boat-like transition state **156** (bβ) to give the 3,8-*trans* lactams **157** (R^1=OTBS, entries 15–19, Table 8). The corresponding *cis* product **158** (R^1=OTBS) resulting from the expected chair-like intermediate **156** (cβ) had only once been isolated as a minor compound

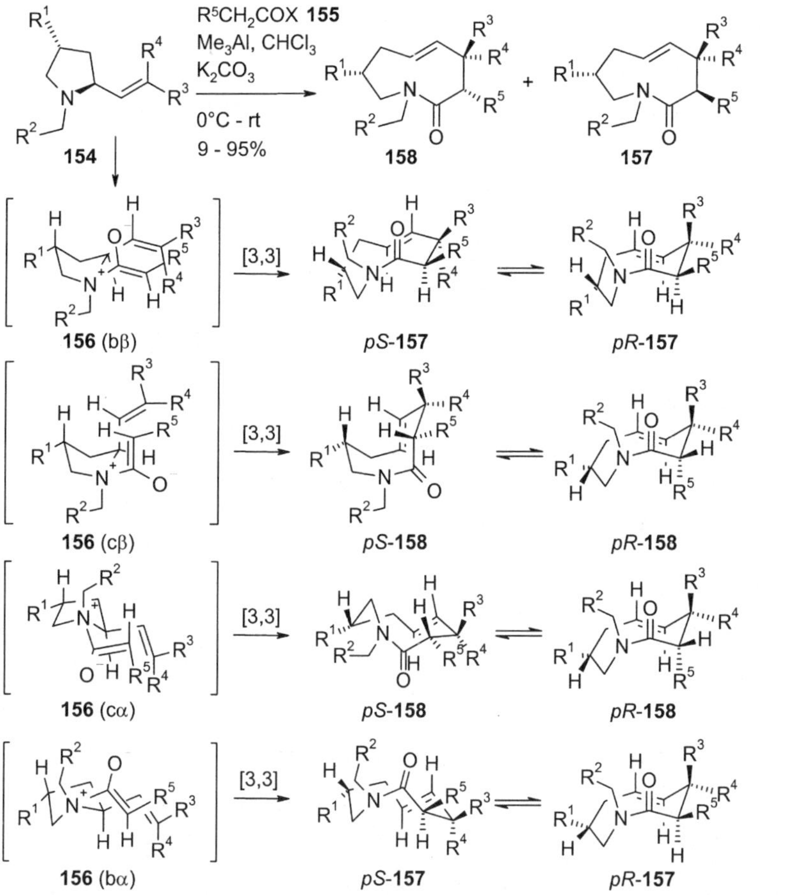

R^1 = H, OTBS, R^2 = H, Ph, R^3 = H, CO_2Et, CH_2OBn, R^4 = H, Me,
R^5 = H, Alkyl, Vinyl, Ph, Cl, OBn, NPht, X = Cl, F (Table 8)

Scheme 30

(entry 17, Table 8). The completeness of the 1,4-chirality transfer should be pointed out [33]. Obviously, the configuration of the intermediately generated stereogenic ammonium center in **156** had to be considered: Rearranging the 2-vinylpyrrolidines **154** (R^1=H), the *N*-acylation should have been directed by the adjacent side chain to the opposite face of the five-membered ring to give **156** (α) (1,2-*anti* induction, as found analyzing appropriate acylammonium salts). Consequently, the rearrangement proceeded via a chair-like transition state **156** (cα), as known for the acyclic 3,3-sigmatropic reaction leading predominantly to lactams **158**. In contrast, the *N*-acylation of the 2,4-*trans* disubstituted pyrrolidines **154** (R^1=OTBS) were directed by the bulky silyl ether to generate a *syn* arrangement of vinyl and acyl group in an intermediate ammonium salt **156** (β) (1,3-*anti* induction, 1,2-*syn*). Then, an appropriate conformation to undergo a

Table 8 [33]

Entry	R^1	R^2	R^3	R^4	R^5	X	Yield (%)	Ratio **158**:**157**	Ref.
1	H	H	CO_2Et	H	H	Cl	70	–	a
		Ph				Cl	60		b
2	OTBS	H	CO_2Et	H	H	Cl	53	–	a
3	OTBS	H	CO_2Et[b]	H	H	Cl	47	–	a
4	H	H	CO_2Et	H	Me	Cl	77	>95:<5	b
		Ph[a]				F	73	>6:<1	d
5	H	Ph[a]	CO_2Et	H	CH_2CH_2Cl	F	51	>6:<1	d
6	H	H	CO_2Et	H	$CH=CH_2$	Cl	80	>95:<5	b
		Ph[a]				F	72	>3:>1	d
7	H	H	CO_2Et	H	Ph	Cl	32[c]	45:55	b
		Ph[a]				F	79	1:2	d
8	H	H	CO_2Et	H	Cl	Cl	72	>95:<5	b
		Ph				Cl	22	90:10	b
		Ph				F	81	90:10	d
9	H	H	CO_2Et	H	OBn	Cl	68	80:20	b
		Ph				Cl	30	80:20	b

[a] Rearrangement with acid chloride failed.
[b] Reaction with 2*R*-vinylpyrrolidine.
[c] Up to 50% of the reactant recovered.
[d] Only *pS*-diastereomers.

Table 8 (continued)

Entry	R^1	R^2	R^3	R^4	R^5	X	Yield (%)	Ratio **158**:**157**	Ref.
10	H	H	CO_2Et	H	NPht	Cl	35[c]	>94:<6	b
11	H	Ph	CH_2OBn	H	Cl	Cl	11	87:13	b
12	H	Ph[a]	H	Me	CH_2CH_2Cl	F	51	>6:1	d
13	H	Ph[a]	H	Me	Ph	F	87	1:>4	d
14	H	Ph	H	Me	Cl	Cl	9	1:1	d
						F	91	3:1	d
15	OTBS	H	H	H	Ph	Cl	17[c]	1:>10	c
		Ph				Cl	26[c]	1:>10	c
		Ph				F	95	1:>10	e
16	OTBS	H[b]	H	H	Cl	Cl	22	1:>10	c
17	OTBS	H	H	H	Cl	Cl	29[c]	1:>10	c
		Ph				Cl	20[c]	1:5	c
		Ph				F	92	1:>10	e
18	OTBS	Ph[a]	H	H	OBn	F	73	1:>10	e
19	OTBS	Ph	H	H	NPht	Cl	17	1:>10	c
20	H	Ph	H	H	Cl	F	77	1:1.4[d]	f

Claisen rearrangement presumably was the boat-like form **156** (bβ) with minimized 1,3 repulsive interactions resulting in the lactams **157**. However, the 2,4-*cis* disubstituted pyrrolidine **159** (R^1=OTBS, R^3, R^4=H) gave the expected lactam diastereomer **158** via a chair-like transition state conformation **160** (entry 16, Table 8) (Scheme 31).

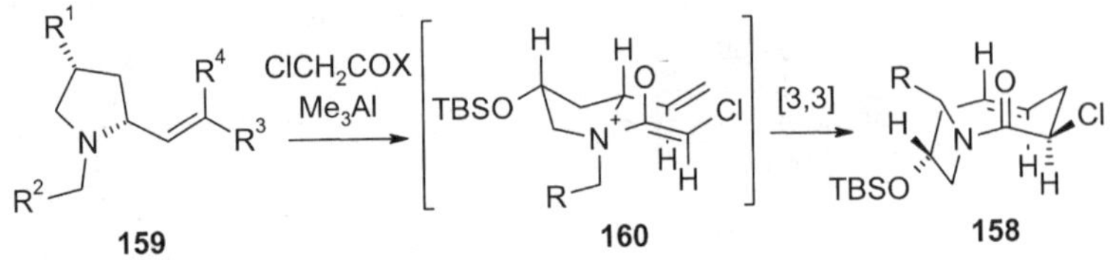

Scheme 31 R^1 = OTBS, R^2, R^3, R^4 = H, R^5 = Cl, X = Cl

The lactam and the olefin unit characterized the heterocyclic cores **157** and **158** as constrained ring systems, the conformations of which were found to be strongly dependent on the substitution pattern and the relative configuration of the stereogenic centers. The planar chiral properties of the medium-sized rings with internal *trans* double bonds had to be taken into account for analyzing the nine-membered rings [35]. The rearrangements of the 2*S*-vinylpyrrolidines **154** passing through a boat-like transition state **156** (b) effected initially the formation of the medium-sized ring with *pS* arrangement of the *E* double bond (*pS*-**157**). This planar diastereomer *pS*-**157** was obviously unstable: NMR and NOE analyses indicated the coexistence of one preferred *pS*-**157** and at least one additional minor conformation as a highly flexible equilibrium of some arrangements of the lactam function. Finally, the epimerization (flipping of the *E* double bond) to give the *pR* arrangement *pR*-**157** of the olefin with respect to the ring generated the most stable and rigid conformation. Preliminary force field calculations of the azoninones **157** and molecular mechanics calculations of the related *E*/*Z*-1,5-nonadiene confirmed these observations [36]. In contrast, the lactams **158** (R^4=H) generated via chair-like zwitterions **156** (c) were found to be generated directly in a stable *pS* arrangement of the *E* double bond *pS*-**158** (Schemes 30, 31). Nevertheless, a high activation barrier had to be passed to achieve the change of the planar chiral information (*pS*-**157**→*pR*-**157**). This fact allowed the isolation and the characterization of the conformers of the nine-membered rings (Schemes 30, 31) [33c,e,f].

The proof of principle gave the aza-Claisen rearrangement of vinylpyrrolidine **154** (R^1, R^3, R^4=H, R^2=Ph) and chloroacetyl fluoride **155** (R^5=Cl) under standard conditions. A mixture of two diastereomers *pS*-**157** and *pS*-**158** was obtained in 77% yield and a ratio of 1.4:1, which could be separated by means of column chromatography and preparative HPLC. On handling these compounds, any warming up to 30–40 °C was avoided to maintain the planar chiral properties (*pS*) of the diastereomers resulting from the ring expansion. For testing the conformational stability of both compounds, the separated diastereomers *pS*-**157** and *pS*-**158** were heated to about 60 °C [37]. After 3 to 10 h a second diastereomer *pR*-**157** occurred starting from *pS*-**157**, indicating the flipping of the double bond with respect to the ring. All spectral data of the new diastereomer *pR*-**157** were identical with those determined for lactam *pS*-**158** except the specific rotation proving the formation of the enantiomer. Lactam *pS*-**158** suffered the analogous process generating *pR*-**158** (enantiomer of *pS*-**157**), but the conversion was found to be incomplete. The relative arrangement of the double bond and the stereogenic center of the diastereomers was proven via NOE analyses. While the lactam *pS*-**158** was characterized by a single set of peaks (almost rigid conformation), the spectral data of lactam *pS*-**157** indicated the coexistence of two conformers (double set of peaks in ^1H and ^{13}C spectra) potentially originating from some mobility of the lactam function (Table 8, entry 20) (Scheme 32) [33f].

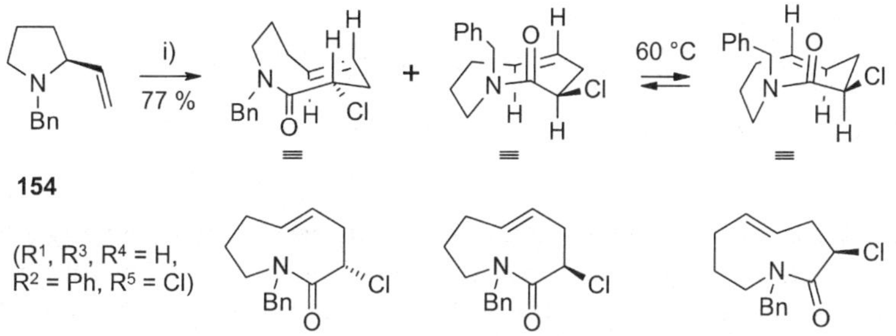

154

pS-**158** *pS*-**157** *pR*-**157**

i) $ClCH_2COF$, Me_3Al, CH_2Cl_2, K_2CO_3, 0°C, 3h

Scheme 32

The azoninones **157** and **158** with defined stereochemical properties served as key compounds in natural product syntheses. Firstly, the planar chiral information was used to generate stereospecifically new stereogenic centers depending on the defined conformation of the nine-membered rings [38]. Upon treatment of p*S*-**158** (R^1, R^4, R^5=H, R^2=Ph, R^3=CO_2Et) with PhSeCl in MeCN, the *anti* addition of $[PhSe]^+$ and the lactam lone pair to the double bond gave an intermediate acylammonium ion **161**, which suffered from immediate von Braun degradation to form benzyl chloride and the indolizidinone **162** as a single regio- and stereoisomer (ring contraction) with 70% yield. Several further steps allowed completion of a total synthesis of (–)-8-*epi*-dendroprimine [38a]. A mixture of p*S*-**157**/p*S*-**158** (R^1, R^3, R^4=H, R^2=Ph, R^5=Cl) was epoxidized by means of MCPBA to give the diastereomeric mixture of epoxy azonanones **163** with defined epoxide configuration (stereospecific cycloaddition). Chlorine and benzyl groups were removed by hydrogenation using Pearlman's catalyst to give a single hydroxyindolizidinone **164** after regio- and stereoselective intramolecular oxirane opening. Several further transformations enabled completion of a (+)-pumiliotoxin 251D synthesis (Scheme 33) [38b,c].

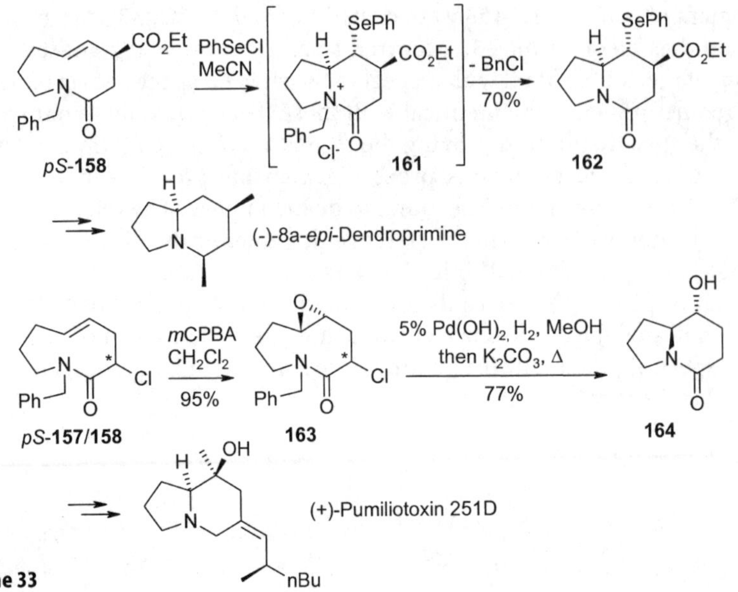

Scheme 33

An alternative pathway using a zwitterionic aza-Claisen rearrangement to generate azoninones was described by Hegedus [39]. 2-Vinylpyrrolidines **165** and chromium carbene complexes **166** underwent photochemical reactions in the presence of a Lewis acid to give the corresponding nine-membered ring lactams **167** bearing *E* double bonds in up to 71% yield. Though reactants and products suffered from some instability against Lewis acids, the presence of the zinc chloride or dimethylaluminum chloride was mandatory to start the rearrangement. In contrast to the classical ketene Claisen process, electron-rich ketene equivalents such as alkoxy or amino ketenes could be used, since the donor substituents stabilized the chromium carbene complex **166**. Furthermore, α,α-disubstituted lactams were synthesized but the stereoselectivity observed was low. The determination of the stereochemical outcome of the reaction proved that the 1,4-chirality transfer was not complete: a Mosher analysis of an appropriate azoninone gave a loss of about 10% of the chiral information. A chiral carbene complex **166** (R^1=oxazolidinyl) was found to have a negligible influence on the stereoselectivity of the rearrangement. Generally, the present variant of the rearrangement was found to be very sensitive to any steric hindrance. Additional substituents in any position (e.g., R^2≠H) led to a severe decrease of the yield and the stereoselectivity. Additionally, one example rearranging a 2-vinylpiperidine **168** was given. The corresponding azecinone **169** was formed in about 33% yield (Scheme 34). Some details are outlined in Table 9. In analogy to Edstrom's experiments [32], the nine- and ten-membered ring lactams **167** and **169** underwent regio- and stereoselective transannular ring contractions to give the corresponding indolizidinones and quinolizidinones, respectively (*vide supra* Scheme 29) (Scheme 34, Table 9).

Table 9 [39]

Entry	n	R[1]	X	R[2]	Lewis acid	Yield (%)	Ratio **167/169**
1	1	Me	OMe	H	$ZnCl_2$	71	–
2	1	Me	OBn	H	$ZnCl_2$	66	62% de[b]
3	1	$-(CH_2)_3-O-$		H	$ZnCl_2$	15	–
					Me_2AlCl	22	
4	1	H	NMe_2	H	Me_2AlCl	9	–
5	1	H	Oxazolidine[a]	H	$ZnCl_2$	19	74% de[b]
6	1	Me	OMe	Me	$ZnCl_2$	20	60% de[c]
7	1	Me	OBn	Me	$ZnCl_2$	40	33% de[c]
8	2	Me	OMe	H	Me_2AlCl	33	–

[a] Chiral oxazolidine.
[b] Determined via Mosher analysis of a derivative.
[c] Mixture of 3,4 diastereomers.

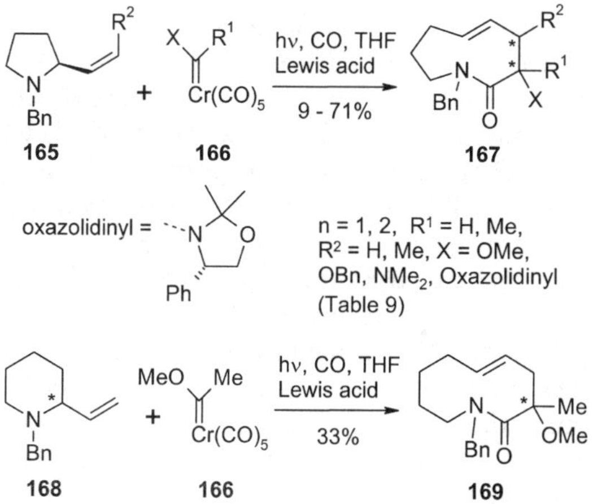

Scheme 34

First systematic efforts to investigate the internal asymmetric induction of ketene Claisen rearrangements have been published by Yu [40]. Some simple *E* and *Z N*-crotylpiperidines **170** were treated with propionyl, methoxyacetyl, and fluoroacetyl chloride **171**. Renouncing Lewis acid support, the reactants were combined in toluene at 0 °C in the presence of solid K_2CO_3 as a proton acceptor (Schotten–Baumann conditions). The piperidides **173** were isolated in 38–61% yield. The formation of side products (von Braun degradation) was not reported. All reactions were found to be highly diastereoselective. The stereochemical outcome could be rationalized by the initial formation of the zwitterion **172** with a defined double bond and a defined *Z*-ammonium enolate geometry because of steric and/or electronic reasons. The passing of a chair-like transition state gave rise to the formation of the *anti* product, *anti*-**173**, starting from *Z*-**170** and the *syn* product, *syn*-**173**, starting from *E*-crotylpiperidine *E*-**170**. Alternatively, the use of the corresponding crotylpyrrolidines and the employment of in situ formed propionyl bromide gave somewhat lower yields and diastereoselectivities (Scheme 35).

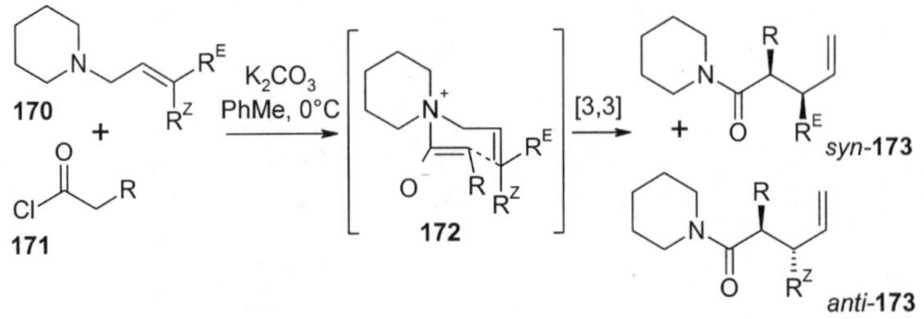

E-**170**: R = Me (41% yield, *syn*/*anti* = 94:6), R = OMe (44%, 92:8), R = F (61%, 95:5)
Z-**170**: R = Me (38% yield, *syn*/*anti* = 3:97), R = OMe (39%, 5:95), R = F (57%, 4:96)

Scheme 35

A more recent systematic investigation of the internal asymmetric induction in ketene Claisen rearrangement was contributed by MacMillan [41]. *E* and *Z* *N*-allylmorpholine derivatives **174** were reacted with acid chlorides **171** in the presence of Hünig's base supported by 5 to 20 mol% of a Lewis acid (Einhorn conditions). The product γ,δ-unsaturated morpholine amides **176** were obtained with 70 to 95% yield. The conditions reported indicate the presence of a high concentration of nucleophilic chloride ions in the reaction medium. Though acid chlorides and tertiary amines tend to a rapid formation of *N*-acylammonium salts, even in the presence of substoichiometric amouts of a Lewis acid, no von Braun degradation-generating allyl chlorides and carboxylic acid morpholine amides were mentioned (Fig. 3). Best results were achieved running the rearrangement in the presence of Yb(OTf)$_3$, AlCl$_3$, and TiCl$_4$. As expected, a very high internal asymmetric induction was found in most experiments. Compared to the *Z* amine *Z*-**174** resulting in *anti*-**176** as the major compound, the corresponding *E*-configured reactants *E*-**174** gave somewhat higher yields and diastereoselectivities forming predominantly the *syn* amides *syn*-**176**. In accordance with previous reports the use of α-alkyloxyacetyl chlorides caused decreased diastereoselectivities of 86:14 to 90:10 [33, 43]. The stereochemical outcome could be rationalized by the intermediate formation of the zwitterion **175** with *Z* enolate geometry, which rearranged passing a chair-like transition state to give the desired product **176**. Furthermore, new quaternary carbon centers could be built up by means of the present protocol. The rearrangement of 3-ethyl-3-methyl-substituted allylamine *E*-**174** allowed generation of the corresponding amide *syn*-**176** with 72% yield and 99:1 *syn* selectivity (Table 10, entry 14). The cyclohexenylamine **177** could be converted into the amide **178** with 75% yield and 99:1 *syn* diastereoselectivity, too (Table 10, entry 15). Detailed information is outlined in Table 10 (Scheme 36).

Table 10 [41a]

Entry	R^E	R^Z	R	Lewis acid	Mol%	Yield (%)	Ratio *syn:anti*
1	Me	H	Me	$Yb(OTf)_3$	10	80	99:1
2	Me	H	Me	$AlCl_3$	10	90	99:1
3	Me	H	Me	$Ti(iOPr)_2Cl_2$	10	76	99:1
4	Me	H	Me	$TiCl_4$·2 THF	5	92	99:1
5	Ph	H	Me	$TiCl_4$·2 THF	10	76	99:1
6	Cl	H	Me	$TiCl_4$·2 THF	10	95	99:1
7	H	H	Me	$TiCl_4$·2 THF	10	95	–
8	H	Me	Me	$TiCl_4$·2 THF	20	74	5:95
9	Me	H	NPht	$TiCl_4$·2 THF	10	77	99:1
10	Me	H	SPh	$TiCl_4$·2 THF	10	81	92:8
11	Me	H	OBn	$TiCl_4$·2 THF	10	91	86:14
12	Cl	H	OBn	$TiCl_4$·2 THF	10	83	90:10
13	H	Cl	OBn	$TiCl_4$·2 THF	10	70	90:10
14	Et	Me	Me	$TiCl_4$·2 THF	10	72	99:1
15	Me	R[a]	Me	$TiCl_4$·2 THF	10	75	5:95

[a] Reactant **177**, product **178**.

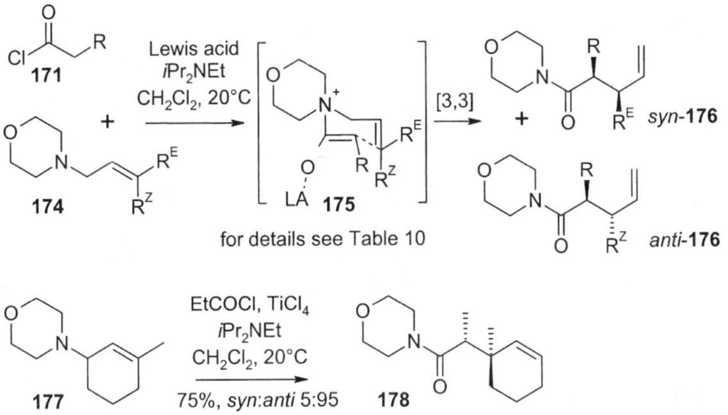

Scheme 36

The present protocol enabled tandem ketene aza-Claisen rearrangements to be run [41b]. The reaction of allyl systems **179** bearing two allylamine fragments gave the corresponding diamides **183** with high yield and high simple diastereoselectivity as well as an excellent 1,3-asymmetric induction. The amount of Lewis acid had to be increased to about 200 to 400 mol%. The process was explained by two consecutive sigmatropic rearrangements. In the first step, the *E*-allylmorpholine moiety of **179** reacted with the ketene fragment from **171** passing the zwitterionic intermediate **180** displaying the well-known

Z-enolate geometry. The chair-like transition state gave rise to the formation of the *syn* intermediate *syn*-**181**. Then, the newly generated allyl morpholine in **181** suffered from a second rearrangement. The addition of another acid chloride **171** led to the zwitterion **182**, which was immediately transformed into the diamide *syn*/*anti*-**183**, passing a chair-like conformation with minimized repulsive interactions. High yields of 71 to 99% were reported; in most runs one major diastereomer could be detected with >92:8 diastereoselectivity. Detailed information is given in Table 11 (Scheme 37).

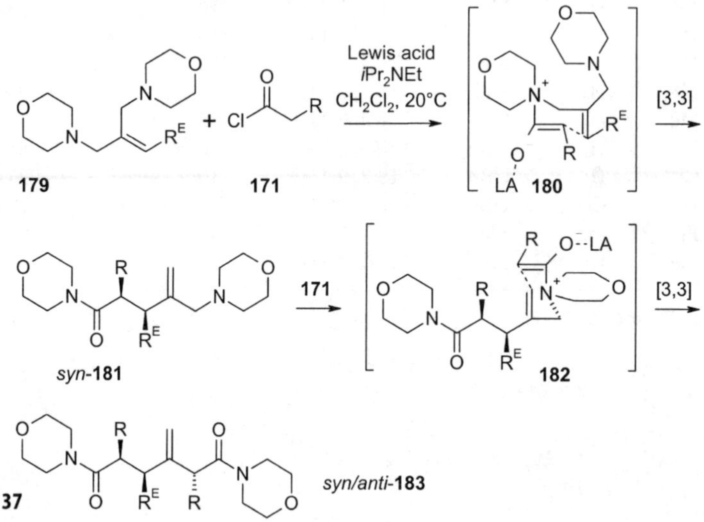

Scheme 37

Table 11 [41b]

Entry	R^E	R	Lewis acid	Mol%	Yield (%)	Ratio *syn*/*anti*:*anti*/*anti*
1	Me	Me	$Yb(OTf)_3$	200	97	98:2
2	Me	Me	$AlCl_3$	200	93	64:36
3[a]	Me	Me	MgI_2	400	70	98:2
4	Me	Me	$TiCl_4 \cdot 2$ THF	200	93	98:2
5	Cl	Me	$Yb(OTf)_3$	200	98	99:1
6	OBz	Me	$Yb(OTf)_3$	200	86	91:9
7	CN	Me	$TiCl_4 \cdot 2$ THF	200	78	97:3
8	SPh	Me	$TiCl_4 \cdot 2$ THF	200	70	93:7
9	Me	Bn	$Yb(OTf)_3$	200	99	92:8
10	Me	NPht	$Yb(OTf)_3$	200	98	95:5
11	Me	OPiv	$TiCl_4 \cdot 2$ THF	200	97	97:3
12	Cl	OPiv	$TiCl_4 \cdot 2$ THF	200	84	95:5
13	OBz	OPiv	$TiCl_4 \cdot 2$ THF	200	71	92:8

[a] Reaction at –20°C.

5.2.2
Stereochemical Results: External Asymmetric Induction/Remote Stereocontrol

The low reaction temperatures of the zwitterionic ketene aza-Claisen rearrangements recommended the process for further testing of the stereodirecting properties. An efficient external chiral induction within the addition rearrangement sequence requires not only a highly ordered transition state of the six core atoms, but also a defined arrangement of the external chiral subunit with respect to the rearrangement framework. Always, a single defined transition-state conformation causes a highly selective reaction [41b]. With the intention of achieving a maximal asymmetric induction via remote stereo control, the chiral information should predominantly be placed next to the nascent sigma bond formed during the course of the rearrangement and carrying the new chiral centers. A stereogenic center in the allylamine moiety adjacent to C3 fulfilled such a prerequisite. Particularly, defined C atom–heteroatom bonds offer the advantage of being potential excellent stereodirecting subunits. Since the electron-rich 1,2-vinyl double bond should attack the allyl system at position 6, an adjacent C–X (nucleophilic X) bond should adopt an *anti* arrangement with respect to the incoming donor. In other words, the extended C–X–σ^* orbital is *syn* coplanar positioned with respect to the attacking vinyl double bond. The transition state is stabilized by an additional delocalizing of some electron density in the empty *anti* binding orbital. Such weak electronic effects have been successfully used in ketene thia-Claisen rearrangements [42].

Allylamines **184** bearing an additional chiral center adjacent to C6 (R^1=**a**–**d**) were efficiently synthesized via short ex-chiral pool sequences starting from D-mannitol (→**c**), L-malic (→**b**), L-lactic acid (→**a**), and L-proline (→**d**) [43]. The treatment of such allylamines **184** with acid chlorides **185** (X=Cl), even in the presence of several Lewis acids (Me_2AlCl, $MeAlCl_2$, $ZnCl_2$, $TiCl_4$, $SnCl_4$, $Yb(OTf)_3$, etc.), suffered from the formation of varying amounts of von Braun degradation products **187** and **188** among the desired γ,δ-unsaturated amides **189**–**192**. The nucleophilic attack of the chloride at the intermediately formed acylammonium salt **186** sometimes predominated. Subjecting the allylamines **184** to α-monosubstituted acid halides **185** in the presence of Me_3Al, the desired rearrangement got the upper hand. Conducting the reaction at 0 °C in a two-phase system of CH_2Cl_2 and K_2CO_3, the degradation could be almost suppressed depending on the substitution pattern in **184** and **185**. Finally, the use of carboxylic acid fluorides **185** (X=F) in the presence of Me_3Al enabled the process to be run in the absence of nucleophiles [33d]. Since then, allyl halides **187** have not been found any more. Analyzing the stereoselection properties of the conversion to **189**–**192**, the 1,2-asymmetric induction was found to be mostly >90:10 in favor of the *syn* product – even in the presence of the nitrogen as directing function. The minor *anti* diastereomer only occurred in appreciable amounts if acetyl chloride **185** (R^2=H) was used as C2 source. Furthermore, the simple diastereoselectivity (internal asymmetric induction) was high, allowing the diastereoselective generation of two new stereogenic centers in a single step bearing a variety of functional groups. The γ,δ-unsaturated amides **189–192** formed represented useful intermediates for natural product total syntheses, as demonstrated by completing the synthesis of (+)-dihydrocanadensolide **193** [43b] and the formal synthesis of (–)-petasinecin **194** (**195**=petasinecin) (Scheme 38) [33d]. Detailed information is summarized in Table 12.

The stereochemical outcome of the reaction could be explained by the passing of a clearly preferred transition state. Generally, the ketene equivalent (from **185**) and the allylamine **184** (R^1: **c**) combined to form a hypothetical intermediate acylammonium enolate with a defined *Z*-enolate geometry (in *b*-**189**, *c*-**190**, *c*-**191**), as is known for amide and acylammonium enolates. Adopting the chair-like conformations *c*, the *anti* arrangement of the attacking enolate and the guiding heteroatom (N and O at C6a) favored *c*-**190** facing *c*-**191**; the *anti/syn* product **190** was isolated as the major compound. Surprisingly, R^2 substituents characterized by extended Π-systems led to the *syn–syn* products **189** selectively. Here, the high remote stereocontrol must involve an alternative boat-like transition state conformation *b*-**189**. In Fig. 4 the hypotheses concerning **184** (R^1: **c**) are outlined. With respect to the inverted configuration of the directing center adjacent to C6 in **184** (R^1: **a, b, d**), **189–192** (R^1: **a, b, d**), the enantiomer stereotriads were formed.

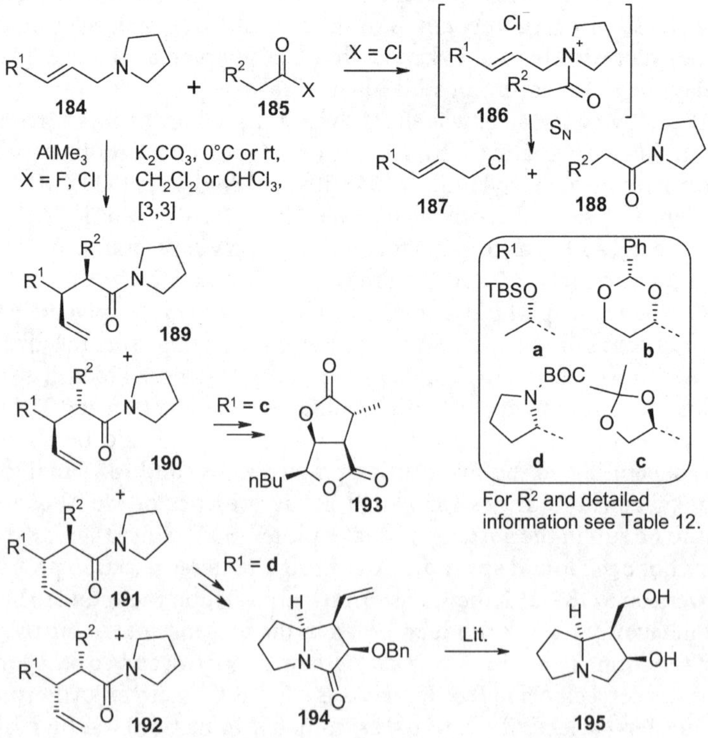

Scheme 38

Table 12 [43]

Entry	R^1	R^2	X	Yield (%)	Ratio **189/190/191/192**	Ref.
1	**a**	H	Cl	84	–:1:2:–	a
2	**a**	Me	Cl	73	<1:<1:>15:<1	a
3	**b**	H	Cl	80	–:4:7:–	a
4	**b**	Me	Cl	74	<1:1:10:<1	a
5	**c**	H	Cl	82	–:3:2:–	a
6	**c**	Me	Cl	77	<1:9:1:<1	a
7	**c**	$ClCH_2CH_2$	Cl	74	3:7:<1:<1	b
8	**c**	*i*Pr	Cl	45	<1:97:<1:<1	b
9	**c**	H_2C=CH	Cl	62	<1:97:<1:<1	b
10	**c**	H_2C=CH–CH=CH–	Cl	60	97:<1:<1:<1	b
11	**c**	Ph	Cl	52	97:<1:<1:<1	b
12	**c**	Cl	Cl	82	2:96:<1:<1	b
13	**c**	OBn	Cl	83	9:86:<1:4	b
14	**d**	Me	F	78	<1:<1:>97:<1	c
15	**d**	Ph	F	85	<1:<1:>97:<1	c
16	**d**	Cl	Cl	24	<1:<1:>97:<1	c
17	**d**	Cl	F	76	<1:<1:>97:<1	c
18	**d**	OBn	F	68	<1:<1:>97:<1	c

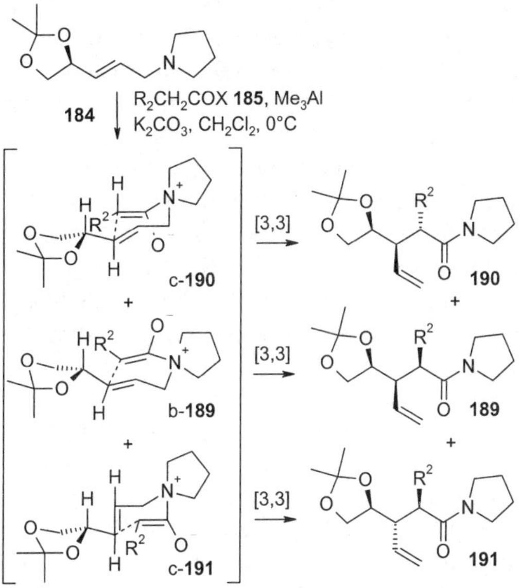

Fig. 4 Formation of enantiomer stereotriads of **184**

5.2.3
Stereochemical Results: External Asymmetric Induction/Auxiliary Control

Auxiliary-controlled 3,3-sigmatropic rearrangements represent an almost classical approach to introducing chiral information into a more complicated rearrangement system. The major advantage is a reliable control of the stereochemical outcome of the reaction. All products are diastereomers, e.g., the separation of minor compounds should be more or less easy and the well-known spectroscopic analyses will always be characterized by defined and reproducible differences. However, the auxiliary strategy requires two additional chemical transformations: the attachment and the removal of the auxiliary must be carried out in two steps. The efficiency of these steps influences the whole sequence. High yields and the avoiding of stereochemical problems are the prerequisites of each step. Furthermore, the synthesis and the recycling of an auxiliary have to be taken into account. Overall, a set of advantages and problems has to be considered before deciding on an auxiliary strategy.

An aza-Claisen rearrangement enables a chiral auxiliary to be attached to the central nitrogen via the third binding valence next to the 3,3-sigmatropic framework. In analogy to the remote stereocontrol prerequisites mentioned above, a low reaction temperature was crucial to guarantee a restrained conformational mobility of the potential transition state. The system should be forced to take a single reaction path for obtaining a high diastereoselectivity.

The zwitterionic aza-Claisen rearrangement seemed to fulfill these prerequisites using L-(–)-proline derivatives **196** as chiral auxiliaries [44]. Several *N*-allyl pyrrolidines **199** were synthesized via a Pd(0)-catalyzed amination of the corresponding allyl mesylates **197** and **198**. Always, the double bond was *E* configured. The treatment with chloro and suitably protected α-amino acetyl fluorides **200** in the presence of solid K_2CO_3 and trimethyl aluminum in $CHCl_3$/0 °C led to the formation of the corresponding γ,δ-unsaturated amides **201** and **202**. Again, the charge neutralization served as an efficient driving force allowing the reactions to be conducted at such low temperatures. Especially, the use of azido acetyl fluoride **200** (R^3=N_3) enabled a subsequent reductive cyclization to generate D-proline/L-proline dipeptides **203**, allowing introduction of varying substituents in the new D-proline moiety (Schemes 39, 40, Table 13).

The removal strategy of the auxiliary should be chosen depending on the auxiliary and the substitution pattern of the amide **201**/**202**. Generally, the iodo lactonization as described by Tsunoda [22] and Metz [45] led to smooth cleavages of all types of amides. In particular, the prolinol auxiliaries (R^1=CH_2OTBS) offered further advantages. In the presence of acid-stable substituents R^2 and R^3, a neighboring group-assisted esterification (R^1=CH_2OTBS) with HCl/MeOH allowed conversion of the amides **201** into the corresponding esters **205**. Alternatively, the auxiliary can be used as a leaving group in an intramolecular metal organic reaction of **201** (R^3=CH_2–Ar–Br) to generate a cyclic ketone **206** without any loss of the chiral information (Scheme 40).

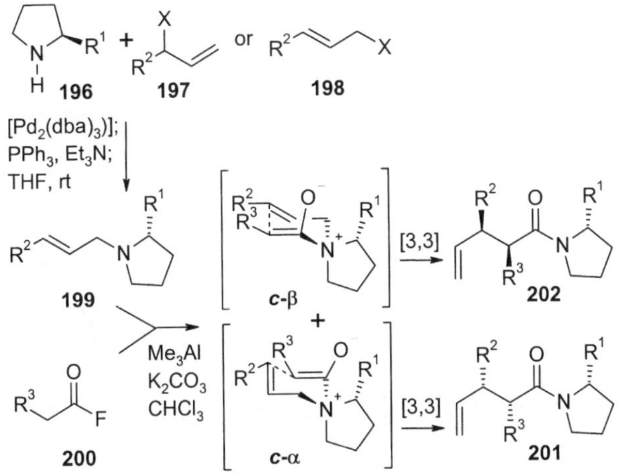

Scheme 39

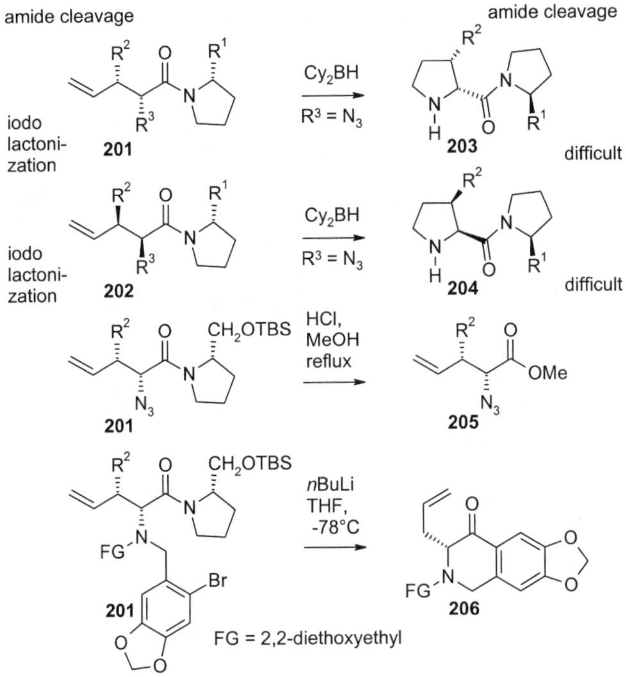

Scheme 40

Discussing the stereochemical outcome of the Claisen rearrangements, two aspects had to be considered. On the one hand, the relative configuration of the new stereogenic centers was found to be exclusively *syn* in **201** and **202**, pointing out the passing of a chair-like transition state ***c*-α** and ***c*-β**, respectively, including a *Z*-acylammonium enolate structure (complete simple diastereoselectivity/internal asymmetric induction).

Table 13 [44]

Entry	R^1	R^2	R^3	*T* (°C)	Yield (%)	Ratio **201/202**	Ref.
1	H	3,4-Methylenedioxyphenyl	Cl	0	36[a]	1:1	a
2	CO_2Me	3,4-Methylenedioxyphenyl	Cl	0	69	2:1	a
3	CO_2tBu	3,4-Methylenedioxyphenyl	Cl	0	50[a]	>95:5	a
4	CH_2OTBS	3,4-Methylenedioxyphenyl	Cl	0	72	>95:5	a
5	CO_2Me	3,4-Methylenedioxyphenyl	N_3	0	77	4:1	a
6	CO_2Me	4-Methoxyphenyl	N_3	0	13[a]	1:1	a
7	CH_2OTBS	3,4-Methylenedioxyphenyl	N_3	0	87	>95:5	a
8	CH_2OTBS	4-Methoxyphenyl	N_3	0	57	>95:5	a
9	CH_2OTBS	Ph	N_3	0	91	>95:5	a
10	CO_2Me	H	NPht	0	74	1:1	b
11	CO_2Me	H	N_3	20	77	4:1	b
12	CO_2Me	H	N_3	0	77	7:1	b
13	CO_2Me	H	N_3	–20	77	9.5:1	b
14	CO_2Me	H	$(EtO)_2CHCH_2NBOC$	20	73	15:1	b
15	CH_2OTBS	H	N_3	0	77	>95:5	b
16	CH_2OTBS	H	HNBOC	0	6	–	b
17	CH_2OTBS	H	$(EtO)_2CHCH_2NBn$	0	0	–	b
18	CH_2OTBS	H	$(EtO)_2CHCH_2NBOC$	0	51[b]	>95:5	b
19	CH_2OTBS	H	$(EtO)_2CHCH_2NCbz$	0	75	>95:5	b
20	CH_2OBn	H	$(EtO)_2CHCH_2NBOC$	0	47[b]	>95:5	b

[a] Not optimized.
[b] Yield including four further steps.

On the other hand, the external asymmetric induction strongly depended on the chiral auxiliary. The careful analysis of the hypothetical zwitterionic intermediates ***c-α*** and ***c-β*** indicated the formation of a stereogenic ammonium center. In terms of the well-known 1,3-chirality transfer of 3,3-sigmatropic rearrangements, the present reaction allowed the chiral information to be shifted from the ammonium center (1) to the enolate C (3). The amide **201/202** *α*-carbon atom had been built up with a defined configuration after passing the above-mentioned chair-like transition state ***c-α/c-β***, including a defined olefin geometry and the equatorial arrangement of the bulky (chain branch) part of the auxiliary. Consequently, the crucial step of the whole process must have been the diastereoselective addition of the ketene equivalent from **200** on generating the zwitterionic intermediates. Thus, employing the auxiliaries bearing the small proline methyl ester substituent (R^1=CO_2Me) in **199**, the reaction with nonhindered acid fluorides **200** gave the corresponding amides **201/202** with low or moderate diastereoselectivity indicating unselective *N*-acylation. In contrast, conversions at lower temperatures or with bulky substituted acid fluorides **200** resulted in significantly higher selectivities (more selective acylation). The use of reactant allylamines **199** bearing the bulky proline *tert*-butyl ester and the OTBS prolinol auxiliaries as R^3 was characterized by a high auxiliary-directed diastereoselectivity, indicating the passing of a defined acylation rearrangement path via ***c-α***. At present, the OTBS prolinol (R^1=CH_2OTBS) is the auxiliary of choice because of the easy introduction, the high auxiliary-directed induction of chirality, the stability against a set of consecutive processes, and the simple cleavage by the neighboring group-assisted amide **201**–ester **205** conversion. Detailed information is summarized in Table 13 (Schemes 39, 40). The zwitterionic aza-Claisen rearrangement has been developed as a reliable method for synthesizing suitably protected nonnatural *α*-amino acid derivatives, e.g., C-allyl glycines type **205** and 3-arylprolines type **203**.

The major disadvantage of the classical auxiliary-controlled 3,3-sigmatropic rearrangements is still the requirement of two additional chemical transformations: the attachment and the removal of the auxiliary had always to be considered. The efficiency of these steps influences the usability of the whole sequence.

Since coordination of the Lewis acid metal salt at the core heteroatoms of the 3,3-sigmatropic system was found to accelerate the process, the proximity of the Lewis acid ligands should allow one to influence the stereochemical outcome of the rearrangement. Hence, the use of chiral ligands should cause an external chiral induction. In conclusion, a Lewis acid carrying chiral ligands should serve as a chiral auxiliary. The separate attachment and the final removal of the auxiliary could be saved, and the enantioselective Claisen rearrangement arose as a more straightforward process. Generally, such a reaction should be run in a catalytic sense, but the increased complexation ability of the product in comparison to the reactants mostly inhibited the release of the Lewis acid right after a rearrangement step until the aqueous cleavage. It is understood that the stereochemical properties of the products had to be carefully analyzed using chiral GC, HPLC, and derivatization techniques.

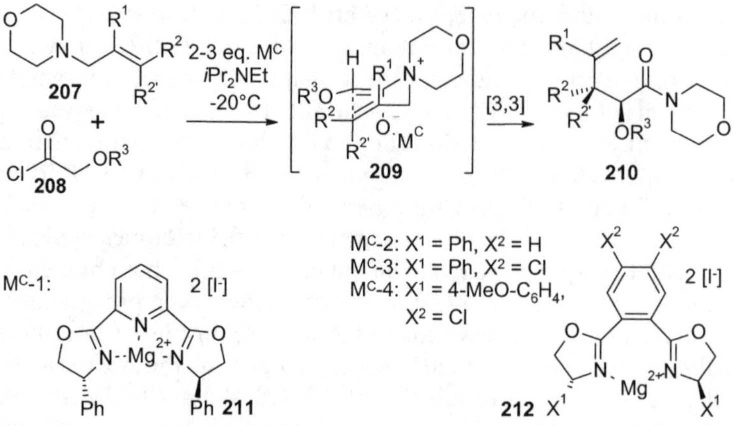

Scheme 41 For detailed information including R^1 - R^3 see Table 14

Table 14 [46]

Entry	R^1	R^2	$R^{2'}$	R^3	Eq. (M^C)	Yield (%)	de (%)	ee (%)
1	H	H	H	Bn	–	42	–	–
2	H	H	H	Bn	2.0 (1)	87	–	56
3	H	H	H	Bn	2.0 (2)	88	–	83
4	H	H	H	Bn	2.0 (3)	65	–	86
5	H	H	H	Bn	0.5 (4)	81	–	42
6	H	H	H	Bn	1.0 (4)	63	–	81
7	H	H	H	Bn	2.0 (4)	80	–	91
8	H	H	H	Ac	2.0 (4)	44	–	37
9	H	H	H	TBS	2.0 (4)	67	–	38
10	H	H	H	4-Cl-Ph	2.0 (4)	59	–	71
11	H	H	H	Ph	2.0 (4)	48	–	78
12	H	H	H	Me	2.0 (4)	28	–	80
13	Me	H	H	Bn	3.0 (4)	78	–	91
14	Ph	H	H	Bn	3.0 (4)	79	–	90
15	H	CH_2OBz	H	Bn	3.0 (4)	86	84	86
16	H	4-NO_2-Ph	H	Bn	3.0 (4)	82	98	97
17	H	CO_2Et	H	Bn	3.0 (4)	84	94	96
18	H	Cl	H	Bn	3.0 (4)	95	96	91
19	H	H	Cl	Bn	3.0 (4)	74	96	91
20	H	CO_2Et	Me	Bn	3.0 (4)	75	88	97

MacMillan [46] investigated the intermolecular aza-Claisen rearrangements treating *N*-allyl morpholines **207** with glycolic acid chlorides **208** in the presence of a chiral-chelated Lewis acid. The so-called magnesium BOX systems **211** and **212** gave the best results concerning yield (up to 95%) and chirality transfer (up to 97% ee) generating the amides **210**. Usually, 2–3 mol equiv of the chiral metal complexes **211** and **212** had to be employed to achieve satisfactory ee values. The glycolic acid framework seemed to play a crucial role in terms of asymmetric induction: A nonchelating α-oxygen substituent R^3 produced amides **210** with only moderate enantiomeric excess. In contrast, the use of benzyloxyacetyl chloride allowed very high ee values to be achieved. It seemed reasonable that this α-oxygen substituent enabled an efficient chelation of the chiral modified Lewis acid in **209** causing the high level of external chirality transfer. The substitution pattern of the allyl morpholine remained variable, and the use of *E* and *Z* olefin led to the defined formation of enantiomer amides **210** with comparable asymmetric induction. The scope and limitations are outlined in Table 14. Until now, the catalytic enantioselective aza-Claisen rearrangement involving substoichiometric amounts of the chiral information remained undiscovered (Scheme 41, Table 14).

5.3
Allene Carbonester Claisen Rearrangement

The third type of zwitterionic aza-Claisen rearrangement can be termed as *N*-allyl ammonium enolate Claisen rearrangement [47]. The first step of this tandem process was a Lewis acid-catalyzed Michael addition of a tertiary allylamine **213** to the β-carbon center of an allene carbonester **214**. The so-formed hypothetical zwitterion **215** must be characterized by a highly resonance-stabilized anion, and additional support should have given the O-coordinated Lewis acid. Then, the allyl vinyl ammonium moiety of **215** underwent a 3,3-sigmatropic rearrangement to give the unsaturated ester **216** bearing two new stereogenic centers in the γ and δ positions. The formation of the vinylogous carbamate in **216** and the charge neutralization served as potential driving forces, allowing such a reaction to be run at 23 °C. Best results were obtained using 5 to 10 mol% of $Zn(OTf)_2$.The products **216** were isolated with 75–97% yield and excellent diastereoselectivity of >98:2. Variation of the allylamine **213** and the allene substitution pattern in **214** gave a first insight into the scope and limitation of the transformation (Table 15). It should be pointed out that the present protocol enabled generation of defined quaternary centers. Furthermore, the double bond geometry of the allylamine **213** moiety allowed prediction of the stereochemical outcome of the reaction: The geranyl derivative *E*-**213** rearranged upon treatment with methyl pentadienoate **214** to give the *syn* product *syn*-**216** (methyl groups) with 94% yield and 98:2 dr. In contrast, the analogous reaction of nerylamine *Z*-**213** delivered the corresponding *anti* derivative *anti*-**216** (methyl groups) with 93% yield and 98:2 dr. The stereochemical outcome could be rationalized by the favored passing of a chair-like transition state in **215**; the vinyl

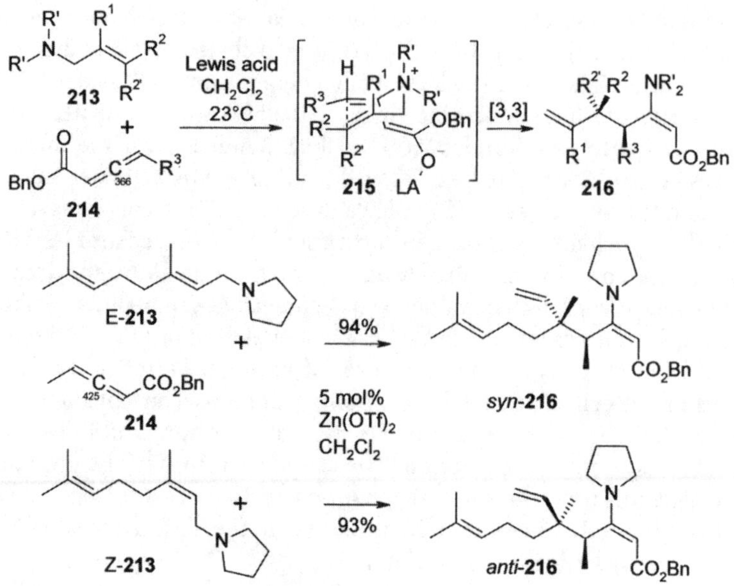

Scheme 42 For detailed information including R^1 - R^3 see Table 15

Table 15 (Reaction with 10 mol% $Zn(OTf)_2$) [47]

Entry	R′	R^1	$R^2/R^{2'}$	R^3	Yield (%)	Ratio *syn:anti*
1	$-(CH_2)_4-$	H	Me/H	Me	95	>98:2
2	$-(CH_2)_4-$	H	H/Me	Me	94	2:>98
3	$-(CH_2)_4-$	H	Ph/H	Me	97	94:6
4	$-(CH_2)_4-$	H	*i*Pr/H	Me	81	>98:2
5	$-(CH_2)_4-$	Me	H/H	Me	80	–
6	Me	H	Ph/H	Me	81	94:6
7	$-(CH_2)_5-$	H	Ph/H	Me	87	94:6
8[a]	$-(CH_2)_4-$	H	Me/H	Ph	86	97:3
9[a]	$-(CH_2)_4-$	H	Ph/H	Ph	94	94:6
10	$-(CH_2)_4-$	H	Ph/H	*i*Pr	94	94:6
11	$-(CH_2)_4-$	H	Ph/H	Cl	84	93:7
12	$-(CH_2)_4-$	H	Ph/H	$-CH_2CH=CH_2$	96	95:5
13	$-(CH_2)_4-$	H	Ph/H	H	84	–
14	$-(CH_2)_4-$	H	Me/H	PhtN	75	91:9
15[b]	$-(CH_2)_4-$	H	R/Me	Me	94	>98:2
16[b]	$-(CH_2)_4-$	H	Me/R	Me	93	2:>98

[a] Methyl ester.

[b] R=4-methyl-3-pentenyl, 5 mol% $Zn(OTf)_2$.

double bond should have been *E* configured because of the arrangement of R^3 and the bulky ammonium center with maximized distance. Enantioselectively catalyzed experiments will be reported in the future (Scheme 42, Table 15).

6
Alkyne Aza-Claisen Rearrangements

Propargylamines could serve as a suitable allyl moiety in aza-Claisen rearrangements. The 3,3-sigmatropic bond reorganization led to allenes, which easily underwent consecutive processes like nucleophile addition and cyclization in a tandem process.

Frey developed a pyrrole **219** synthesis starting from vinyl dibromides **217** and enamines **218** [48]. In the presence of a strong base (KO*t*Bu) an initial dehydrobromination of **217** led to an alkynyl bromide **220**. A consecutive equilibration was found to be crucial. Involving activating aryl substituents Ar (Ph, naphthyl), a reversible base-induced H shift should have formed the corresponding allene **221**. Without such a substituent, no cyclization took place. Then, the nucleophilic attack of the enamine **218** nitrogen proceeded to give the propargyl vinyl framework in **222** ready for the sigmatropic reaction. At 23–65 °C, the aza-Claisen rearrangement generated the β,γ-allenylimine **223** which underwent a final 5-exo-trig cyclization to produce the pyrrole **219**. The present procedure allowed 1,2,3,5 tetrasubstituted pyrroles to be built up with 32 to 50% yield overall including annulated bicyclic structures (Table 16, Scheme 43).

A related amination/rearrangement/cyclization tandem sequence had been introduced by Cossy [49]. Starting from cyclic epoxyketones **224** the reaction with propargylamines **225** caused an oxirane-opening condensation process to generate the enaminoketones **226**. Upon heating in toluene to reflux, aza-Claisen rearrangement delivered the intermediate allenyl imines **227**, which

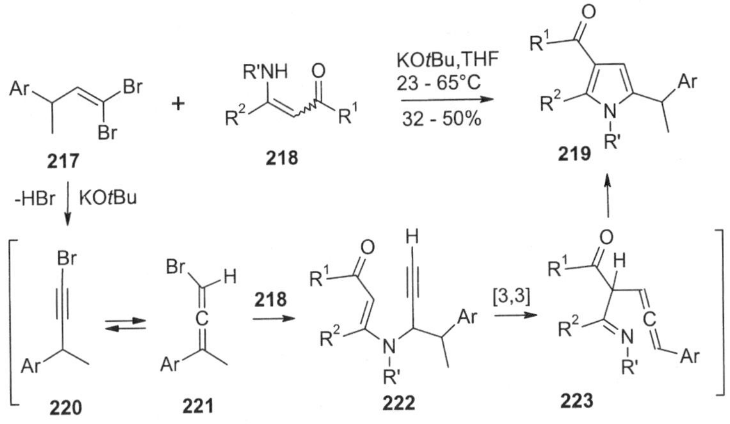

Scheme 43

Table 16 [48]

Entry	Ar	R′	R^1	R^2	Yield (%)
1	Ph	H	OMe	Me	45
2	Ph	Me	OEt	Me	32
3	Ph	Ph	OMe	Me	35
4	Ph	Bn	OEt	Me	50
5	Ph	Bn	Me	Me	35
6	Ph	Ph	$-(CH_2)_3-$		33
7	Ph	Ph	$-[CH_2C(Me)_2CH_2]-$		34

suffered from keto–enol tautomerism to **228** and a final cyclization to give the annulated pyrroles **229** with 33 to 80% yield overall. In contrast to all other reactions, the conversion of the *N*-benzyl propargylamine **224** (R=Bn) proceeded at ambient temperature; a best yield of 80% was obtained. The whole process could be run as a one-pot reaction without any heating. Furthermore, cycloketones **230** and propargylamines **225** gave rise to the formation of simple *N*-propargyl enamines **231**. The aza-Claisen rearrangements of these systems required significantly prolonged reaction times to achieve about 60 to 70% conversion of **231**. However, the corresponding pyrroles **232** were isolated in 50 to 60% yield recommending the procedure for further investigation (Scheme 44).

An uncatalyzed amination/aza-Claisen rearrangement/cyclization cascade described by Majumdar et al. was terminated by a final six-membered ring for-

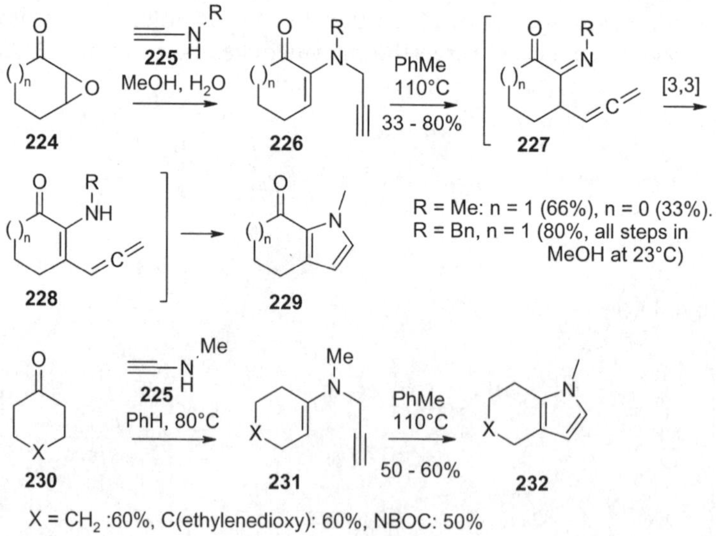

Scheme 44

mation [50]. *N*-propargylenamines **235** were generated from vinylogous acid chlorides **233** and propargylamines **234** by means of a Michael addition–elimination process with 68–77% yield (X=O, coumarins) and 80–90% yield (X=S). Upon heating the propargyl vinyl amines in *o*-dichlorobenzene to about 180 °C, a cascade of rearrangement and cyclization steps allowed generation of the tricyclic products **236** and **237** with 56–72% yield (X=O) and 60–90% yield (X=S). Generally, the exocyclic olefin **236** was obtained as the major compound. If any, the material characterized by an endocyclic double bond **237** was isolated as a side product, which could be converted into **236** by prolonged heating (1,3-H shift). Though propargylamine **235** displayed a propargyl vinyl amine as well as a propargyl vinyl (aryl) ether subunit, the aza-Claisen rearrangement proceeded. The ether system remained untouched despite a broad variation of the aryl system. The reaction path was rationalized by starting with an aza-Claisen rearrangement to produce allene imine **238**. Imine–enamide tautomerism led to the vinylogous amide **239**, which suffered from a 1,5-H shift to build up 1-azahexatriene **240**. Then, an electrocyclic ring closure formed the dehydropiperidine **237**, which finally underwent double bond migration to give the coumarin derivative **236** as the major compound (Scheme 45).

Additionally, dimedone derivative **241** and propargylamine **234** could be combined to give the alkynyl vinyl amine **242**. The rearrangement/cyclization cascade could be induced upon heating until reflux in chlorobenzene. The an-

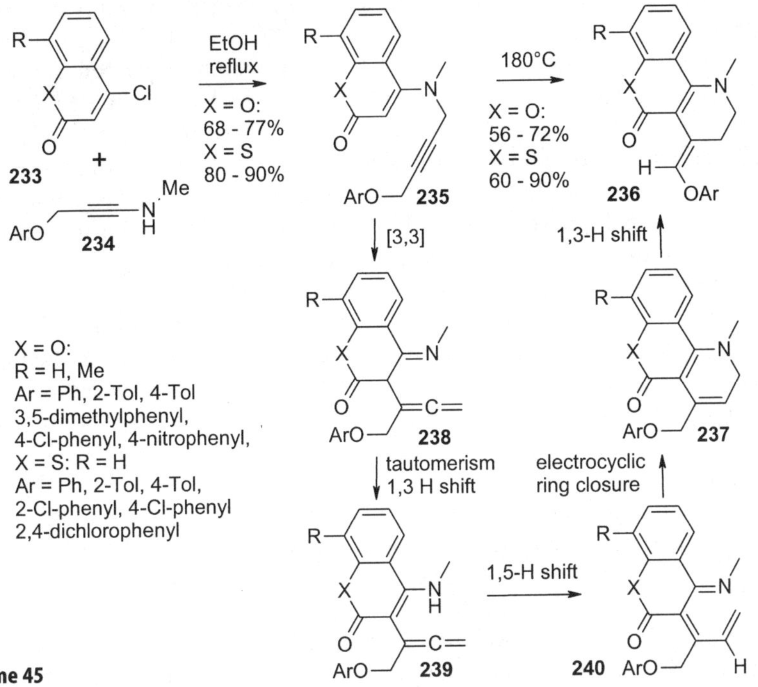

Scheme 45

nulated piperidines **243** and **244** were isolated with 75 to 80% yield. In analogy to the coumarin series, the product **243** displaying the exocyclic double bond was formed as the major product; the endocyclic olefin **244** was obtained as the side product (up to about 20%) (Scheme 46).

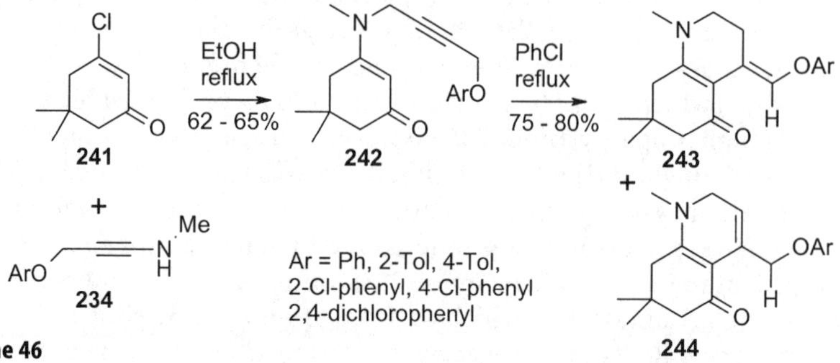

Scheme 46

7 Iminoketene Claisen Rearrangements

The iminoketene Claisen rearrangement has been investigated by Walters et al. [51]. Motivated by an early publication from Brannock and Burpitt in 1965 [52], *N*-allyl amides **245** were activated by means of a strong water-removing reagent like Ph_3PBr_2 and PPh_3/CCl_4. The dehydration at 20 °C led to a highly active hypothetical intermediate *N*-allyl iminoketene **246**, which underwent immediate aza-Claisen rearrangement to generate the product γ,δ-unsaturated nitrile **247**. The low reaction temperature of the present protocol recommended the process for further investigation. Extensive variation of the water-removing reagent and the conditions showed that the originally introduced activated triphenylphosphine produced the best results. Additionally, the combination of trimethylphosphite/iodine and Et_3N was found to be useful in reacting α-heteroatom-substituted amides (R^1=OBn, NPht, etc.). Quaternary centers in the α-position to the nascent nitrile functions ($R^1, R^{1'} \neq H$) were generated smoothly. Generally, alkyl OH groups were converted into the corresponding halides during the course of the reaction. In most cases oxygen substituents placed anywhere in the reactant resulted in moderate yields because of some side reactions, presumably caused by an oxygen–phosphorus interaction. For detailed information see Table 17.

The rearrangement of *E* and *Z* *N*-crotylamines **245** (R^3, *E* or *Z*=Me) gave the corresponding nitriles **247** with 82 and 68% yield, respectively. Disappointingly, the product was obtained as an inseparable mixture of *syn*/*anti* diastereomers **247** indicating a low simple diastereoselectivity. Obviously, the intermediate ketene imine fitted neither a chair- nor a boat-like conformation. Hence, a low axis-to-center chirality induction was operative, and *E* and *Z* reactants gave a

Table 17 [51]

Entry	R^1	$R^{1'}$	R^2	R^{3Z}	R^{3E}	Method	Yield (%)
1	Ph	H	H	H	H	b	89
2	Ph	H	H	H	H	a	94
3	Bn	H	H	H	H	a	67
4	Ph	Me	H	H	H	a	60
5	Ph	Ph	H	H	H	a	75
6	p-$MeOC_6H_4$	H	H	H	H	a	85
7	p-$F_3CC_6H_4$	H	H	H	H	a	73
8	MeO_2CCH_2	H	H	H	H	a	69
9	Et	H	H	H	H	a	36
10	BnO	H	H	H	H	c	40
11	Br-$(CH_2)_3$	H	H	H	H	b	30
12	Br-$(CH_2)_4$	H	H	H	H	b	45
13	MOMO	H	H	H	H	c	59
14	PhtN	H	H	H	H	b	78
15	PhtN	H	H	H	H	c	66
16	Me	H	–$(CH_2)_3$–		H	a	47
17	Ph	H	H	H	Me	c	82
18	Ph	H	H	Me	H	c	68
19[a]	o-$HOCH_2C_6H_4$	H	H	H	H	b	46

[a] Product: R^1=o-$BrCH_2C_6H_4$.

1.1:1 and a 1.6:1 ratio in favor of the major compound (isomer not determined, Table 17) (Scheme 47).

Further information concerning the stereochemical properties of the rearrangement were evaluated by submitting rigid cyclohexane derivatives **254/255** to the reaction conditions. In 1975, House described the allylation of a cyclohexyl cyanide **248** [53]. The initial deprotonation with LDA led to a ketene imine anion **249**, which was then treated with allyl bromide. Two potential paths rationalized the outcome: an *N*-allylation generated the intermediate ketene imines **250/251**, which underwent aza-Claisen rearrangement to deliver the nitriles **252/253**; alternatively, the direct C-allylation of **249** produced the nitriles,

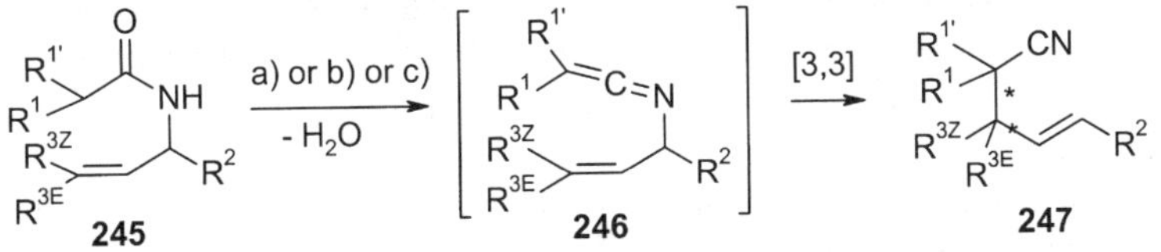

a) Ph_3P, CCl_4, Et_3N, MeCN, 23°C. b) Ph_3PBr_2, Et_3N, CH_2Cl_2, 23°C. c) $(MeO)_3P$, I_2, Et_3N, CH_2Cl_2, 23°C. For details see Table 17

Scheme 47

with the ratio of 88:12 in favor of the axial nitrile **252**. Walters presumed that the aza-Claisen rearrangement of the allyl amides **254** and **255** should have given the same nitriles **252/253** with comparable dr after passing the ketene imines **250/251**. In fact, the reaction of the axial amide **254** led to the corresponding nitriles with 41 to 46% yield and 75:25 ratio in favor of the axial nitrile **252**. In contrast, the equatorial amide **255** was converted into the imidate **256** and the azadiene **257**, and only traces of the nitriles **252/253** were found. It seemed reasonable that the iminoketene Claisen rearrangement was sensitive to sterically encumbered situations. The formation of the ketene imines **250/251** starting from axial amide **254** represented a sterically favored process leading to the nitriles **252/253**. In contrast, the formation of the ketene imines **250/251** starting from equatorial amide **255** must have been disfavored and the system gave rise to the formation of the competing products **256/257** (Scheme 48).

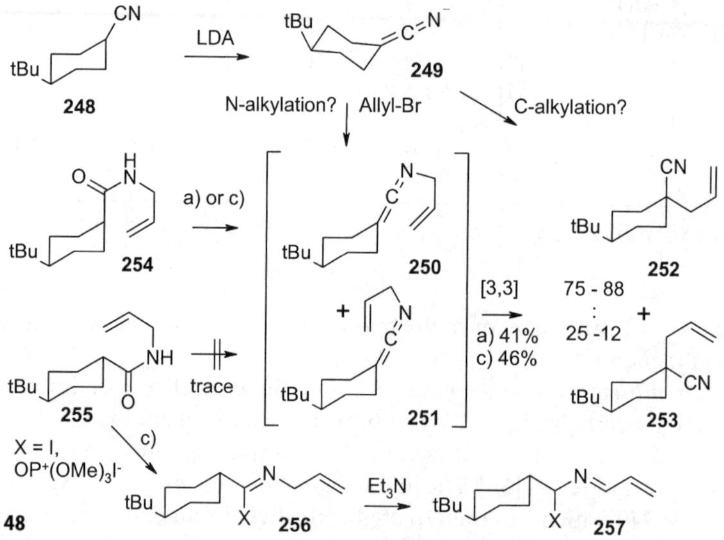

Scheme 48

Preliminary investigations were undertaken rearranging propargylamides **258**. In the presence of an alkyl substituent R[1] (R^1=Bn, Et), the use of standard reaction conditions caused dehydration to give intermediate **259**. The final aza-Claisen rearrangement delivered allenylnitrile **260** with moderate yield. The reaction cascade of the phenyl derivative **258** (R^1=Ph) suffered from a final double bond migration to give the $\alpha,\beta,\gamma,\delta$-unsaturated nitrile **261** (56% yield) (Scheme 49).

In 1993, a first application of the Walters protocol in natural product syntheses was reported [54]: *N*-allylamide **262** could be converted into a 1:1 mixture of the diastereomer nitriles **263** with 56% yield. Despite the mild reaction conditions, no external 1,2-asymmetric induction (remote stereo control) was operative when conducting such a rearrangement. The diastereomers were sep-

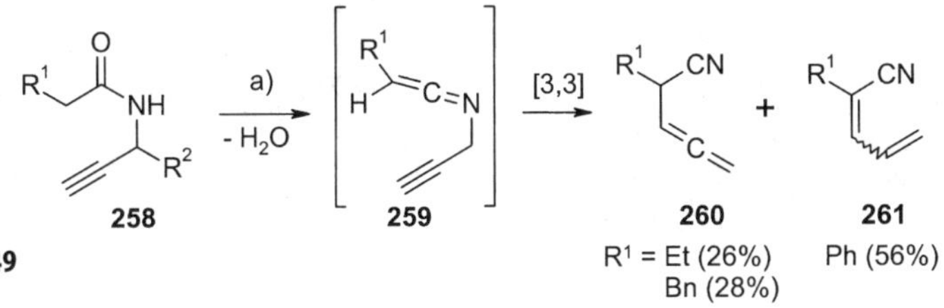

Scheme 49

arated and the cleavage of the nitriles allowed the lactones **264** to be built up. The *syn* lactone *syn*-**264** was involved as a key intermediate in a (+)-canadensolide total synthesis. The *anti* lactone *anti*-**264** enabled completion of the total syntheses of (+)-santolinolide A and (–)-santolinolide B via several further steps (Scheme 50).

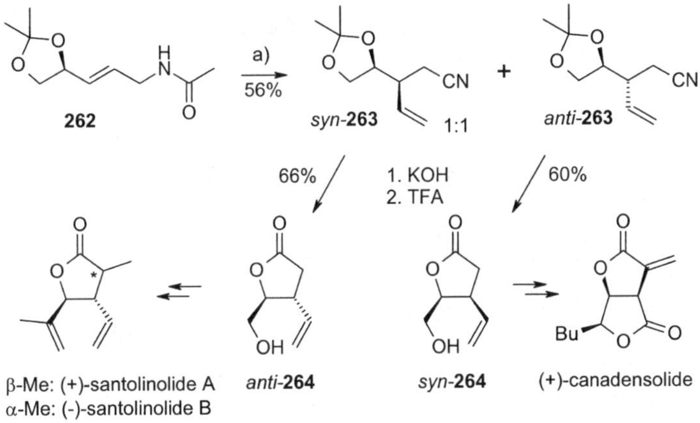

Scheme 50

At the same time as Walters' publications concerning iminoketene Claisen rearrangements, Molina reported on a related process [55]. *N*-allyl azides **265** were subjected to a Staudinger reaction to generate phosphine imines **266**. Then, the addition of stable ketenes **267** (synthesized separately) caused an aza-Wittig reaction to give iminoketenes **268**, which underwent immediate aza-Claisen rearrangement to produce the γ,δ-unsaturated nitriles **269** (method A). The driving force of the cascade was high enough to generate two adjacent quaternary carbon centers. The diastereoselectivities observed on generating the nitriles varied between 1:1 and about 4:1; the configuration of the major compound was not determined.

Alternatively, phosphine imines **266** were treated with various phenylacetyl chlorides **270** (method B). Surprisingly, phosphonium salts **271** were isolated with 25 to 97% yield, which could be deprotonated by means of a base to build up the corresponding phosphoranes **272** (66–89% yield). Upon heating to

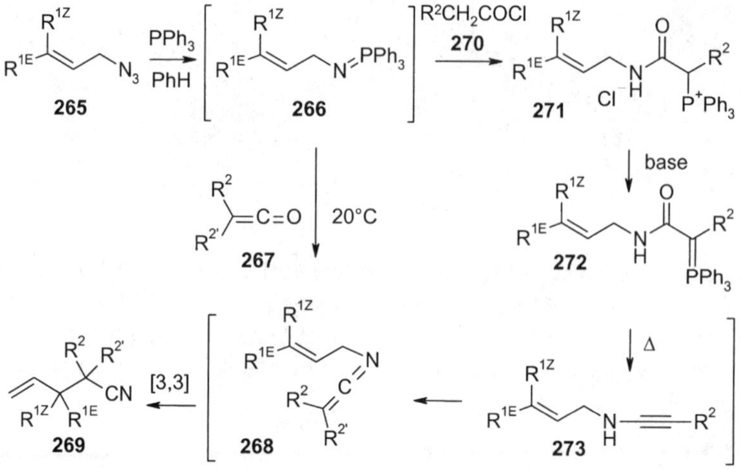

Scheme 51 For detailed information see Table 18

Table 18 [55]

Entry	R^{1E}	R^{1Z}	R^2	$R^{2'}$	Method	T (°C)	Yield (%)
1	Ph	H	Ph	Ph	A	20	51
2	Ph	H	Ph	Et	A	20	48
3	Ph	H	*p*-Tolyl	Ph	A	20	44
4	Me	H	Ph	Et	A	20	55
5	Me	Me	Ph	Ph	A	20	47
6	$H_2C{=}CH$	H	Ph	Ph	A	20	60
7	$H_2C{=}CH$	H	Ph	Et	A	20	41
8	Me	Me	Ph	H	B	140	55
9	$H_2C{=}CH$	H	Ph	H	B	130	57
10	$H_2C{=}CH$	H	*p*-Cl-C_6H_4	H	B	115	29
11	$H_2C{=}CH$	H	*p*-F-C_6H_4	H	B	120	69
12	Ph	H	Ph	H	B	125	60
13	Ph	H	*p*-Cl-C_6H_4	H	B	90	25
14	Ph	H	*p*-F-C_6H_4	H	B	130	59

90–130 °C nitriles **269** were formed in 25 to 60% yield. This outcome was explained by an initial extrusion of $Ph_3P{=}O$ to generate ynamines **273**. The consecutive isomerization delivered iminoketenes **268**, which underwent the usual iminoketene Claisen rearrangement to produce the nitriles **269**. Detailed information is given in Table 18 (Scheme 51).

References

1. Cope AC, Hardy EM (1940) J Am Chem Soc 62:441
2. Claisen L (1912) Ber Dtsch Chem Ges 45:1423
3. Claisen rearrangement recent reviews: (a) Frauenrath H (1995) Houben-Weyl (Methods of organic synthesis), Helmchen G, Hoffmann RW, Mulzer J, Schaumann E (eds) Stereoselective synthesis E21d, Thieme, Stuttgart, p 3301; (b)Enders D, Knopp M, Schiffers R (1996) Tetrahedron Asymmetry 7:1847; (c) Ito H, Taguchi T (1999) Chem Soc Rev 28:43; (d) Allin SM, Baird RD, (2001) Curr Org Chem 395; (e) Hiersemann M, Abraham L (2002) Eur J Org Chem 1461; (f) Chai Y, Hong SP, Lindsay HA, McFarland C, McIntosh MC (2002) Tetrahedron 58:2905; (g) Nubbemeyer U (2003) Synthesis 961. Aza-Claisen rearrangements: (h) Majumdar KC, Bhattacharyya T (2002) Ind J Chem 79:112
4. Hill RK, Gilman W (1967) Tetrahedron Lett 8:1421
5. (a) Cook GR, Stille JR (1991) J Org Chem 56:5578; (b) Cook GR, Barta NS, Stille JR (1992) J Org Chem 57:461; (c) Barta NS, Cook GR, Landis MS, Stille JR (1992) J Org Chem 57:7188; (d) Cook GR, Stille JR (1994) Tetrahedron 50:4105. For some early asymmetric aza-Claisen rearrangements see: (e) Kurth MJ, Brown EG (1988) Synthesis 362; (f) Kurth MJ, Soares CJ (1987) Tetrahedron Lett 28:1031 and references cited therein
6. Overman LE (1984) Angew Chem Int Ed Engl 23:579, Angew Chem 96:565
7. Welch JT, De Korte B, De Kimpe N (1990) J Org Chem 55:4981
8. Matzanke N, Gregg RJ, Weinreb SM (1997) J Org Chem 62:1920
9. McComsey DF, Maryanoff BE (2000) J Org Chem 65:4938
10. (a) Winter RF, Hornung FM (1997) Organometallics 16:4248; (b) Winter RF, Klinkhammer KW (2001) Organometallics 20:1317
11. Winter RF, Rauhut G (2002) Chem Eur J 8:641
12. Beholz LG, Stille JR (1993) J Org Chem 58:5095
13. Cooper MA, Lucas MA, Taylor JM, Ward D, Williamson NM (2001) Synthesis 621
14. (a) Majumdar KC, De RN, Saha S (1990) Tetrahedron Lett 31:1207; (b) Majumdar KC, Das U (1996) Can J Chem 74:1592
15. (a) Lai G, Anderson WK (1993) Tetrahedron Lett 34:6849; (b) Anderson WK, Lai G (1995) Synthesis 1287; (c) Lai G, Anderson WK (2000) Tetrahedron 56:2583; (d) Roe JM, Webster RAB, Ganesan A (2003) Org Lett 5:2825
16. (a) Yadav JS, Subba Reddy BV, Abdul Rasheed M, Sampath Kumar HM (2000) Synlett 487; (b) Sreekumar R, Padmakumar R (1996) Tetrahedron Lett 37:5281
17. (a) Irie K, Koizumi F, Iwata Y, Ishii T, Yanai Y, Nakamura Y, Ohigashi H, Wender PA (1995) Bioorg Med Chem Lett 5:453; (b) Irie K, Isaka T, Iwata Y, Yanai Y, Nakamura Y, Koizumi F, Ohigashi H, Wender PA, Satomi Y, Nishino H (1996) J Am Chem Soc 118:10733; (c) Irie K, Yanai Y, Oie K, Ishizawa J, Nakagawa Y, Ohigashi H, Wender PA, Kikkawa U (1997) Bioorg Med Chem 5:1725
18. Hegedus LS, Allen GF, Bozell JJ, Waterman EL (1978) J Am Chem Soc 100:5800.
19. Benali O, Miranda MA, Tormos R, Gil S (2002) J Org Chem 67:7915
20. Majumdar KC, Jana NK (2001) Monatsh Chem 132:633
21. Viallon L, Reinaud O, Capdevielle P, Maumy P (1995) Tetrahedron Lett 36:4787
22. (a) Tsunoda T, Sasaki O, Itô S (1990) Tetrahedron Lett 31:727; (b) Itô S, Tsunoda T (1990) Pure Appl Chem 62:1405; (c) Tsunoda T, Sasaki M, Sasaki O,Sako, Y, Hondo Y, Itô S (1992) Tetrahedron Lett 33:1651; (d) Tsunoda T, Tatsuki S, Shiraishi Y, Akasaka M, Itô S (1993) Tetrahedron Lett 34:3297; (e) Tsunoda T, Tatsuki S, Kataoka K, Itô S (1994) Chem Lett 543; (f) Itô S, Tsunoda T (1994) Pure Appl Chem 66:2071; (g) Tsunoda T, Ozaki F, Shirakata N, Tamaoka Y, Yamamoto H, Itô S (1996) Tetrahedron Lett 37:2463; (h) Tsunoda T, Nishii T, Yoshizuka M, Suzuki T, Itô S (2000) Tetrahedron Lett 41:7667; (i) Davies SG, Garner AC, Nicholson RL, Osborne J, Savory ED, Smith AD (2003) J Chem Soc Chem Commun 2134

23. (a) Suh YG, Lee JY, Kim SA, Jung JK (1996) Synth Commun 26:1675; (b) Suh YG, Kim SA, Jung JK, Shin DY, Min KH, Koo BA, Kim HS (1999) Angew Chem 111:3753, Angew Chem Int Ed 38:3545
24. (a) Lindstroem UM, Somfai P (1997) J Am Chem Soc 119:8385; (b) Lindstroem UM, Somfai P (1998) Synthesis 109; (c) Lindstroem UM, Somfai P (2001) Chem Eur J 7:94
25. Neuschütz K, Simone J-M, Thyrann T, Neier R (2000) Helv Chim Acta 83:2712
26. Overman LE, Clizbe LA, Freerks RL, Marlowe CK (1981) J Am Chem Soc 103:2807
27. Originally, ketene Claisen rearrangements were described by Bellus and Malherbe in the reaction of allyl sulfides and dichloroketene: (a) Malherbe R, Bellus D (1978) Helv Chim Acta 61:1768; (b) Malherbe R, Rist G, Bellus D (1983) J Org Chem 48:860
28. (a) Baxter AW, Labaree D, Ammon HL, Mariano PS (1990) J Am Chem Soc 112:7682; for some preceding publications see (b) Baxter AW, Labaree D, Chao S, Mariano PS (1989) J Org Chem 54:2893; (c) Chao S, Kunng FA, Gu JM, Ammon L, Mariano PS (1984) J Org Chem 49:2708; (d) Kunng FA, Gu JM, Chao S, Chen Y, Mariano PS (1983) J Org Chem 48:4262; (e) Mariano PS, Dunaway-Mariano D, Huesmann PL (1979) J Org Chem 44:124; (f) Mariano PS, Dunaway-Mariano D, Huesmann PL, Beamer RL (1977) Tetrahedron Lett 4299
29. Vedejs E, Gingras M (1994) J Am Chem Soc 116:579
30. Roberts SM, Smith C, Thomas RJ (1990) J Chem Soc Perkin Trans I 1493
31. (a) Cid MM, Eggnauer U, Weber HP, Pombo-Villar E (1991) Tetrahedron Lett 32:7233; (b) Cid MM, Pombo-Villar E (1993) Helv Chim Acta 76:1591
32. (a) Edstrom ED (1991) J Am Chem Soc 113:6690; (b) Edstrom ED (1991) Tetrahedron Lett 32:5709
33. (a) Diederich M, Nubbemeyer U (1995) Angew Chem Int Ed Engl 34:1026; (b) Diederich M, Nubbemeyer U (1996) Chem Eur J 2:894; (c) Sudau A, Nubbemeyer U (1998) Angew Chem Int Ed Engl 37:1140; (d) Laabs S, Scherrmann A, Sudau A, Diederich M, Kierig C, Nubbemeyer U (1999) Synlett 25; (e) Sudau A, Münch W, Nubbemeyer U, Bats JW (2000) J Org Chem 65:1710; (f) Sudau A, Münch W, Nubbemeyer U, Bats JW (2002) Eur J Org Chem 3304
34. (a) von Braun J (1907) Chem Ber 40:3914; (b) Cooley JH, Evain EJ (1989) Synthesis 1
35. (a) The Schlögl nomenclature is used for terming the planar chiral properties: Schlögl K (1984) Top Curr Chem 125:27; (b) alternatively, use of P (pR) and M (pS): Prelog V, Helmchen G (1982) Angew Chem 94:614, Angew Chem Int Ed Engl 21:567
36. (a) White DNJ, Bovill MJ (1977) J Chem Soc Perkin Trans II 1610; (b) Guella G, Chiasera G, N'Diaye I, Pietra F (1994) Helv Chim Acta 77:1203; (c) Deiters A, Mück-Lichtenfeld C, Fröhlich R, Hoppe D (2000) Org Lett 2:2415
37. Preliminary kinetic investigations of the conversion of pS-157 into pR-157; half-life period at 43 °C: 347 min, half-life period at 55 °C: 84 min, half-life period at 65 °C: 26.4 min, half-life period at 75 °C: 10.5 min; activation energy (gG*) of the epimerization: about 24 kcal mol^{-1} (±3)
38. (a) Diederich M, Nubbemeyer U (1999) Synthesis 286; (b) Sudau A, Münch W, Nubbemeyer U, Bats JW (2001) Chem Eur J 7:611; (c) Sudau A, Münch W, Nubbemeyer U, Bats JW (2002) Eur J Org Chem 3315
39. Deur CJ, Miller MW, Hegedus LS (1996) J Org Chem 61:2871
40. Yu C-M, Choi H-S, Lee J, Jung W-H, Kim H-J (1996) J Chem Soc Perkin Trans I 115
41. (a) Yoon TP, Dong VM, MacMillan DWC (1999) J Am Chem Soc 121:9726; (b) Yoon TP, MacMillan DWC (2001) J Am Chem Soc 123:2911; (c) Dong VM, MacMillan DWC (2001) J Am Chem Soc 123:2448
42. (a) Nubbemeyer U, Öhrlein R, Gonda J; Ernst B, Bellus D (1991) Angew Chem 103:1533, Angew Chem Int Ed Engl 30:1465; (b) Ernst B, Gonda J, Jeschke R, Nubbemeyer U, Öhrlein R, Bellus D (1997) Helv Chim Acta 80:876

43. (a) Nubbemeyer U (1995) J Org Chem 60:3773; (b) Nubbemeyer U (1996) J Org Chem 61:3677
44. (a) Laabs S, Münch W, Bats JW, Nubbemeyer U (2002) Tetrahedron 58:1317; (b) Zhang N, Nubbemeyer U (2002) Synthesis 242. For a preliminary result see ref. 33d
45. Metz P (1993) Tetrahedron 49:6367
46. Yoon TP, MacMillan DWC (2001) J Am Chem Soc 123:2911
47. Lambert TH, MacMillan DWC (2002) J Am Chem Soc 124:13646
48. (a) Frey H (1994) Synlett 1007; (b) for an early work see Schmidt G, Winterfeldt E (1971) Chem Ber 104:2483
49. Cossy J, Poitevina C, Sallé L, Gomez Pardo D (1996) Tetrahedron Lett 37:6709
50. (a) Majumdar KC, Gosh S (2001) Tetrahedron 57:1589; (b) Majumdar KC, Samanta SK (2001) Tetrahedron Lett 42:4231; (c) Majumdar KC, Samanta SK (2001) Tetrahedron 57:4955; (d) Majumdar KC, Samanta SK (2002) Synthesis 121
51. (a) Walters MA, McDonough CS, Brown PS, Hoem AB (1991) Tetrahedron Lett 32:179; (b) Walters MA, Hoem AB, Arcand HR, Hegeman AD, McDonough CS (1993) Tetrahedron Lett 34:1453; (c) Walters MA, Hoem AB (1994) J Org Chem 59:2645; (d) Walters MA (1994) J Am Chem Soc 116:11618; (e) Walters MA, Hoem AB, McDonough CS (1996) J Org Chem 61:55
52. Brannock KC, Burpitt RD (1965) J Org Chem 30:2564
53. House HO, Lubinkowski J, Good JJ (1975) J Org Chem 40:862
54. Nubbemeyer U (1993) Synthesis 1120
55. (a) Molina P, Alajarin M, Lopez-Leonardo C (1991) Tetrahedron Lett 32:4041; (b) Molina P, Alajarin M, Lopez-Leonardo C, Alcantara J (1993) Tetrahedron 49:5153

Top Curr Chem (2005) 244: 215–249
DOI 10.1007/b96892

Synthetic Studies on the Pamamycin Macrodiolides

Peter Metz (✉)

Institut für Organische Chemie, Technische Universität Dresden,
Bergstrasse 66, 01069 Dresden, Germany
peter.metz@chemie.tu-dresden.de

1 Introduction 215

2 Total Syntheses 217

3 Synthetic Approaches to the Hydroxy Acid Fragments 230

4 Concluding Remarks 248

References 248

Abstract The pamamycins are structurally intriguing 16-membered macrodiolides displaying a wide range of interesting biological activities. A comprehensive survey of the total syntheses reported in this area so far and the various synthetic approaches to the hydroxy acid constituents of the pamamycins described to date is presented.

Keywords Pamamycins · Macrodiolides · Antibiotics · Synthesis · Hydroxy acids

1 Introduction

The pamamycins (**1**) are a class of 16-membered macrodiolide homologs that have been isolated from various *Streptomyces* species (Fig. 1) [1, 2]. Next to displaying pronounced autoregulatory, anionophoric, and antifungal activities, several members of this family have been shown to be highly active against Gram-positive bacteria including multiple antibiotic-resistant strains of *Mycobacterium tuberculosis* [1c].

More recent investigations on the antimycobacterial activity of pamamycin-607 (**1b**) on 25 independent *M. tuberculosis* clinical isolates (either susceptible, mono-, or multiresistant to the first line antituberculous drugs) established minimum inhibitory concentrations MIC_{100} in the range of 1.5–2.0 µg/ml, while the MIC_{100} of **1b** for a bioluminescent laboratory strain of *M. tuberculosis* (H37Rv) was determined as 0.55 µg/ml [3a]. Parallel studies on the effect of **1b** on the cell cycle distribution of human (HL-60) cells by flow cytometry indicated no

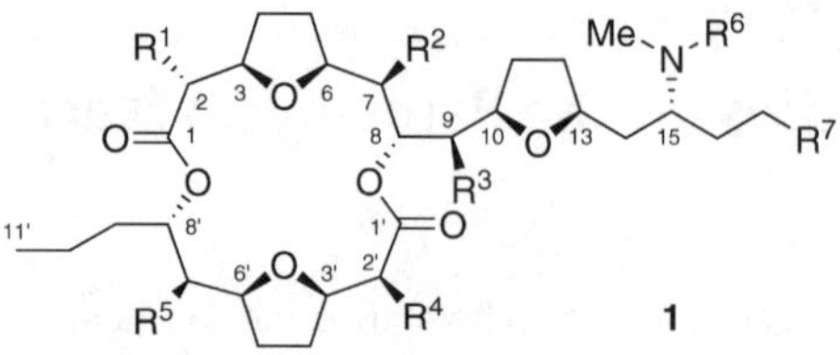

	R^1	R^2	R^3	R^4	R^5	R^6	R^7	pamamycin
1a	Me	H	Me	Me	H	Me	Me	593
1b	Me	Me	Me	Me	H	Me	Me	607
1c	Me	Me	Me	Me	Me	Me	Me	621A
1d	Et	H	Me	Me	Me	Me	Me	621B
1e	Me	Me	Et	Me	H	Me	Me	621C
1f	Et	Me	Me	Me	H	Me	Me	621D
1g	Me	Me	Et	Me	Me	Me	Me	635A
1h	Me	Me	Me	Et	Me	Me	Me	635B
1i	Et	Me	Me	Me	Me	Me	Me	635C
1j	Et	Me	Me	Et	H	Me	Me	635D
1k	Me	Me	Et	Et	H	Me	Me	635E
1l	Et	Me	Et	Me	H	Me	Me	635F
1m	Et	Me	Et	Et	H	Me	Me	649A
1n	Et	Me	Et	Me	Me	Me	Me	649B
1o	Me	H	Me	Me	H	H	Me	De-*N*-methyl-579
1p	Me	Me	Me	Me	H	H	Me	De-*N*-methyl-593A
1q	Et	H	Me	Me	H	H	Me	De-*N*-methyl-593B
1r	Me	Me	Me	Me	H	Me	Pr	MS-282a
1s	Me	Me	Et	Me	H	Me	Et	MS-282b

Fig. 1 Pamamycin homologs

(cell cycle) or only small effects (apoptosis) at the latter concentration [3b]. Thus, **1b** might emerge as a promising lead molecule for the development of novel antituberculous drugs.

Due to the biological properties of these macrodiolides and challenged by their unique structure, many groups have initiated programs aimed at the total synthesis of pamamycin homologs with a special focus on pamamycin-607 (**1b**) as the target. This account covers the total syntheses reported in this area so far, as well as the various synthetic approaches to the hydroxy acid constituents **2** ("larger fragment") and **3** ("smaller fragment") of the pamamycins (Scheme 1) described to date.

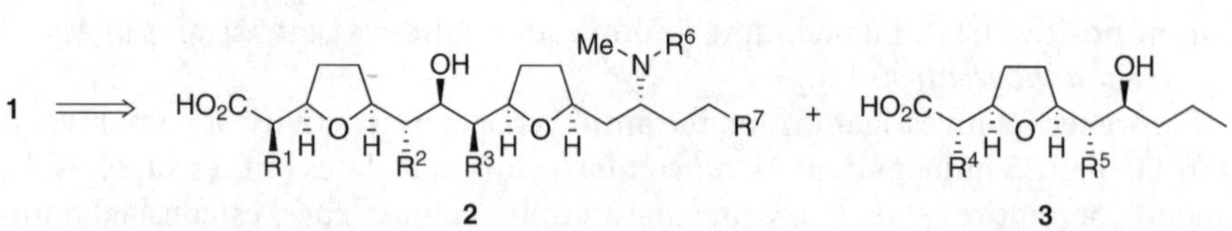

Scheme 1 Hydroxy acid constituents **2** ("larger fragment") and **3** ("smaller fragment") of the pamamycins **1**

2
Total Syntheses

So far, four total syntheses of the homolog pamamycin-607 (**1b**) have been reported [4–7]. This section will detail the different routes and strategies that were utilized to access this macrodiolide.

The first total synthesis of **1b** was published in 2001 by Thomas [4]. Central to his approach was the application of reagent-controlled remote asymmetric induction in the preparation of homoallylic alcohols by tin(IV) halide-promoted addition of chiral 5-alkoxypent-2-enylstannanes to aldehydes [8], followed by a stereoselective phenylselenenyl-induced cyclization to give 2,5-*cis*-disubstituted tetrahydrofurans. Synthesis of the larger fragment of **1b** commenced with addition of the lithium amide from **5** to enoate **4** and subsequent hydrogenolysis to furnish the *β*-amino ester **7** corresponding to the C(13)-C(18) moiety of **1b** with excellent enantiomeric excess (Scheme 2). *N*-Methylation of the derived tosylamide and reduction/oxidation provided aldehyde **9** that was subjected to the first crucial allylation reaction, with the allyltin trichloride generated in situ by transmetallation of allyl stannane **10** with tin(IV) chloride. Treatment of the resultant homoallylic alcohol **11** formed with at least 96% diastereoselectivity with phenylselenenyl chloride and catalytic amounts of tin(IV) chloride effected a completely stereoselective cyclization to give the 2,5-*cis*-disubstituted tetra-

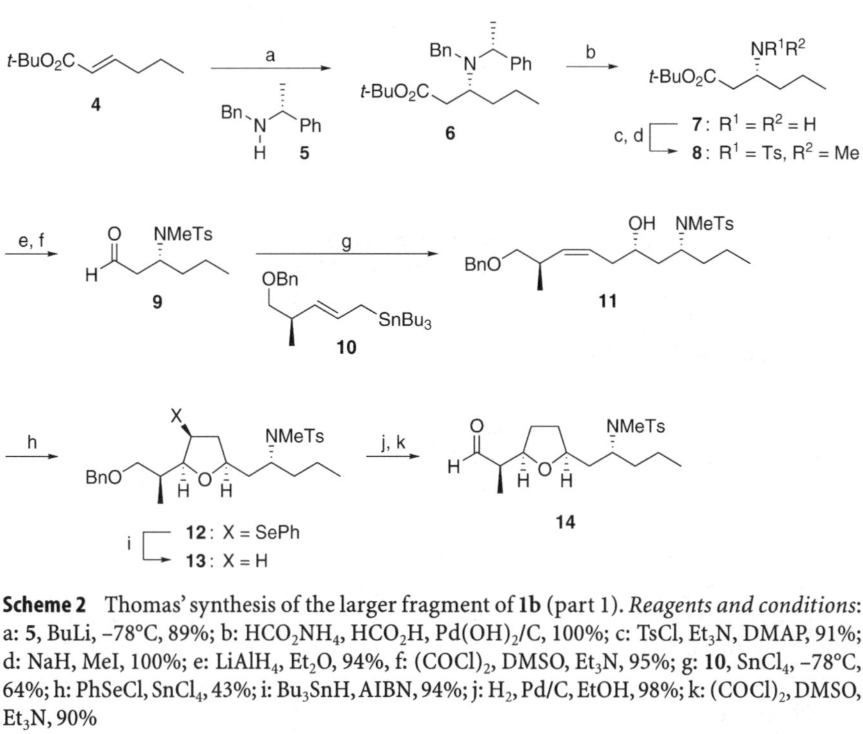

Scheme 2 Thomas' synthesis of the larger fragment of **1b** (part 1). *Reagents and conditions*: a: **5**, BuLi, –78°C, 89%; b: HCO_2NH_4, HCO_2H, $Pd(OH)_2/C$, 100%; c: TsCl, Et_3N, DMAP, 91%; d: NaH, MeI, 100%; e: $LiAlH_4$, Et_2O, 94%, f: $(COCl)_2$, DMSO, Et_3N, 95%; g: **10**, $SnCl_4$, –78°C, 64%; h: PhSeCl, $SnCl_4$, 43%; i: Bu_3SnH, AIBN, 94%; j: H_2, Pd/C, EtOH, 98%; k: $(COCl)_2$, DMSO, Et_3N, 90%

hydrofuran **12** via a selenonium ion intermediate. After reductive removal of the phenylselanyl group, hydrogenolysis, and oxidation, aldehyde **14** was obtained.

Anti aldol reaction of aldehyde **14** with the lithium enolate derived from propionate **15** and silyl protection yielded the Felkin–Anh product **16** as the major stereoisomer (Scheme 3). Conversion of ester **16** to aldehyde **17** set the scene for the second key allylation using the chiral stannane **10** and tin(IV) chloride to give homoallylic alcohol **18**, which in turn was cyclized to the bis-tetrahydrofuran **19** on exposure to *N*-phenylselenenyl phthalimide in the presence of zinc(II) chloride. Reductive deselenylation, exchange of *N*-tosyl against *N*-Boc protection, and conversion of the benzyl ether to a carboxyl group eventually delivered the larger fragment surrogate **21**.

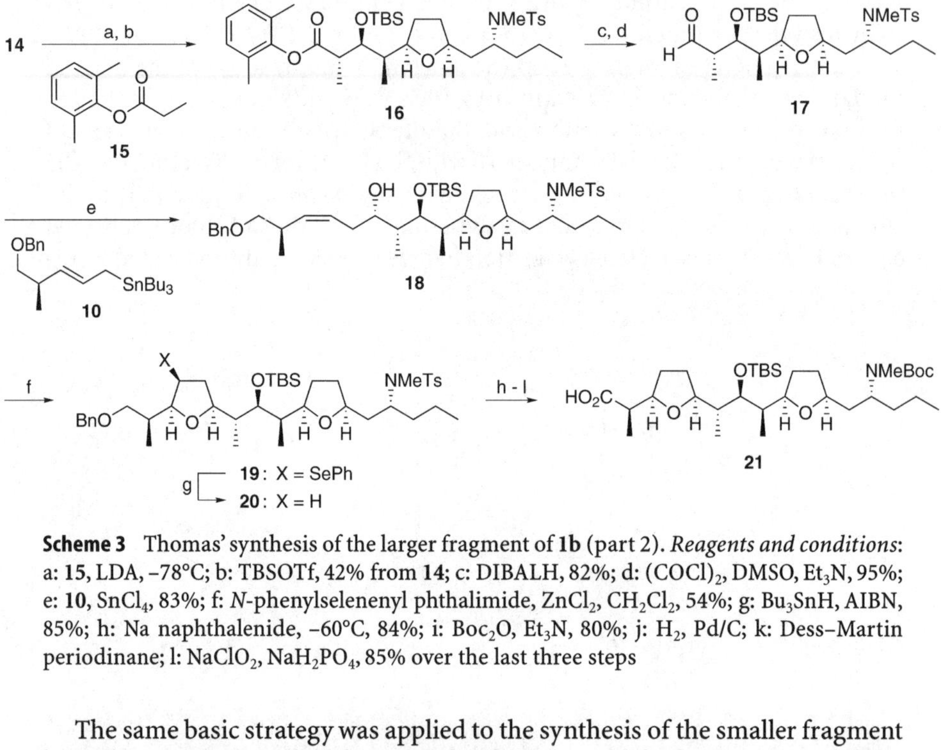

Scheme 3 Thomas' synthesis of the larger fragment of **1b** (part 2). *Reagents and conditions*: a: **15**, LDA, –78°C; b: TBSOTf, 42% from **14**; c: DIBALH, 82%; d: $(COCl)_2$, DMSO, Et_3N, 95%; e: **10**, $SnCl_4$, 83%; f: *N*-phenylselenenyl phthalimide, $ZnCl_2$, CH_2Cl_2, 54%; g: Bu_3SnH, AIBN, 85%; h: Na naphthalenide, –60°C, 84%; i: Boc_2O, Et_3N, 80%; j: H_2, Pd/C; k: Dess–Martin periodinane; l: $NaClO_2$, NaH_2PO_4, 85% over the last three steps

The same basic strategy was applied to the synthesis of the smaller fragment benzyl ester **28** as well (Scheme 4). In this case, aldehyde **22** prepared from (*S*)-2-hydroxypentanoic acid [9] was allylated with *ent*-**10** and tin(IV) chloride, and the resulting alcohol **23** was converted to epimer **24** via Mitsunobu inversion prior to phenylselenenyl-induced tetrahydrofuran formation. Reductive cleavage of the phenylselanyl group, hydrogenolysis of the benzyl ether, oxidation, carboxylate benzylation, and desilylation then furnished ester **28**.

With both building blocks **21** and **28** secured, the total synthesis of **1b** was completed as depicted in Scheme 5. Intermolecular Yamaguchi esterification to give **29** and subsequent acidic desilylation, which required reinstatement of the

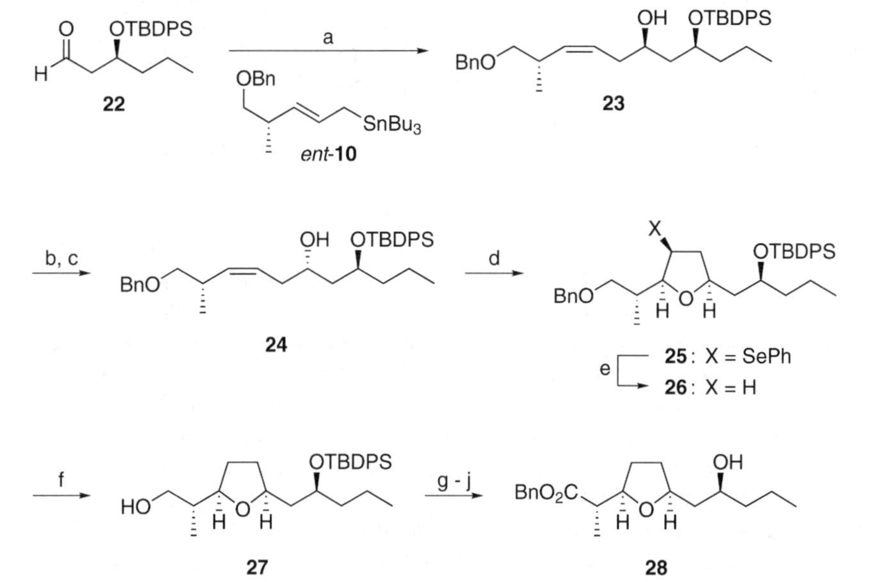

Scheme 4 Thomas' synthesis of the smaller fragment of **1b**. *Reagents and conditions*: a: *ent*-**10**, $SnCl_4$, 80%; b: *p*-nitrobenzoic acid, DEAD, Ph_3P, 68%; c: NaOH, 94%; d: *N*-phenylselenenyl phthalimide, $SnCl_4$, 60%; e: Bu_3SnH, AIBN, 89%; f: H_2, 10% Pd/C, 70%; g: Dess–Martin periodinane; h: $NaClO_2$, NaH_2PO_4; i: *i*-Pr_2NEt, BnBr, 81% from **27**; j: conc. aq. HCl, MeOH, 49%

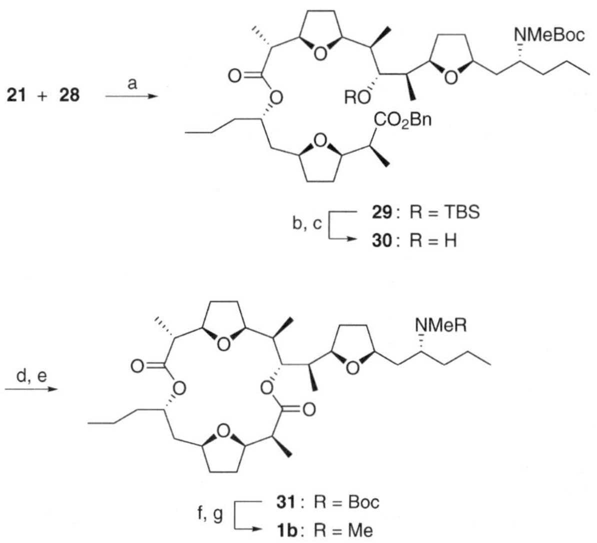

Scheme 5 Completion of Thomas' synthesis of **1b**. *Reagents and conditions*: a: 2,4,6-trichlorobenzoyl chloride, DMAP, CH_2Cl_2, rt, 63%; b: HCl, EtOH, 40 to 50°C; c: Boc_2O, Et_3N, 80% from **29**, d: H_2, Pd/C, EtOH; e: (i) 2,4,6-trichlorobenzoyl chloride, Et_3N, rt, (ii) DMAP, rt, 25% from **30**; f: TFA, CH_2Cl_2, rt, 82%; g: CH_2O, $NaBH_3CN$, HOAc, 60%

N-Boc group, was followed by reductive debenzylation of **30** and Yamaguchi lactonization of the resultant hydroxy acid to provide macrodiolide **31** in 25% yield accompanied by a dimer. Finally, removal of the *N*-Boc group and reductive *N*-methylation yielded pamamycin-607 (**1b**). In total, ca. 40 steps were required to access the target from ester **4**, aldehyde **22**, and allyl stannanes **10** and *ent*-**10**.

A further total synthesis of pamamycin-607 (**1b**) was first described in 2001 by Lee [5]. His approach relied on the stereoselective radical cyclization of *β*-alkoxyvinyl ketone and *β*-alkoxymethacrylate substrates to give 2,5-*cis*-disubstituted tetrahydrofurans with additional control of an exocyclic stereogenic center in the latter case as the key transformation [10]. Synthesis of the larger fragment of **1b** began with generation of the central stereopentade C(6)-C(10) by application of Evans' aldol [11] and Keck's allylation [12] methodologies (Scheme 6). Thus, addition of the boron enolate derived from **32** to aldehyde **33** followed by Weinreb amide formation gave rise to **34**, which was converted to aldehyde **35** by BOM protection and DIBALH reduction. Another Evans aldol reaction, this time with the boron enolate of *ent*-**32**, and subsequent reductive cleavage of the chiral auxiliary yielded diol **36**, which was elaborated to aldehyde

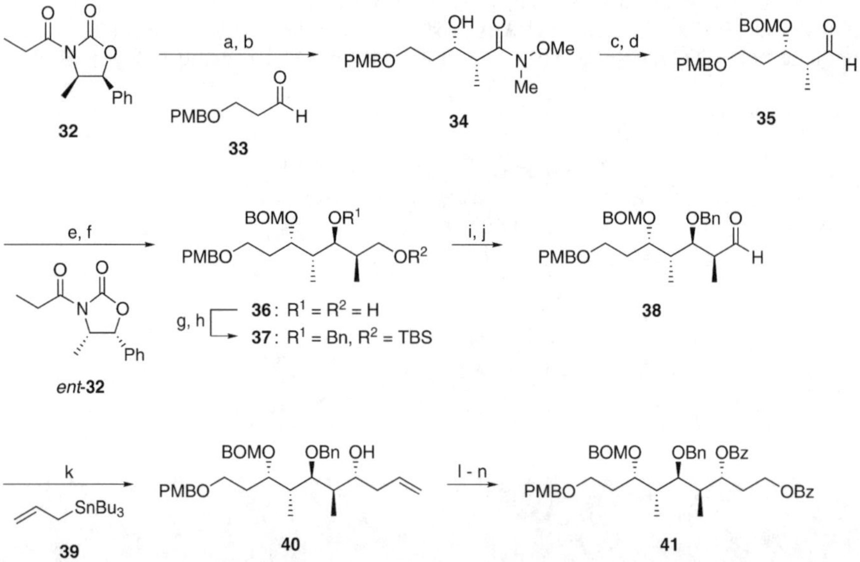

Scheme 6 Lee's synthesis of the larger fragment of **1b** (part 1). *Reagents and conditions*: a: (i) Bu_2BOTf, Et_3N, CH_2Cl_2, –40 to 0°C, (ii) **33**, –78 to 0°C, 88%; b: MeONHMe·HCl, Me_3Al, THF, –20°C to rt, 93%; c: BOMCl, *i*-Pr_2NEt, Bu_4NI, CH_2Cl_2, rt, 85%; d: DIBALH, THF, –78°C, 98%; e: (i) *ent*-**32**, Bu_2BOTf, Et_3N, CH_2Cl_2, –40 to 0°C, (ii) **35**, –78 to 0°C, 97%; f: $NaBH_4$, THF, H_2O, rt, 92%; g: TBSCl, imidazole, CH_2Cl_2, 0°C, 100%; h: BnBr, NaHMDS, THF, DMF, 0°C, 93%; i: Bu_4NF, THF, rt, 93%; j: SO_3·pyridine, Et_3N, DMSO, CH_2Cl_2, rt, 96%; k: **39**, $MgBr_2·Et_2O$, CH_2Cl_2, rt, 99%; l: (i) OsO_4, NMO, acetone, H_2O, rt, (ii) $NaIO_4$, rt; m: $NaBH_4$, EtOH, rt, 88% from **40**; n: BzCl, pyridine, DMAP, CH_2Cl_2, rt, 93%

38 by chemoselective protecting group operations and oxidation. The remaining stereogenic center of the stereopentade was installed via Keck allylation using allyl stannane **39**. Oxidative olefin scission in the resulting homoallyl alcohol **40**, reduction, and benzoylation provided the C(4)-C(12) segment **41**.

Protecting group manipulations transformed **41** to alcohol **42**, which was condensed with *rac*-**43** followed by iodide substitution to give *β*-alkoxymethacrylate **44** (Scheme 7). Low-temperature radical cyclization initiated by triethylborane/air yielded a 10.8:1 diastereomeric mixture of tetrahydrofurans featuring **45** as the major stereoisomer. Benzoate deblocking and mono-tosylation furnished alcohol **46**, condensation of which with acetal **47** and subsequent iodide substitution to give **48** set the stage for the second key radical cyclization. Standard high-dilution conditions with AIBN initiation effected clean formation of the second tetrahydrofuran subunit, and the minor impurity resulting from the incomplete diastereoselectivity of the first radical ring closure could be removed after this step. Reductive debenzylation then afforded the larger fragment surrogate **49**.

This radical cyclization strategy was utilized for the synthesis of the smaller fragment silyl ether **54** as well (Scheme 8). Evans aldol reaction of the boron enolate derived from *ent*-**32** with aldehyde **33**, samarium(III)-mediated imide methyl ester conversion, and protecting group exchange led to tosylate **51**. Elaboration of **51** to ketone **53** was achieved under the conditions used for construction of the second tetrahydrofuran moiety of **49** from **46**. A highly diastereoselective reduc-

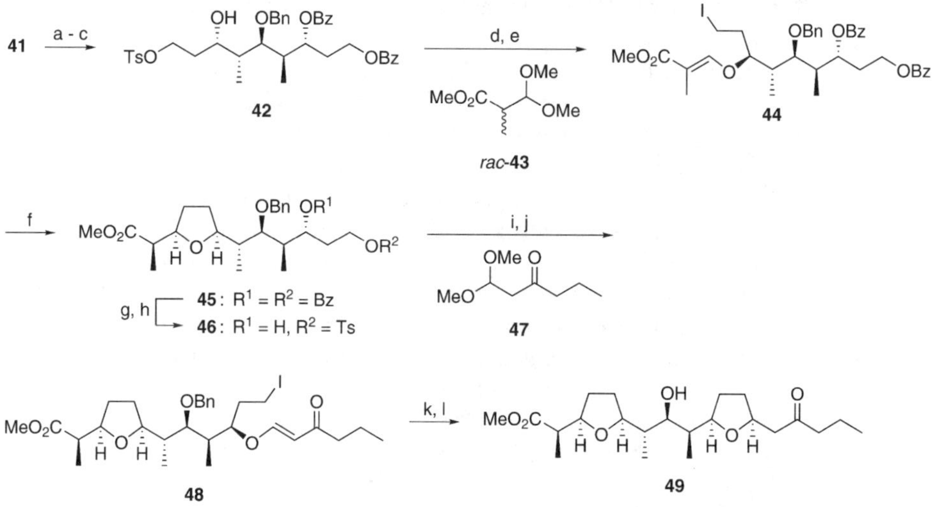

Scheme 7 Lee's synthesis of the larger fragment of **1b** (part 2). *Reagents and conditions*: a: CAN, MeCN, THF, rt, 88%; b: TsCl, Et_3N, CH_2Cl_2, 0°C, 99%; c: conc. HCl, MeCN, rt, 99%; d: *rac*-**43**, TsOH, $CHCl_3$, reflux, 68%; e: NaI, acetone, reflux, 99%; f: Bu_3SnH, Et_3B, air, toluene, –78°C, 96%; g: K_2CO_3, MeOH, rt, 98%; h: TsCl, Et_3N, CH_2Cl_2, 0°C, 98%; i: **47**, TFA, benzene, reflux, 84%; j: NaI, acetone, reflux, 97%; k: Bu_3SnH, AIBN, benzene, reflux, 89%; l: H_2, Pd/C, MeOH, rt, 98%

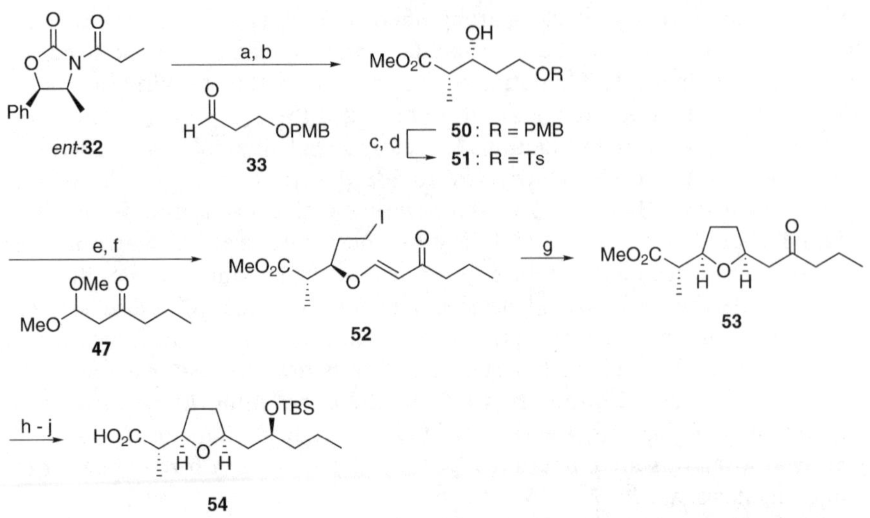

Scheme 8 Lee's synthesis of the smaller fragment of **1b**. *Reagents and conditions*: a: (i) Bu_2BOTf, Et_3N, CH_2Cl_2, –40 to 0°C, (ii) **33**, –78 to 0°C, 82%; b: $Sm(OTf)_3$, MeOH, THF, rt, 92%; c: H_2, Pd/C, MeOH, rt, 98%; d: TsCl, Et_3N, CH_2Cl_2, 0°C, 83%; e: **47**, TFA, benzene, reflux, 94%; f: NaI, acetone, reflux, 95%; g: Bu_3SnH, AIBN, benzene, reflux, 97%; h: SmI_2, MeOH, THF, 0°C to rt, 70%; i: TBSCl, imidazole, DMF, rt, 98%; j: NaOH, MeOH, H_2O, rt, 97%

tion of **53** with samarium(II) iodide/methanol (dr of crude product=8.5:1), silylation, and saponification provided the carboxylic acid **54**.

Completion of the total synthesis of **1b** by coupling of the partially protected hydroxy acid building blocks **49** and **54** is illustrated in Scheme 9. Intermolecular Yamaguchi esterification of **54** with **49** followed by stereoselective reductive amination of the ketone, *N*-Boc protection, and desilylation afforded hydroxy methyl ester **56**. Chemoselective saponification of the methyl ester succeeded without epimerization to give hydroxy acid **57**. The presence of the NHBoc moiety in **56** is obviously pivotal to the success of this transformation, since methyl ester hydrolysis of a compound very similar to **56** (NMe_2 instead of NHBoc) could not be effected without extensive epimerization at C(2′) [6]. Whereas attempted Yamaguchi lactonization of **57** only met with failure due to epimerization at C(2) under these conditions, which was also noted by our own group in parallel studies [6], a 1,3-dicyclohexylcarbodiimide-based method accomplished the desired cyclization to give macrodiolide **58** in 56% yield. Finally, cleavage of the *N*-Boc protecting group and subsequent twofold reductive *N*-methylation yielded pamamycin-607 (**1b**). The total synthesis was achieved in 44 steps from oxazolidinones **32** and *ent*-**32**, aldehyde **33**, and acetals *rac*-**43** and **47**.

Another total synthesis of pamamycin-607 (**1b**) was reported in 2001 by our own group [6] at about the same time as the Lee synthesis. Here, the approach was based on the stereoselective intramolecular Diels–Alder reaction of vinylsulfonates and novel methods for elaboration of the resulting sultones [13, 14].

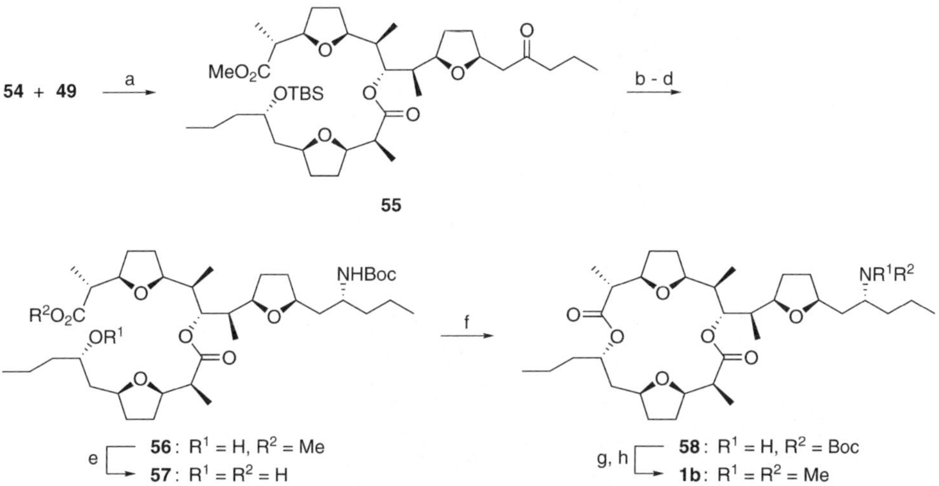

Scheme 9 Completion of Lee's synthesis of **1b**. *Reagents and conditions*: a: (i) **54**, 2,4,6-trichlorobenzoyl chloride, Et_3N, THF, rt, (ii) **49**, DMAP, benzene, rt, 90%; b: NH_4OAc, $NaBH_3CN$, 4 Å sieves, *i*-PrOH, 0°C; c: Boc_2O, Et_3N, CH_2Cl_2, rt; d: conc. aq. HCl, MeOH, rt, 79% **56** from **55**; e: LiOH, MeOH, H_2O, rt, 81%; f: DCC, PPTS, pyridine, $ClCH_2CH_2Cl$, reflux, 56%; g: TFA, CH_2Cl_2, rt; h: aq. CH_2O, HOAc, 3.5 bar H_2, Pd/C, MeOH, rt, 79% **1b** from **58**

Furan and (*S*)-1,2-epoxypentane (**59**), prepared by hydrolytic kinetic resolution of the racemic epoxide using Jacobsen's (salen)cobalt(III) catalyst [15], served as the starting points for the synthesis of the larger fragment of **1b** (Scheme 10) [16, 17]. Alkylation of furan with **59** provided the enantiomerically pure (>99% ee) alcohol **60**, which reacted with vinylsulfonyl chloride (**61**) to give sultone **62** by a domino esterification/cycloaddition with complete diastereoselectivity. Subsequent treatment of **62** with 2 equivalents of methyllithium induced a domino elimination/alkoxide-directed 1,6-addition to yield the bicyclic compounds **63–65**. Ozonolysis of this mixture followed by eliminative workup afforded two diastereomeric hemiacetals **66**. Only the trisubstituted olefins **63** and **64** were attacked by ozone, while the vinylic sultone **65** could easily be separated. A Lewis acid-catalyzed exchange of the hydroxyl group in **66** against a phenylthio group in **67** then set the stage for a domino reductive elimination/hydrogenation with Raney nickel under hydrogen pressure to give **68** via a single 2,3-dihydrofuran with high diastereoselectivity (dr of crude product=18:1). Silylation followed by ester reduction and iodide substitution then delivered the iodide **71** [18].

Halogen–lithium exchange of iodide **71** and subsequent addition of 2-acetylfuran (**72**) to the resultant organolithium intermediate yielded two diastereomeric tertiary alcohols (dr=1:1), which were converted to (*E*)-olefin **73** with complete diastereoselectivity upon brief exposure to catalytic amounts of concentrated aqueous hydrogen chloride (Scheme 11) [18]. Diastereoselective hydroboration/oxidation of **73** gave largely the desired stereoisomer **74** due to

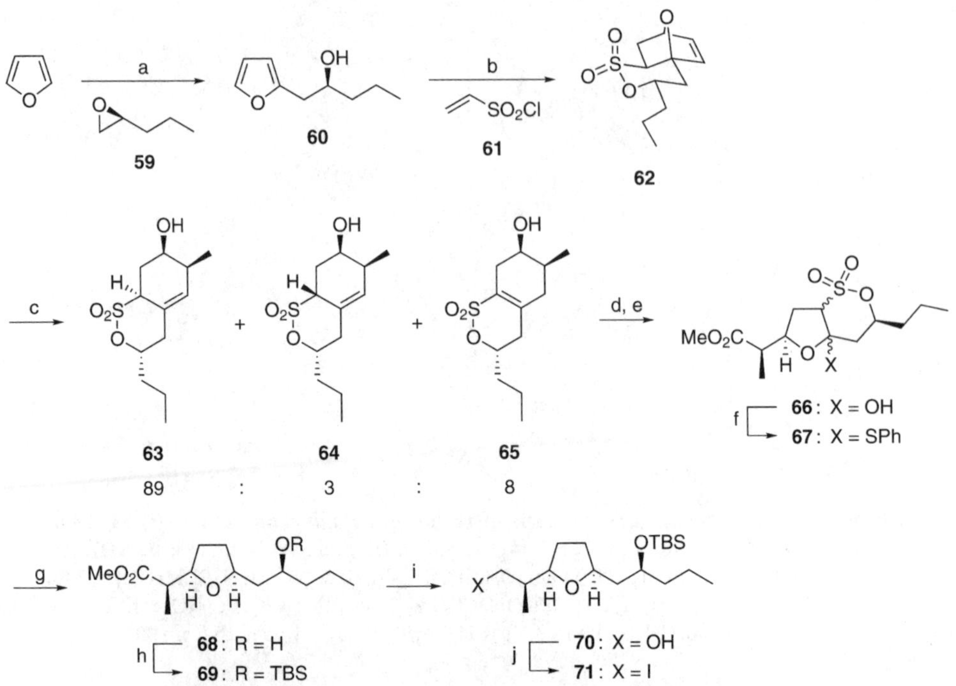

Scheme 10 Metz' synthesis of the larger fragment of **1b** (part 1). *Reagents and conditions*: a: (i) BuLi, THF, –78 to –15°C, (ii) **59**, –15°C to rt, 84%; b: **61**, Et_3N, THF, 0°C to rt, 88%; c: (i) MeLi, THF, –78 to 0°C; (ii) NH_4Cl, H_2O, –78°C to rt, 66%; d: O_3, $NaHCO_3$, CH_2Cl_2, MeOH, –78°C; e: Ac_2O, pyridine, CH_2Cl_2, rt, 83% from **63-65**; f: PhSH, $BF_3{\cdot}Et_2O$, CH_2Cl_2, rt, 82%; g: Raney Ni (W2), 50 bar H_2, EtOH, rt, 54%; h: TBSCl, imidazole, DMAP, DMF, rt, 100%; i: $LiAlH_4$, Et_2O, 0°C to rt, 96%; j: I_2, Ph_3P, imidazole, Et_2O, MeCN, rt, 99%

minimization of allylic 1,3-strain [19]. The second iterative cycle of our sultone route commenced with another domino esterification/cycloaddition by reacting hydroxyalkylfuran **74** with vinylsulfonyl chloride (**61**) to produce a single sultone **75**. Following domino elimination/alkoxide-directed 1,6-addition by treatment of **75** with two equivalents of methyllithium, ozonolysis with eliminative workup delivered hemiacetal **76** as a single stereoisomer [20]. Subjecting **76** to thiophenol in the presence of trifluoroborane not only effected lactol *S*,*O*-acetal interchange, but simultaneously cleaved the silyl ether on the side chain to give alcohol **77**. Mitsunobu reaction of **77** with hydrazoic acid yielded azide **78** with clean inversion of configuration. This set the scene for a one-pot reaction cascade leading to hydroxy ester **79**. In the event, treatment of **78** with Raney nickel under hydrogen pressure followed by addition of an aqueous formaldehyde solution to the reaction mixture caused desulfurization via domino reductive elimination/hydrogenation, azide reduction, and twofold reductive *N*-methylation to provide **79** with complete diastereoselectivity. Silylation of **79** followed by mild saponification then yielded the larger fragment coupling component **80** [6].

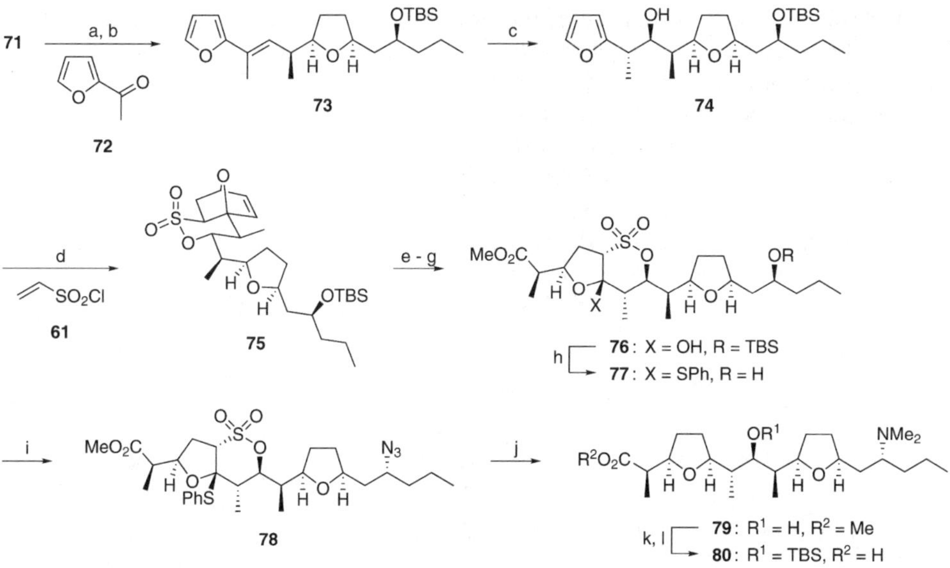

Scheme 11 Metz' synthesis of the larger fragment of **1b** (part 2). *Reagents and conditions*: a: (i) *t*-BuLi, Et_2O, –78°C, (ii) **72**, –78°C to rt; b: conc. aq. HCl, $CHCl_3$, rt, 69% from **71**; c: (i) BH_3·THF, THF, 0°C to rt, (ii) 30% aq. H_2O_2, NaOH, 0°C to rt, 65%; d: **61**, Et_3N, THF, 0°C to rt, 93%; e: (i) MeLi, THF, –78°C to rt; (ii) NH_4Cl, H_2O, –78°C to rt; f: O_3, $NaHCO_3$, CH_2Cl_2, MeOH, –78°C; g: Ac_2O, pyridine, CH_2Cl_2, rt to reflux, 53% from **75**; h: PhSH, BF_3·Et_2O, CH_2Cl_2, rt, 78%; i: HN_3, DEAD, Ph_3P, toluene, 0°C to rt, 97%; j: (i) Raney Ni (W2), 50 bar H_2, EtOH, rt, (ii) aq. CH_2O, rt, 56% from **78**; k: TBSOTf, 2,6-lutidine, CH_2Cl_2, rt, 90%; l: LiOH, THF, MeOH, rt, 97%

A similar sultone-based strategy was first applied to the synthesis of the smaller fragment methyl ester of **1b** as well (see Scheme 26) [16, 17]. However, the three-step sequence depicted in Scheme 12 allowed a considerable shortcut to the smaller fragment of **1b**. Saponification of methyl ester **68**, which we used as a key intermediate for the synthesis of the larger hydroxy acid constituent of **1b** (see Scheme 10), and subsequent Yamaguchi lactonization of the resulting hydroxy acid **81** yielded the C(2′) epimerized lactone **82** as the major product [21]. Since we faced severe problems with epimerization during attempted methyl ester cleavage in a first approach to **1b** involving a final

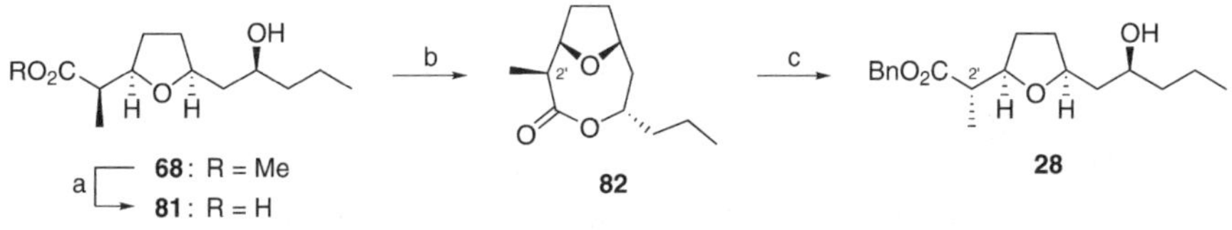

Scheme 12 Metz' synthesis of the smaller fragment of **1b**. *Reagents and conditions*: a: NaOH, rt, 100%; b: (i) 2,4,6-trichlorobenzoyl chloride, Et_3N, THF, rt, (ii) DMAP, toluene, reflux, 71%; c: BnOLi, BnOH, THF, rt, 83%

lactonization with formation of the ester linkage between C(1) and the C(8′) oxygen [6], we chose benzyl ester **28** as the smaller fragment surrogate. Upon treatment of lactone **82** with the lithium alkoxide of benzyl alcohol, coupling component **28** was readily obtained.

The total synthesis of **1b** was first completed using the two hydroxy acid building blocks **80** and **28** (Scheme 13) [6]. Intermolecular Yamaguchi esterification of **80** with **28** provided coupling product **83**. Desilylation of **83** and reductive debenzylation of the resulting benzyl ester proceeded uneventfully to give the *seco* acid **84**. Finally, modified Yamaguchi cyclization [22] of **84** afforded pamamycin-607 (**1b**) as a single stereoisomer in 78% yield.

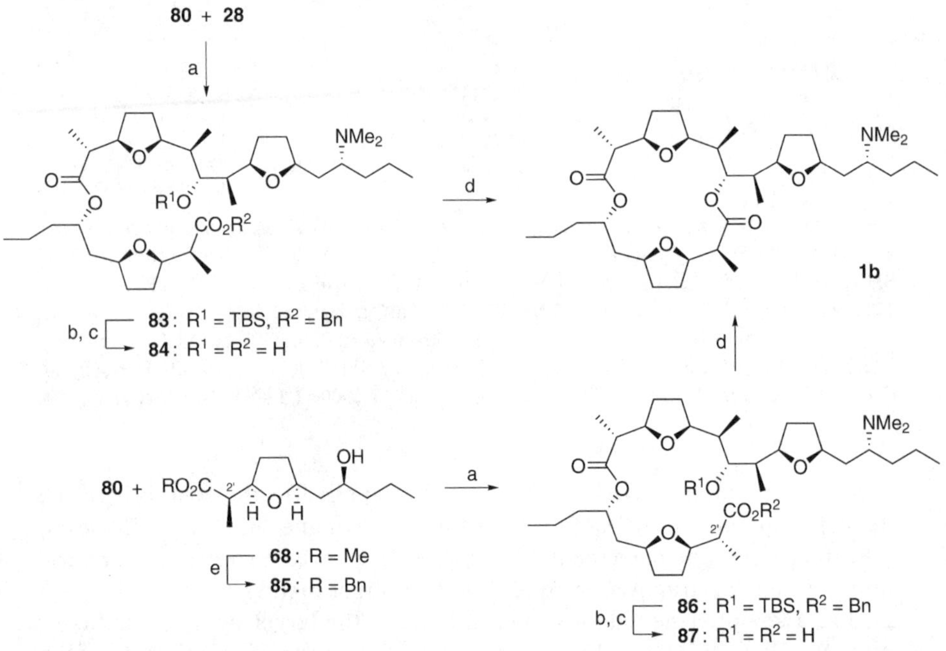

Scheme 13 Completion of Metz' synthesis of **1b**. *Reagents and conditions*: a: (i) **80**+2,4,6-trichlorobenzoyl chloride, Et_3N, THF, rt, (ii) **28**, DMAP, toluene, rt, 88% **83** or (ii) **85**, DMAP, toluene, rt, 84% **86**; b: aq. HF, MeCN, rt; c: H_2, 10% Pd/C, THF, rt, 93% **84** from **83**, 88% **87** from **86**; d: 2,4,6-trichlorobenzoyl chloride, DMAP, 4 Å sieves, CH_2Cl_2, rt, 78% **1b** from **84**, 65% **1b** from **87**; e: BnOH, $Ti(OEt)_4$, 187 mbar, 80°C, 87%

Recently, we found that this synthesis could be further streamlined by entirely omitting a selective synthesis of the smaller fragment [23]. Inspired by our observation of complete epimerization at C(2) during an attempted Yamaguchi lactonization to **1b** involving a ring closure between C(1) and the C(8′) oxygen [6] and the almost C_2-symmetric nature of the macrodiolide, we anticipated that epimerization at C(2′) in the desired sense might occur as well during a Yamaguchi lactonization of *seco* acid **87** featuring the "wrong" configuration at C(2′). In order to test this hypothesis, acid **80** and alcohol **85** prepared by titanate-mediated transesterification [24] of **68** with benzyl alcohol were coupled to give ester **86**, which was converted to *seco* acid **87** by desilylation and reductive debenzylation. Indeed, modified Yamaguchi macrolactonization of **87** again afforded pamamycin-607 (**1b**) as the single macrodiolide product in 65% yield. With this additional shortcut, only 27 steps were needed to access **1b** from furan, (*S*)-1,2-epoxypentane (**59**), and vinylsulfonyl chloride (**61**).

A fourth total synthesis of pamamycin-607 (**1b**) was published in 2002 by Kang [7]. Key operations of his approach were iodoetherifications of γ-triethylsilyloxy alkenes to generate the three 2,5-*cis*-disubstituted tetrahydrofuran units with complete diastereoselectivity. Synthesis of the larger fragment of **1b** started from diol **88** [25], which was transformed to aldehyde **90** by benzylidene formation, regioselective reduction to give benzyl ether **89**, and oxidation (Scheme 14). Julia–Kociénski olefination [26] of **90** with sulfone **91** followed by desilylation and oxidation provided the C(10)-C(18) segment **92**. A highly *anti*-selective reagent-controlled Paterson aldol reaction [27] of ketone **93** [28] with aldehyde **92** yielded β-hydroxy ketone **94**, which was subjected to Evans *anti* reduction [29] and subsequent oxidative deblocking to afford triol **95** with excellent stereoselectivity. Chemoselective oxidation of **95** to a δ-lactone followed by Weinreb amide formation and chemoselective silylation of the less hindered secondary alcohol then gave rise to β-hydroxy amide **96**.

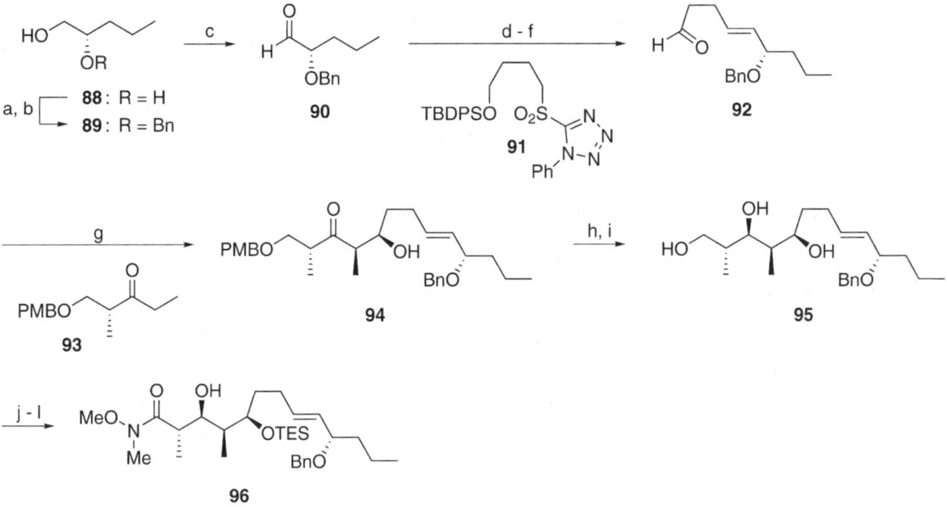

Scheme 14 Kang's synthesis of the larger fragment of **1b** (part 1). *Reagents and conditions*: a: TsOH, PhCHO, toluene, reflux, 94%; b: DIBALH, toluene, 0°C, 88%; c: Swern oxidation; d: **91**, KHMDS, DME, –78°C to rt; e: Bu_4NF, THF, rt, 72% from **90**; f: Swern oxidation; g: (i) **93**, $(cyclohexyl)_2BCl$, Et_3N, Et_2O, 0°C, (ii) **92**, –78 to –20°C, 85%; h: $Me_4NBH(OAc)_3$, HOAc, MeCN, –30 to –20°C, 88%; i: CAN, H_2O, MeCN, 0°C, 88%; j: TEMPO, NCS, Bu_4NCl, aq. $NaHCO_3$, K_2CO_3 (pH 8.6), CH_2Cl_2, rt, 97%; k: MeONHMe·HCl, Me_3Al, CH_2Cl_2, –78°C, 92%; l: TESCl, imidazole, CH_2Cl_2, –40°C, 99%

Addition of the organolithium intermediate derived from iodide *rac*-**97** by halogen–lithium exchange to Weinreb amide **96** and *anti* reduction of the resulting *β*-hydroxy ketones afforded diols **98**, double silylation of which set the stage for the crucial iodoetherification process (Scheme 15). In the event, treatment of **99** with iodine in the presence of silver carbonate effected the two-directional generation of both tetrahydrofuran moieties of **100** with complete stereocontrol. Elaboration to epoxides **101** was performed by successive acidic acetal cleavage, base-induced epoxide formation, reductive deiodination, and silylation. Due to the bulky protected tertiary carbinol in **101**, epoxide opening with lithium dimethylcuprate occurred only adjacent to the tetrahydrofuran unit, and tetraols **102** were obtained after reductive debenzylation/desilylation. Chemoselective Mitsunobu reaction of the least hindered alcohol with hydrazoic acid followed by 1,2-diol cleavage and aldehyde oxidation then delivered the larger fragment surrogate **103**.

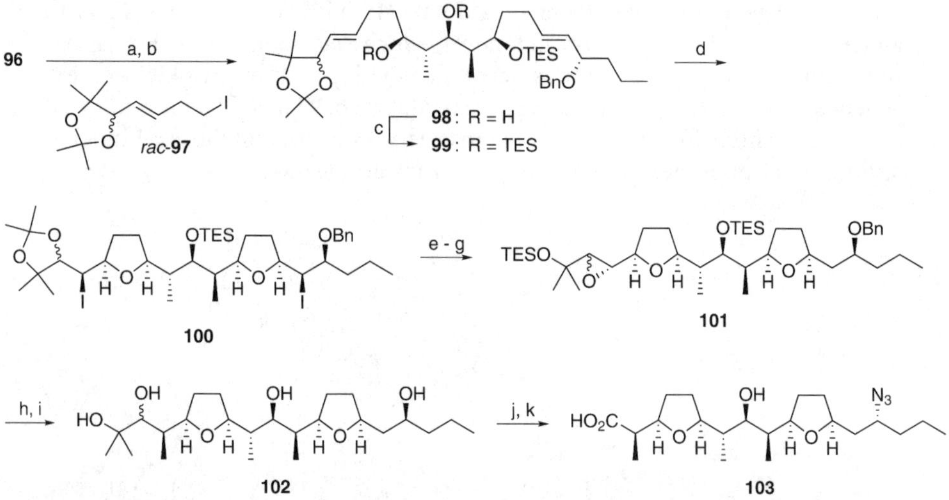

Scheme 15 Kang's synthesis of the larger fragment of **1b** (part 2). *Reagents and conditions*: a: (i) *rac*-**97**, *t*-BuLi, Et_2O, –78 to –20°C, (ii) **96**, –50 to –20°C, 82%; b: $Me_4NBH(OAc)_3$, HOAc, MeCN, –20°C, 92%; c: TESOTf, Et_3N, CH_2Cl_2, 0°C, 98%; d: I_2, Ag_2CO_3, Et_2O, rt, 81%; e: (i) HCl, MeOH, reflux, (ii) K_2CO_3, rt, 86%; f: Ph_3SnH, Et_3B, THF, 0°C, 96%; g TESOTf, Et_3N, CH_2Cl_2, –20°C, 89%; h: Me_2CuLi, Et_2O, 5°C; i: H_2, $Pd(OH)_2/C$, EtOH, rt, 88% from **101**; j: HN_3, DEAD, Ph_3P, benzene, 0°C, 97%; k: (i) $NaIO_4$, *t*-BuOH, H_2O, rt, (ii) NaH_2PO_4, $KMnO_4$, rt, 87%

The iodoetherification strategy was applied to the synthesis of the smaller fragment coupling component **109** as well (Scheme 16). Silylation of alcohol **104** [30] (76% de) allowed the separation of the pure desired diastereomer, which in turn was subjected to hydroboration/oxidation, sulfide formation with thiol **105**, and oxidation to give sulfone **106**. The requisite *γ*-triethylsilyloxy alkene functionality in **107** was constructed as a diastereomeric (*E*):(*Z*)=1.2:1 mixture by another sulfone-based olefination of aldehyde **90** with **106**. Treatment of **106** with

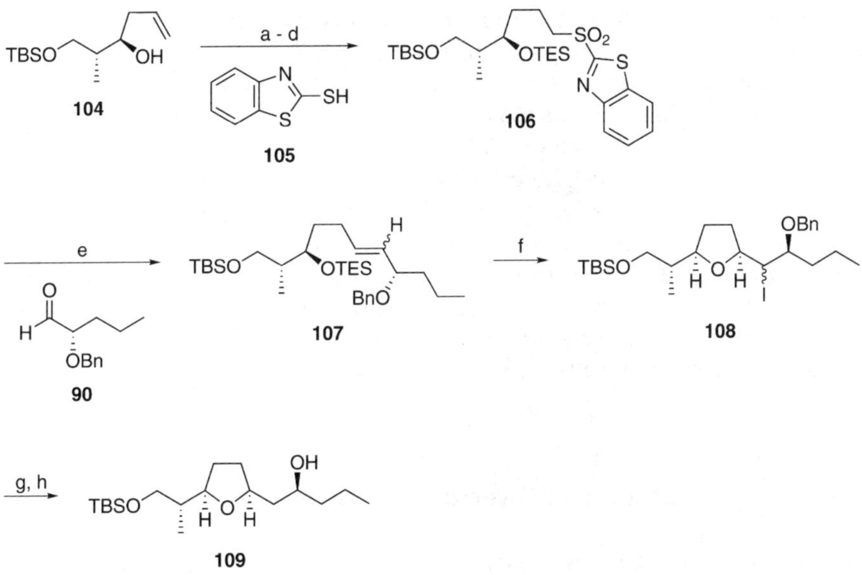

Scheme 16 Kang's synthesis of the smaller fragment of **1b**. *Reagents and conditions*: a: TESCl, imidazole, DMF, rt, 83%; b: (i) $BH_3{\cdot}Me_2S$, THF, rt, (ii) aq. H_2O_2, NaOH, rt, 85%; c: **105**, DEAD, Ph_3P, THF, 0°C, rt, 87%; d: *m*-CPBA, CH_2Cl_2, rt, 90%; e: (i) LHMDS, THF, –78°C, (ii) **90**, rt, 80%; f: I_2, Ag_2CO_3, Et_2O, rt, 92%; g: Ph_3SnH, Et_3B, THF, 0°C, 90%; h: H_2, 10% Pd/C, MeOH, rt, 99%

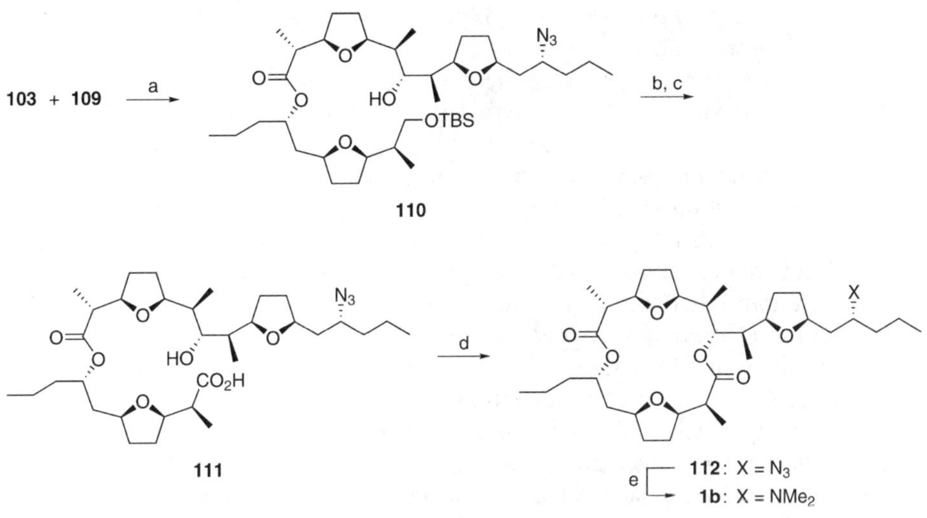

Scheme 17 Completion of Kang's synthesis of **1b**. *Reagents and conditions*: a: (i) **103**, 2,4,6-trichlorobenzoyl chloride, Et_3N, THF, rt, (ii) **109**, DMAP, benzene, rt, 90%; b: Bu_4NF, THF, rt, 94%; c: TEMPO, $NaClO_2$, NaH_2PO_4 buffer (pH 6.7), NaOCl, MeCN, rt, 91%; d: (i) 2,2′-dipyridyl disulfide, Ph_3P, MeCN, (ii) $CuBr_2$, MeCN, rt, 62%; e: (i) H_2, 10% Pd/C, MeOH, rt, (ii) aq. CH_2O, HOAc, rt, 89%

iodine/silver carbonate led to exclusive 2,5-*cis*-disubstitution of the tetrahydrofuran moiety in the resulting 1.2:1 mixture of iodides **108**. Successive reductive deiodination and debenzylation afforded alcohol **109** as a single stereoisomer.

With both building blocks **103** and **109** in hand, the total synthesis of **1b** was completed as shown in Scheme 17. Coupling of acid **103** and alcohol **109** under Yamaguchi conditions to give ester **110** and subsequent desilylation followed by chemoselective oxidation provided hydroxy acid **111**. Lactonization of the 2-thiopyridyl ester derived from **111** in the presence of cupric bromide produced the macrodiolide **112** in 62% yield, which was finally converted to pamamycin-607 (**1b**) via one-pot azide reduction/double reductive *N*-methylation. In summary, 36 steps were necessary to accomplish the synthesis of **1b** from alcohols **88** and **104**, sulfone **91**, ketone **93**, and iodide *rac*-**97**.

3
Synthetic Approaches to the Hydroxy Acid Fragments

Next to the total syntheses discussed in the previous section, many synthetic studies toward the hydroxy acid constituents **2** and **3** of the pamamycins **1** have been reported. This section will give a comprehensive survey of the various approaches published to date.

Pioneering studies in this area that culminated in the first enantioselective syntheses of all four known smaller fragments of the pamamycins **1**, as well as a first enantioselective synthesis of a highly advanced segment of the larger fragment of **1b** (and thus also of **1c**, **1h**; cf. Fig. 1), were performed by Walkup [31–34]. Central to his approach was the stereoselective generation of 2,5-*cis*-disubstituted tetrahydrofurans by one-pot intramolecular oxymercuration/palladium(II)-mediated methoxycarbonylation of γ-silyloxy allenes. Following a proof of principle in the racemic series [31], improved conditions for this key step were later implemented in a general enantioselective access to methyl esters **144–147** of the smaller fragments **2** (Schemes 18–20) [32, 33].

Scheme 18 illustrates Walkup's methodology as applied to precursors **122** and **123** of smaller fragments **2** with R^5=H. In these cases, control of absolute configuration relied on lipase-catalyzed hydrolytic kinetic resolution of acetates *rac*-**116** and *rac*-**117** prepared from allenic aldehydes **113** and *rac*-**114**, respectively, by aldol reaction with lithium enolate **115** and subsequent acetylation. In general, diastereomers due to axial chirality of the allene moiety could not be distinguished by NMR or chromatography. However, this did not obstruct the resolution step. Thus, using Amano lipase AK, acetates **116** and **117** were isolated with 93–95% ee. Elaboration of these acetates to γ-trimethylsilyloxy allenes **118** and **119** via reduction, chemoselective PMB protection, and silylation then set the scene for the crucial metal-assisted cyclization event to give tetrahydrofurans **120** and **121** with dr≥95:5. The exocyclic stereogenic center of the saturated derivatives was established with good diastereoselectivity (from **120**: dr of crude product=81:19, from **121**: dr of crude product=87:13) by chelation-

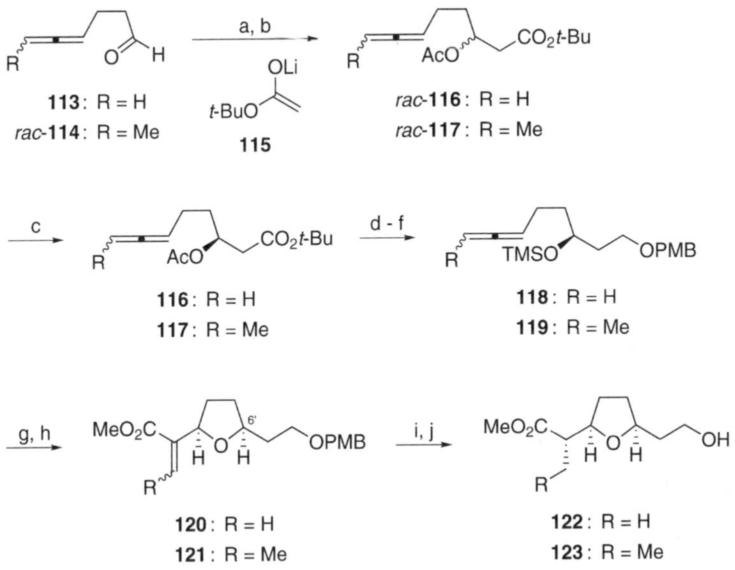

Scheme 18 Walkup's synthesis of all four known smaller fragments of the pamamycins (part 1). *Reagents and conditions:* a: **115**, THF, –78°C to rt; b: Ac_2O, DMAP, Et_3N, 0°C to rt, 64% *rac*-**116** from **113**, 74% *rac*-**117** from *rac*-**114**; c: Amano lipase AK, pH 7 phosphate buffer, rt, 45% **116**, 41% **117**; d: $LiAlH_4$, Et_2O, 0°C to rt; e: (i) NaH, DMF, 0°C to rt, (ii) PMBCl, 0°C to rt; f: TMSCl, Et_3N, THF, rt, 71% **118** from **116**, 72% **119** from **117**; g: $Hg(OCOCF_3)_2$, CH_2Cl_2, rt; h: 1 bar CO, MeOH, 10 mol% $PdCl_2$, $CuCl_2$, $MeC(OEt)_3$, propylene oxide, rt, 57% **120** from **118**, 57% **121** from **119**; i: Mg, MeOH, 0°C to rt; j: DDQ, CH_2Cl_2, pH 7 buffer, 0°C to rt, 73% **122** from **120**, 70% **123** from **121**

controlled conjugate reduction of enoates **120** and **121** with magnesium in methanol. After purification of the major products by preparative HPLC, oxidative PMB deblocking delivered the pure diastereomers **122** and **123**.

Access to the corresponding enantiopure hydroxy esters **133** and **134** of smaller fragments **2** with R^5=Me employed a highly stereoselective (ds>95%) Evans aldol reaction of allenic aldehydes **113** and *rac*-**114** with boron enolate **124** followed by silylation to arrive at the γ-trimethylsilyloxy allene substrates **125** and **126**, respectively, for the crucial oxymercuration/methoxycarbonylation process (Scheme 19). Again, this operation provided the desired tetrahydrofurans **127** and **128** with excellent diastereoselectivity (dr=95:5). Chemoselective hydrolytic cleavage of the chiral auxiliary, chemoselective carboxylic acid reduction, and subsequent diastereoselective chelation-controlled enoate reduction (**133**: dr of crude product=80:20, **134**: dr of crude product=84:16) eventually provided the pure stereoisomers **133** and **134** after preparative HPLC.

The synthesis of the smaller fragment methyl esters **144–147** was completed as depicted in Scheme 20. Chelation-controlled allylation of aldehydes **135–138** prepared by chromium(VI) oxidation of alcohols **122, 123, 133,** and **134** with allyltrimethylsilane (**139**) in the presence of titanium(IV) chloride proceeded

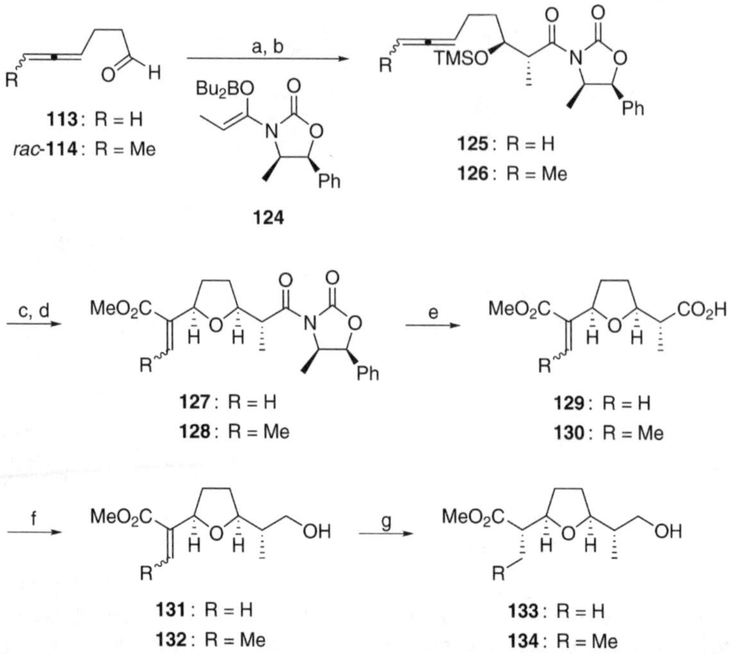

Scheme 19 Walkup's synthesis of all four known smaller fragments of the pamamycins (part 2). *Reagents and conditions*: a: **124**, CH_2Cl_2, –78°C to rt, 75% from **113**, 80% from *rac*-**114**; b: TMSCl, Et_3N, THF, rt, 92% **125**, 84% **126**; c: $Hg(OCOCF_3)_2$, CH_2Cl_2, rt; d: 1 bar CO, MeOH, 10 mol% $PdCl_2$, $CuCl_2$, $MeC(OEt)_3$, propylene oxide, rt, 85% **127** from **125**, 83% **128** from **126**; e: LiOH, 30% aq. H_2O_2, THF, H_2O, rt, 90% **129**, 81% **130**; f: BH_3·THF, THF, 0°C to rt, 80% **131**, 79% **132**; g: Mg, MeOH, 0°C to rt, 58% **133**, 74% **134**

with good to excellent stereocontrol (**140:** dr of crude product=85:15, **141:** dr of crude product=86:14, **142**, **143:** dr>99:1). Finally, hydrogenation of the major allylation products purified by preparative HPLC in the case of **140** and **141** afforded the smaller fragment derivatives **144–147**.

In a similar fashion, Walkup synthesized the enantiomerically pure C(1)-C(14) segment of the larger fragment of **1b** (and **1c**, **1h**) (Scheme 21) [34]. A rather unusual facet of his approach was an apparent kinetic resolution of the racemic aldehyde *rac*-**153** via the Evans aldol reaction. Preparation of *rac*-**153** began with a highly diastereoselective (dr=95:5) Heathcock *anti* aldol reaction [35] of 4,5-hexadienal (**113**) with lithium enolate **148** to furnish pure *rac*-**149** in the indicated yield. Silylation followed by one-pot intramolecular oxymercuration/palladium(II)-mediated methoxycarbonylation of γ-silyloxy allene *rac*-**150** generated tetrahydrofuran *rac*-**151** with excellent stereocontrol (dr of crude product=98:2). Stereoselective chelation-controlled conjugate reduction yielded *rac*-**152** (dr of crude product=84:16), which was elaborated to *rac*-**153** via chemoselective methyl ester reduction and subsequent oxidation. Treatment of aldehyde *rac*-**153** with 3 equivalents of boron enolate **154** gave rise to

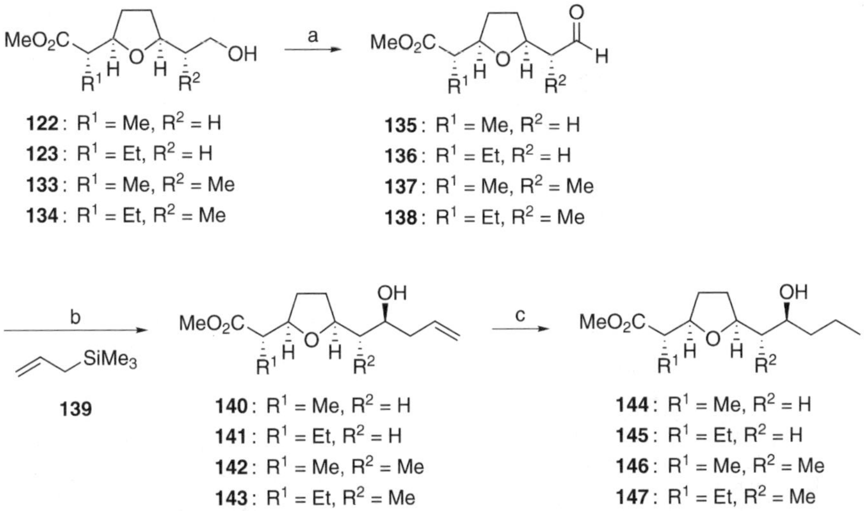

Scheme 20 Walkup's synthesis of all four known smaller fragments of the pamamycins (part 3). *Reagents and conditions*: a: PCC, $MgSO_4$, KOAc, CH_2Cl_2, rt, 70% **135**, 60% **136**, 90% **137**, 91% **138**; b: **139**, $TiCl_4$, CH_2Cl_2, –78°C, 66% **140**, 63% **141**, 84% **142**, 84% **143**; c: 1 bar H_2, Pd/C, EtOH, rt, 71% **144**, 90% **145**, 71% **146**, 75% **147**

syn aldol adduct **155** as the major product, whereas no *syn* aldol diastereomer with intact tetrahydrofuran moiety from addition of **164** to the (2*S*,3*S*,6*R*,7*S*) enantiomer of aldehyde **153** was detected. Conversion of **155** to aldehyde **156** succeeded by chemoselective imide methyl ester exchange, silylation, another chemoselective methyl ester reduction, and oxidation. Finally, 3-butenylmagnesium bromide addition to **156** provided 21% of the C(1)-C(14) segment **157** next to 49% of the undesired C(10) epimer. Additionally, a "reduced" analog of C(1)-C(10) segment **155** featuring a *tert*-butyldimethylsilyl ether at C(1) was synthesized from **113** along similar lines, however, using established asymmetric aldol methodology [34].

Perlmutter used an oxymercuration/demercuration of a γ-hydroxy alkene as the key transformation in an enantioselective synthesis of the C(8′) epimeric smaller fragment of **1b** (and many more pamamycin homologs; cf. Fig. 1) [36]. Preparation of substrate **164** for the crucial cyclization event commenced with silylation and reduction of hydroxy ester **158** (85–89% ee) [37] to give aldehyde **159**, which was converted to alkenal **162** by (*Z*)-selective olefination with ylide **160** (dr=89:11) and another diisobutylaluminum hydride reduction (Scheme 22). An Oppolzer aldol reaction with boron enolate **163** then provided **164** as the major product. Upon successive treatment of **164** with mercury(II) acetate and sodium chloride, organomercurial compound **165** and a second minor diastereomer (dr=6:1) were formed, which could be easily separated. Reductive demercuration, hydrolytic cleavage of the chiral auxiliary, methyl ester formation, and desilylation eventually led to **166**, the C(8′) epimer of the

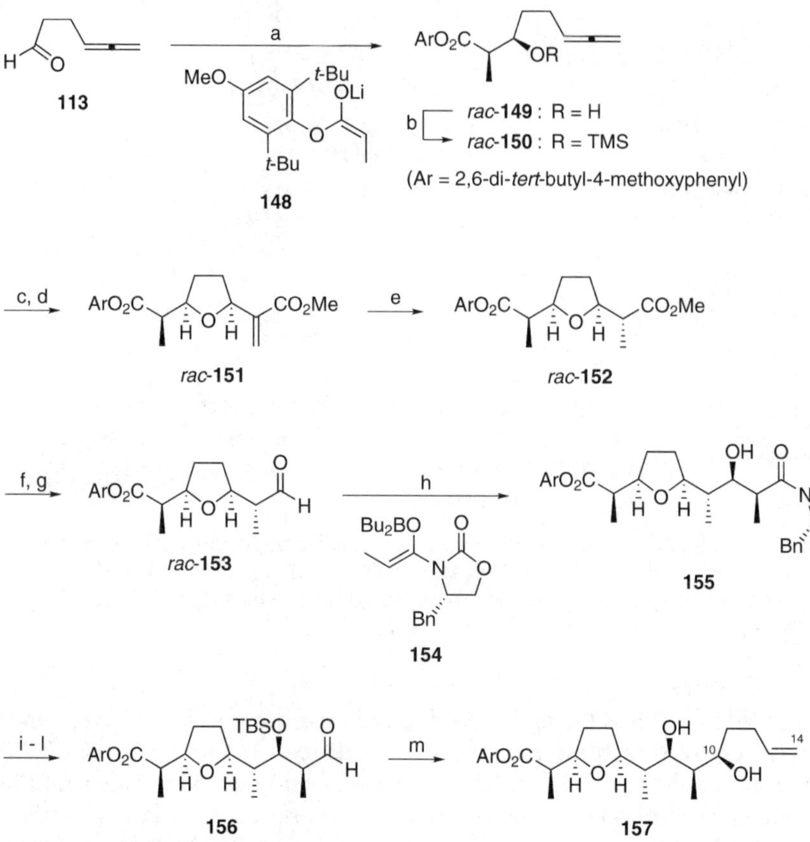

Scheme 21 Walkup's synthesis of a C(1)-C(14) segment of the larger fragment of **1b**. *Reagents and conditions*: a: **148**, 52%; b: TMSCl, Et_3N, 100%; c: $Hg(OCOCF_3)_2$; d: CO, MeOH, $PdCl_2$, $CuCl_2$, 98% from *rac*-**150**; e: Mg, MeOH, 84%; f: $LiAlH_4$, Et_2O, 0°C, 97%; g: (i) $Me_2SCl^+Cl^-$, (ii) Et_3N, 100%; h: **154**, 41%; i: NaOMe, 80%; j: TBSOTf, 98%; k: DIBALH, 92%; l: Swern oxidation, 91%; m: CH_2=CH–CH_2–CH_2–MgBr, 21%

smaller fragment methyl ester **144**. Synthetic utilization of **166** might be possible by coupling with the larger fragment of **1b** via a process that would involve inversion at C(8′).

Prior to the successful completion of his total synthesis of pamamycin-607 (**1b**), Thomas reported the reaction sequence illustrated in Scheme 23 for the construction of a C(1)-C(8) segment of the larger fragment of **1b** and other pamamycins [38]. Allylation of **167** with the allyltin trichloride generated in situ by transmetallation of allyl stannane **10** with tin(IV) chloride followed by hydroxyl-directed epoxidation of the resulting homoallylic alcohol **168** yielded epoxide **169**. Ring opening of **169** with sodium phenylselenide afforded a mixture of hydroxy selenides, which were cyclized upon exposure to catalytic amounts of perchloric acid to give the tetrahydrofuran **170** with excellent over-

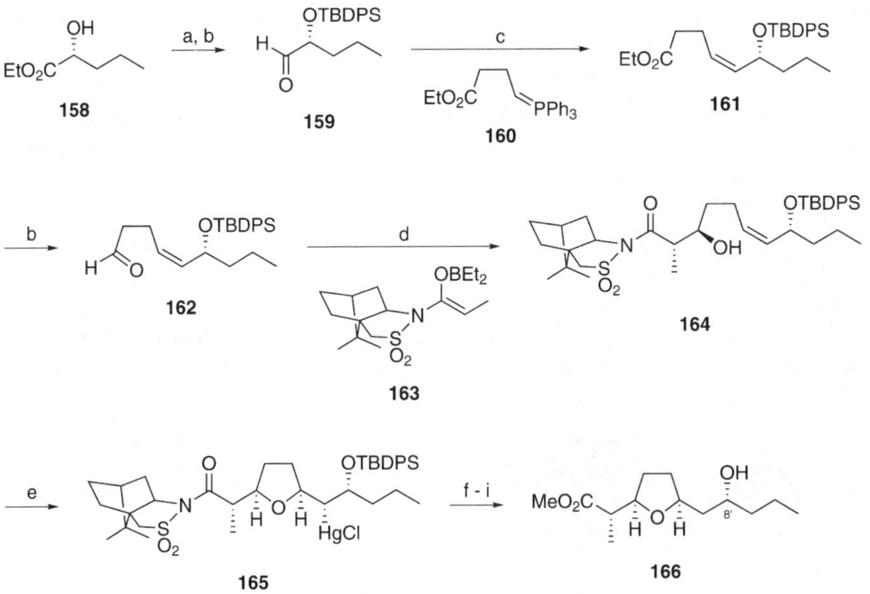

Scheme 22 Perlmutter's synthesis of the C(8′) epimeric smaller fragment of **1b**. *Reagents and conditions*: a: TBDPSCl, DMAP, imidazole, DMF, rt, 96%; b: DIBALH, toluene, –78°C, 95% **159**, 97% **162**; c: **160**, THF, 0°C, 51%; d: **163**, 54%; e: (i) $Hg(OAc)_2$, MeCN, rt, (ii) NaCl, H_2O; f: Bu_3SnH, AIBN, toluene, 93%; g: 30% aq. H_2O_2, LiOH, THF, H_2O; h: CH_2N_2, Et_2O, 46% over the last two steps; i: Bu_4NF, THF, rt, 56%

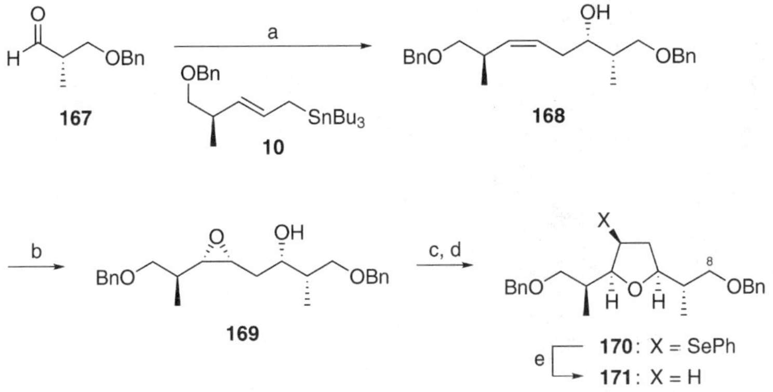

Scheme 23 Thomas' synthesis of a C(1)-C(8) segment of the larger fragment of **1b**. *Reagents and conditions*: a: **10**, $SnCl_4$, –78°C, 75–78%; b: *t*-BuOOH, $VO(acac)_2$, CH_2Cl_2, 73–77%; c: $(PhSe)_2$, $NaBH_4$, EtOH, 75–84%; d: $HClO_4$, CH_2Cl_2, 30%; e: Bu_3SnH, 88–90%

all stereoselectivity, presumably via an intermediate selenonium ion. Reductive deselenylation then furnished the pseudo C_2-symmetric benzyl ether **171**.

Bloch utilized the diastereoselective formation of 2,5-*cis*-disubstituted tetrahydrofurans with additional control over an exocyclic stereogenic center adjacent to the heterocycle via intramolecular Michael reactions of ε-hydroxy-methacrylate substrates to achieve an enantioselective synthesis of the smaller fragment [39] and a C(8)-C(18) segment [40] of the larger fragment of **1b** (Schemes 24 and 25). Scheme 24 illustrates the route to the smaller fragment methyl ester **144**. Acetate **172** prepared by enzymatic transesterification [41] was converted to aldehyde **174** by protecting group operations and subsequent oxidation. Aldol reaction of **174** with lithium enolate **175** proceeded with high substrate-induced diastereoselectivity (dr of crude product=90:10) to provide

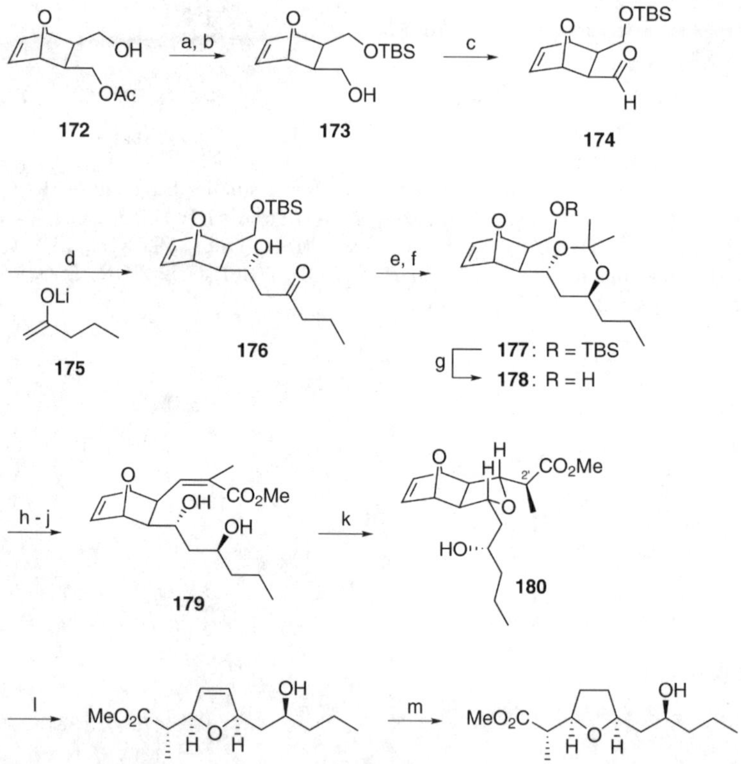

172
173
174
175
176
179
180
181
144

Scheme 24 Bloch's synthesis of the smaller fragment of **1b**. *Reagents and conditions*: a: TBSCl, imidazole, DMF, rt, 95%; b: KOH, MeOH, rt, 99%; c: Dess–Martin periodinane, CH_2Cl_2, 0°C, 86%; d: **175**, THF, hexane, –78°C, 65%; e: $Me_4NBH(OAc)_3$, HOAc, MeCN, –40°C, 68%; f: 2,2-dimethoxypropane, PPTS, CH_2Cl_2, rt, 94%; g: Bu_4NF, THF, 0°C to rt, 95%; h: Dess–Martin periodinane, pyridine, CH_2Cl_2, 0°C, 73%; i: $(PhO)_2POCNa(CH_3)CO_2Me$, THF, –78°C to rt, 80%; j: aq. HCl, THF, 0°C, 94%; k: Bu_4NF, THF, 0°C, 93%; l: 400°C, 1.3 mbar, 80%; m: 1 bar H_2, 5% Pt/C, EtOAc, rt, 79%

hydroxy ketone **176**, which was subjected to Evans *anti* reduction (dr of crude product=80:20) followed by acetonide formation. After desilylation of the resulting acetal **177**, oxidation, (*Z*)-selective Horner–Wadsworth–Emmons olefination, and acetal cleavage, cyclization substrate **179** was isolated as a (*Z*):(*E*)=90:10 mixture. Treatment of **179** with tetrabutylammonium fluoride effected the key ring closure to **180** with complete *cis* selectivity regarding the tetrahydrofuran 2,5-disubstitution and excellent stereocontrol (dr of crude product=95:5) over the configuration at C(2′). Flash vacuum thermolysis of **180** induced a retro Diels–Alder reaction to give dihydrofuran **181**, which was hydrogenated to afford methyl ester **144**.

Synthesis of a C(8)-C(18) segment of the larger fragment of **1b** using the same basic strategy is depicted in Scheme 25. Here, hydroxy ketone **176** was subjected to *syn*-selective (dr of crude product=90:10) reductive amination [42] with sodium cyanoborohydride and benzylamine followed by tetrahydro-oxazine formation using aqueous formaldehyde. The resulting heterocycle **182** was then converted to unsaturated ester **184** by successive desilylation, oxidation, and entirely (*Z*)-selective Horner–Wadsworth–Emmons olefination. Re-

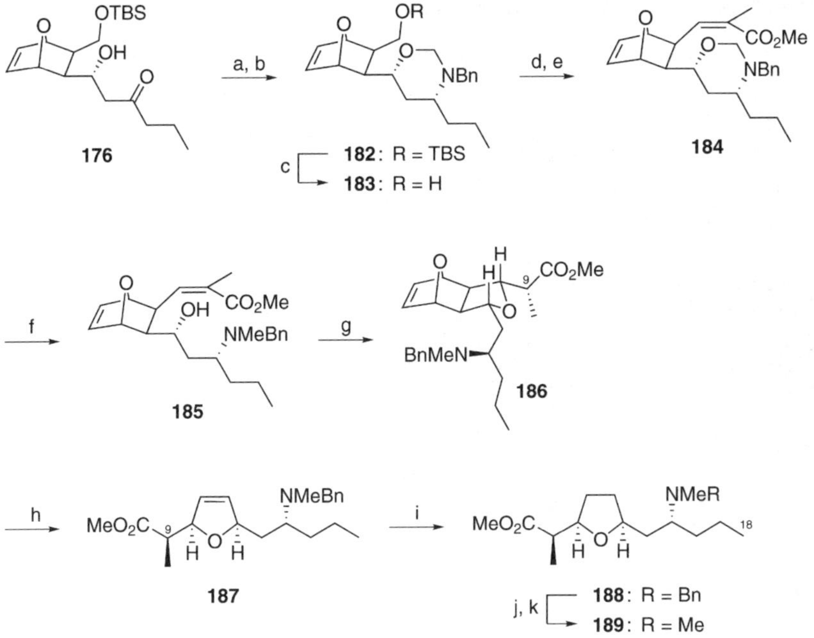

Scheme 25 Bloch's synthesis of a C(8)-C(18) segment of the larger fragment of **1b**. *Reagents and conditions*: a: (i) $BnNH_2$, 4 Å sieves, HOAc, THF, 0°C, (ii) $NaBH_3CN$, –15°C, 70%; b: aq. CH_2O, MeOH, rt, 78%; c: Bu_4NF, THF, 0°C to rt, 86%; d: Dess–Martin periodinane, pyridine, CH_2Cl_2, 0°C; e: $(PhO)_2POCNa(CH_3)CO_2Me$, THF, –78 to 0°C, 56% from **183**; f: $NaBH_3CN$, TFA, MeOH, rt, 96%; g: NaH, Et_2O, rt, 90% **186**+C(9) epimer (60:40); h: 400°C, 1.3 mbar, 75% **187**+C(9) epimer; i: 1 bar H_2, 5% Pt/C, EtOAc, rt, 41%; j: 1 bar H_2, 10% Pd/C, MeOH, rt, 74%; k: aq. CH_2O, HOAc, $NaBH_3CN$, MeCN, rt, 81%

ductive cleavage of the tetrahydrooxazine **184** with simultaneous *N*-methylation gave rise to cyclization substrate **185**. Treatment of **185** with sodium hydride established clean 2,5-*cis*-disubstitution of the resulting tetrahydrofuran. However, control over the relative configuration at C(9) was much less pronounced compared to the corresponding cyclization of **179**. Chromatographic separation of the desired stereoisomer was accomplished after cycloreversion and subsequent hydrogenation at the stage of tetrahydrofuran **188**. Finally, reductive debenzylation and reductive *N*-methylation provided methyl ester **189** as a single stereoisomer.

Before developing the shortcut to benzyl ester **28** of the smaller hydroxy acid constituent of **1b** illustrated in Scheme 12, our own group first employed a sultone-based strategy to the stereoselective synthesis of the smaller fragment methyl ester **144** (Scheme 26) [16, 17]. To this end, sultone **192** was efficiently generated with complete diastereoselectivity via intramolecular Diels–Alder reaction of the vinylsulfonate derived from alcohol **191**, which in turn was prepared by halogen–lithium exchange of 2-bromo-4-methylfuran (**190**) [43] followed by alkylation with epoxide **59**. Upon subjecting **192** to a domino elimination/alkoxide-directed 1,6-hydride addition, the bicyclic compounds **193–195** with the required *trans* relationship of the hydroxyl and the methyl

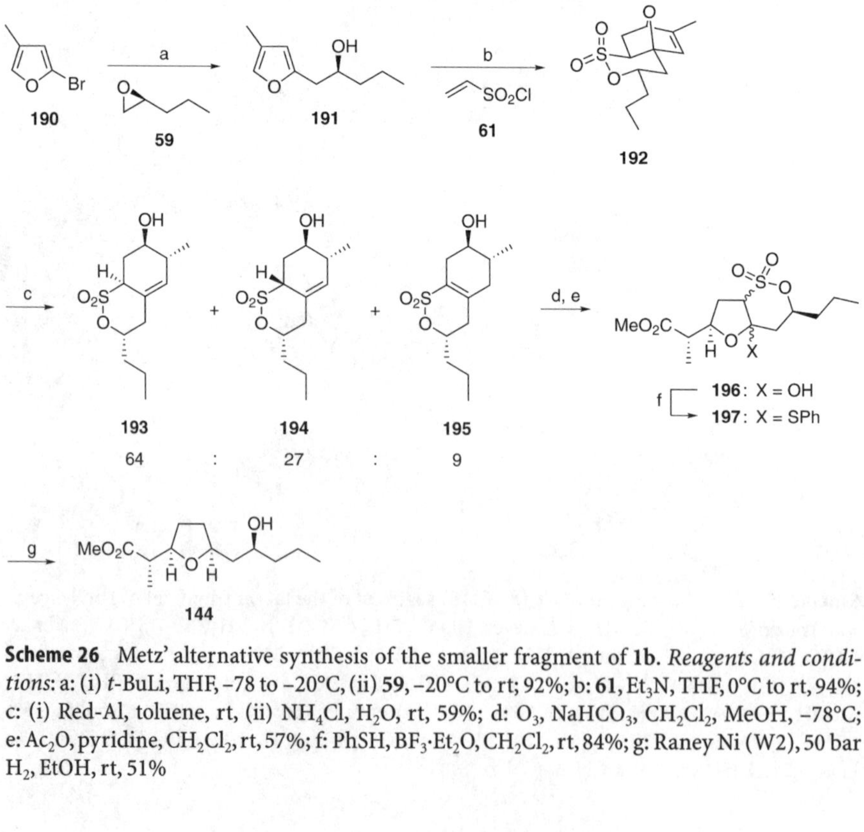

Scheme 26 Metz' alternative synthesis of the smaller fragment of **1b**. *Reagents and conditions*: a: (i) *t*-BuLi, THF, –78 to –20°C, (ii) **59**, –20°C to rt; 92%; b: **61**, Et_3N, THF, 0°C to rt, 94%; c: (i) Red-Al, toluene, rt, (ii) NH_4Cl, H_2O, rt, 59%; d: O_3, $NaHCO_3$, CH_2Cl_2, MeOH, –78°C; e: Ac_2O, pyridine, CH_2Cl_2, rt, 57%; f: PhSH, $BF_3 \cdot Et_2O$, CH_2Cl_2, rt, 84%; g: Raney Ni (W2), 50 bar H_2, EtOH, rt, 51%

substituent were isolated. Application of the reaction sequence domino ozonolysis/cyclization, Lewis acid-catalyzed hydroxyl/phenylthio exchange, and domino reductive elimination/hydrogenation already discussed above (see Schemes 10 and 11) eventually led to methyl ester **144** with high diastereoselectivity (dr of crude product=17:1) for the final step.

Additionally, we also found proper conditions for a direct C(2′) epimerization [21] of the larger fragment precursor **68** to give **144** but again, the three-step sequence depicted in Scheme 12 afforded a higher total yield of a more suitable smaller fragment surrogate of **1b**.

The use of a chiral sulfoxide group as the stereoinducing element is at the center of Solladié's approach to the smaller fragment [44] and a C(8)-C(18) segment [44,45] of the larger fragment of **1b** (Scheme 27). *β*-Ketosulfoxide **200** was obtained from *β*-ketoester **198** via carbonyl protection and condensation with chiral sulfoxide **199**. Two completely diastereoselective reductions of **200**

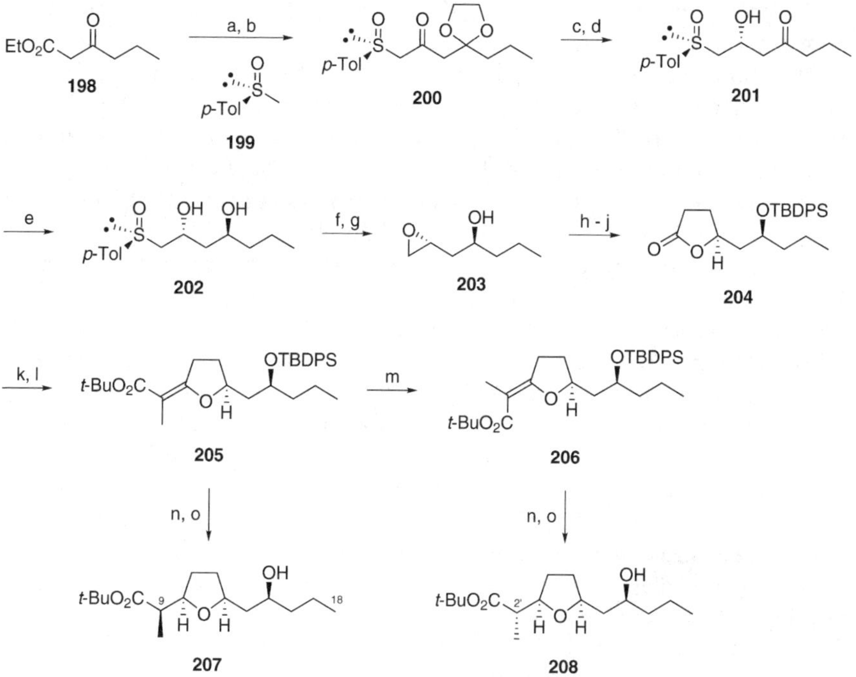

Scheme 27 Solladié's synthesis of the smaller fragment and a C(8)-C(18) segment of the larger fragment of **1b**. *Reagents and conditions*: a: $HOCH_2CH_2OH$, TMSCl, CH_2Cl_2; b: **199**, LDA, 60% from **198**; c: DIBALH; d: $(CO_2H)_2$, THF, H_2O, 80% from **200**; e: $Me_4NBH(OAc)_3$, HOAc, 97%; f: *t*-BuBr, $CHCl_3$; g: (i) Me_3OBF_4, CH_2Cl_2, (ii) K_2CO_3, 75% from **202**; h: TBDPSCl, imidazole, DMF; i: $CH_2(CO_2Et)_2$, EtONa; j: $MgCl_2{\cdot}6H_2O$, DMF, reflux, 73% from **203**; k: $EtCO_2$*t*-Bu, LDA: l: $(CO_2H)_2$, CH_2Cl_2, 75% from **204**; m: (i) LDA, LiCl, Et_2O, –78°C, (ii) EtOH, –78°C, 70%; n: Bu_4NF, THF; o: 4 bar H_2, Rh, Al_2O_3, MeOH, 73% **207** from **205**, 67% **208** from **206**

and the deketalized ketone **201** led to diol **202**, which was converted to epoxide **203** through sulfoxide reduction followed by intramolecular nucleophilic substitution of an intermediate sulfonium salt. Silylation and subsequent regioselective epoxide opening with sodium diethylmalonate afforded γ-lactone **204** after dealkoxycarbonylation. The crucial intermediate **205** was then constructed from **204** as the thermodynamically more stable (*E*) isomer via addition of the lithium enolate of *tert*-butyl propionate and acid-catalyzed dehydration of the resulting hemiacetal. Desilylation to the corresponding alcohol was required for a smooth and highly diastereoselective hydrogenation (85% ds of crude product) to give the C(8)-C(18) segment **207**. On the other hand, a chelation-controlled (*E*) to (*Z*) isomerization by deprotonation of **205** in the presence of lithium chloride followed by protonation of the resulting extended enolate to (*Z*) isomer **206** at low temperature (dr of crude product=95:5) paved the way to the smaller fragment *tert*-butyl ester **208** as well. Thus, desilylation of **206** and subsequent diastereoselective hydrogenation as before (85% ds of crude product) set up the desired (*S*) configuration at C(2′) of **208**.

A convergent synthesis of a C(1′)-C(10′) segment of the smaller fragment of pamamycin-621A (**1c**) and other pamamycin homologs using a catalytic, asymmetric dimerization of methylketene was described by Calter [46]. Methylketene generation from racemic 2-bromopropionyl bromide (*rac*-**209**) with zinc followed by enantioselective [2+2] cycloaddition in the presence of a catalytic amount of quinidine provided β-lactone **210** of 98% ee, which served as a branching point for accessing the two building blocks **212** and **217** (Scheme 28). Vinylketone **212** was derived from **210** by Weinreb amide formation via nucleophilic lactone opening, trapping of the resulting enolate as the silyl enol ether **211**, and subsequent palladium(II)-mediated oxidation. Preparation of the coupling partner **217** from **210** commenced with *N,O*-dimethylhydroxylamine addition to give **213** and highly *anti*-selective reduction of this β-ketoamide to afford the β-hydroxyamide **214** (>95% de, 99% ee). Successive silylation, reduction, diastereoselective addition of tributylstannyllithium to the resultant aldehyde **215** in the Felkin–Anh sense, and MOM protection furnished TBS ether **216**. In order to preclude the possibility of silyl migration during the next step, **216** was converted to PMB ether **217**. The crucial coupling of the two subunits **217** and **212** was accomplished by first transmetallating **217** to a mixed cuprate, conjugate addition of which to vinylketone **212** yielded the desired product **218**. Another *anti*-selective β-ketoamide reduction followed by mesylation of the resulting alcohol to give **219** and acidic removal of both ether protecting groups finally set the scene for a chemoselective base-induced cyclization to deliver the tetrahydrofuran **220**. Whereas the requisite C(11′) carbon is still missing in **220**, synthesis of a complete C(1′)-C(11′) fragment of **1c** along these lines might be possible by conjugate addition of a suitable methylmetal species to vinylketone **212** to provide the propyl homolog of ethylketone **213**.

Kiyota reported an enantioselective access to the smaller fragment [47, 48] and a C(8)-C(18) segment [48] of the larger fragment of **1b** using iodoetheri-

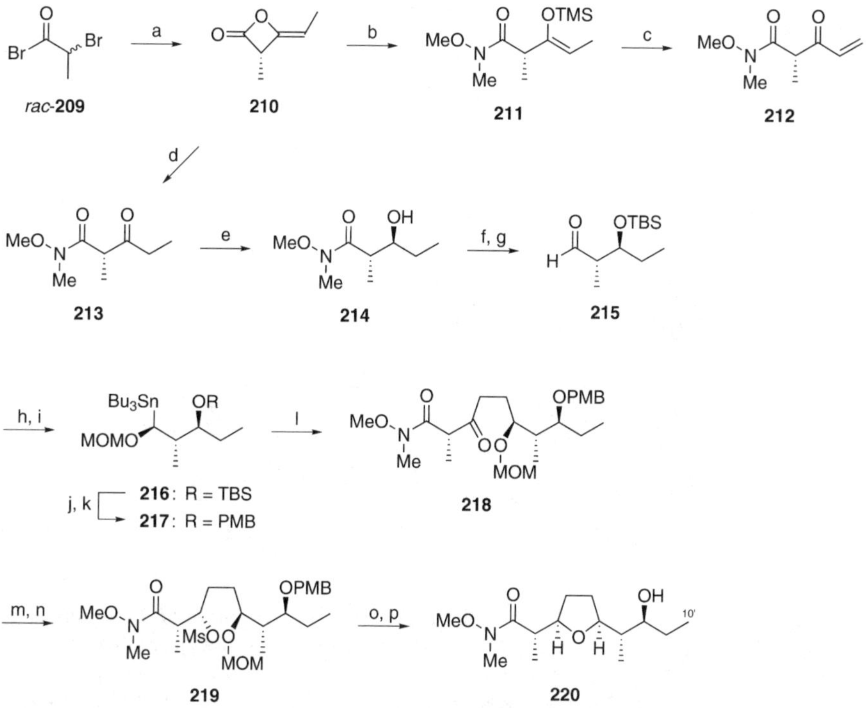

Scheme 28 Calter's synthesis of a C(1′)-C(10′) segment of the smaller fragment of **1c**. *Reagents and conditions*: a: Zn, 0.3 mol% quinidine, 55%; b: (i) MeONLiMe, (ii) TMSCl; c: $Pd(OAc)_2$, 40% from *rac*-**209**; d: MeONHMe, 10 mol% pyridone; e: $KBEt_3H$, 40% from *rac*-**209**; f: TBSCl, imidazole; g: DIBALH, 79% from **214**; h: Bu_3SnLi; i: MOMCl, 31% from **215**; j: Bu_4NF, 91%; k: NaH, PMBBr, 61%; l: (i) BuLi, (ii) [2-thienyl-Cu-CN]$^-$ Li$^+$, (iii) **212**, 45%; m: $KBEt_3H$; n: MsCl, 79% from **218**; o: HCl, MeOH, 62%; p: NaH, 99%

fications of γ-*tert*-butoxy alkenes as the key transformations (Schemes 29 and 30). Similar to Kang's iodoetherification approach published later (see Schemes 14–17, 31, and 32), these cyclofunctionalizations proceed with exclusive formation of 2,5-*cis*-disubstituted tetrahydrofurans. His synthesis of the smaller fragment methyl ester **144** is shown in Scheme 29. Evans aldol reaction of the boron enolate from oxazolidinone **221** with aldehyde **222** gave rise to a single stereoisomer, which was derivatized to *tert*-butyl ether **223**. After hydrolytic cleavage of the chiral auxiliary, esterification with diazomethane, and reductive debenzylation, conversion of hydroxy ester **224** to cyclization substrate **226** was achieved by oxidation and subsequent entirely (*E*)-selective Horner–Wadsworth–Emmons olefination. Upon subjecting γ-*tert*-butoxy alkene **226** to iodine with moderate heating, tetrahydrofuran **227** was formed in a completely diastereoselective fashion. Following reductive deiodination, sodium borohydride reduction of **53** yielded the desired methyl ester **144** as the major product next to 23% of the C(8′) epimer **166**.

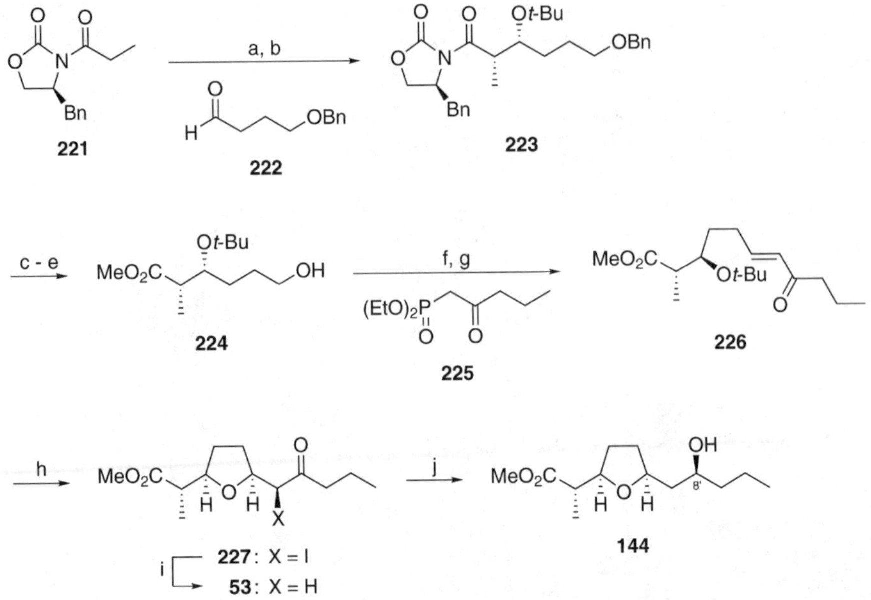

Scheme 29 Kiyota's synthesis of the smaller fragment of **1b**. *Reagents and conditions*: a: (i) Bu_2BOTf, *i*-Pr_2NEt, CH_2Cl_2, 0°C, (ii) **222**, –78°C to rt, 87%; b: isobutene, amberlyst H-15, hexane, –78°C to rt, 72%; c: LiOH, 30% aq. H_2O_2, THF, H_2O, 0°C; d: CH_2N_2, Et_2O, rt, 75% from **223**; e: 1 bar H_2, 10% Pd/C, MeOH, rt, 97%; f: Dess–Martin periodinane, Et_2O, 0°C to rt; g: **225**, LiBr, Et_3N, THF, rt, 63% from **224**; h: I_2, $NaHCO_3$, MeCN, 40°C, 85%; i: Bu_3SnH, AIBN, toluene, rt, 98%; j: $NaBH_4$, MeOH, 0°C, 40%

Synthesis of a C(8)-C(18) segment of the larger fragment of **1b** employing the same key cyclization methodology is illustrated in Scheme 30. Terminal epoxide **229** was ultimately derived from enantiomerically pure epoxide **228** [49]. Elaboration of **229** to sulfone **233** began with copper-catalyzed regioselective epoxide opening using allylmagnesium bromide and subsequent *tert*-butyl ether formation. Two-step oxidative olefin scission and reduction of the resulting aldehyde afforded alcohol **232**, which was transformed to **233** by sulfide formation and oxidation. Julia olefination of aldehyde **234** [50] with sulfone **233** furnished γ-*tert*-butoxy alkene **235** as the pure (*E*) isomer. Under the same conditions, which operated well for the cyclization of **226**, exclusive production of tetrahydrofuran **236** from **235** was observed as well. Finally, reductive deiodination provided the C(8)-C(18) segment **237**.

In addition to his total synthesis of pamamycin-607 (**1b**) (see Schemes 14–17), Kang also communicated an alternative synthesis of a larger fragment surrogate of **1b** [51]. Again, iodoetherifications of γ-triethylsilyloxy alkenes were utilized as key transformations to control the 2,5-*cis*-disubstitution of the two tetrahydrofuran moieties. However, whereas his total synthesis of **1b** involved a two-directional formation of both heterocycles in a single operation, the alternative route depicted in Schemes 31 and 32 is characterized by sequential

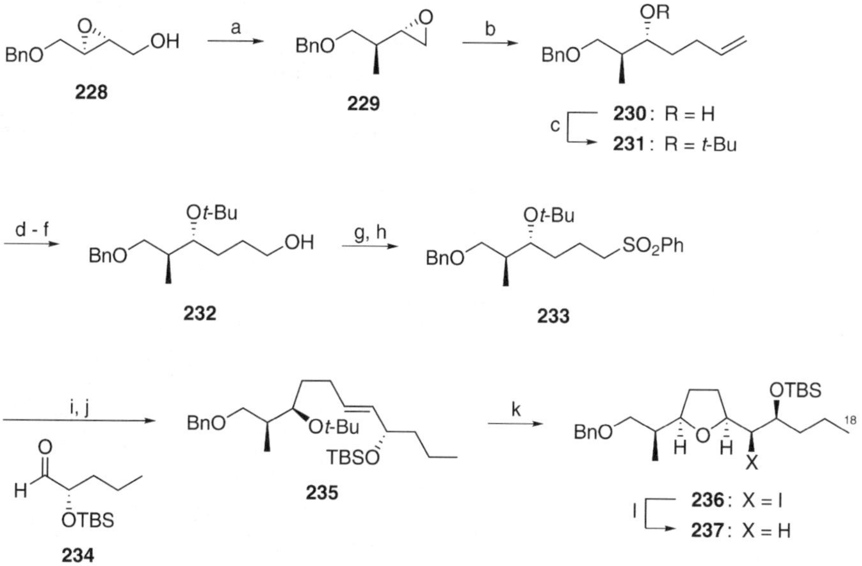

Scheme 30 Kiyota's synthesis of a C(8)-C(18) segment of the larger fragment of **1b**. *Reagents and conditions*: a: see ref. [49]; b: CH_2=CH–CH_2–MgBr, CuI, Et_2O, –30°C to rt, 98%; c: isobutene, amberlyst H-15, hexane, –78°C to rt, 78%; d: OsO_4, NMO, THF, H_2O, 0°C to rt; e: $NaIO_4$, THF, H_2O, rt; f: $LiAlH_4$, Et_2O, –78°C, 85% from **231**; g: Bu_3P, $(PhS)_2$, THF, 0°C to rt; h: *m*-CPBA, $NaHCO_3$, CH_2Cl_2, 0 to 10°C, 69% from **232**; i: (i) BuLi, THF, –78°C, (ii) **234**, –78°C, (iii) PhCOCl, –78°C to rt; j: Na(Hg), MeOH, THF, –78°C, 34% from **233**; k: I_2, $NaHCO_3$, MeCN, 40°C, 76%; l: Bu_3SnH, AIBN, toluene, rt, 26%

construction of the two five-membered rings. Moreover, instead of starting from the C(18) end of the C(1)-C(18) subunit, as has been done for the synthesis of the larger fragment substitute **103**, preparation of **257**, the PMB ether of alcohol **103**, began at the C(1) end (Scheme 31). Synthesis of the first crucial intermediate **241** commenced with oxidation of alcohol **238** [52] and subsequent enantioselective Roush crotylation [53] with boronate **239** to give homoallylic alcohol **240** of 80% ee. Further enantiomeric enrichment was achieved hand in hand with the intended elaboration of **240** by esterification with an (S)-proline derivative followed by two-step oxidative olefin cleavage and aldehyde reduction to arrive at **241**. At this stage, the desired major diastereomer of 100% ee could be separated, which then served as a building block for the C(1)-C(6) segment **243**, as well as the C(7)-C(13) segment **246**. Ester hydrolysis, acetonide formation, reductive debenzylation, and iodination yielded iodide **242**, conversion of which to phosphonium salt **243** was carried out by acidic deprotection, one-pot chemoselective sequential bis-silylation of the resulting diol, and substitution with triphenylphosphane. Aldehyde **246** was derived from **241** via oxidation, vinyl Grignard addition, and saponification to give diastereoisomer **245** as the major product, which in turn was transformed to **246** by acetonide protection followed by ozonolysis of the alkene with reductive workup. The

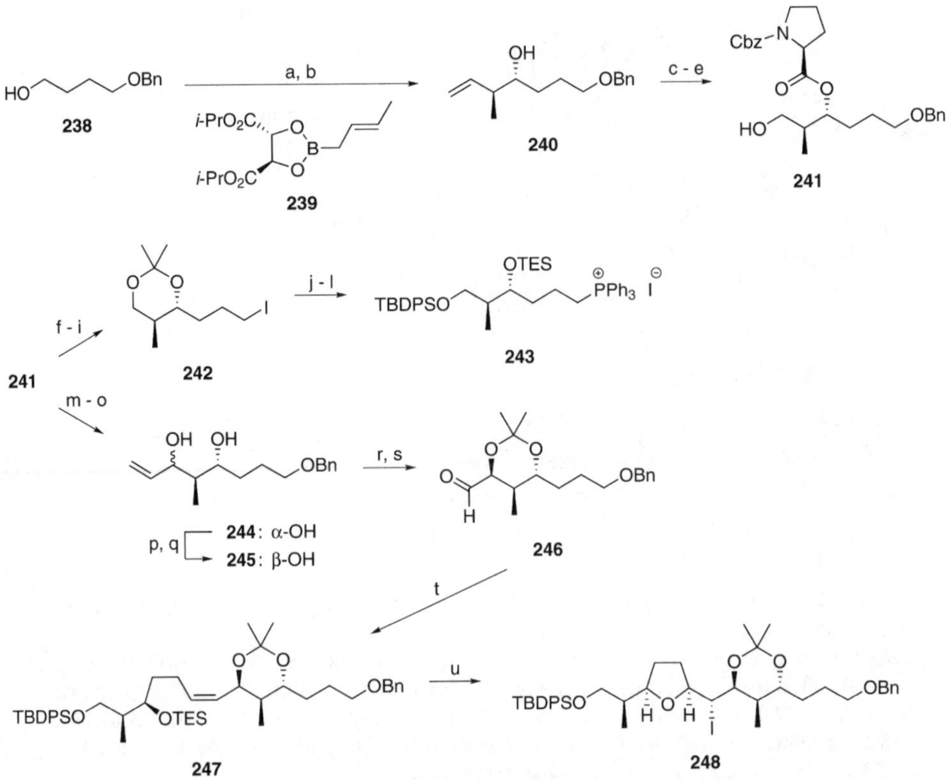

Scheme 31 Kang's alternative synthesis of the larger fragment of **1b** (part 1). *Reagents and conditions*: a: Swern oxidation; b: **239**, 4 Å sieves, toluene, –78°C, 87% from **238**; c: *N*-Cbz-L-proline, DCC, DMAP, CH_2Cl_2, rt; d: OsO_4, NMO, acetone, H_2O, rt; e: (i) $NaIO_4$, THF, H_2O, rt, (ii) $NaBH_4$, 0°C, 78% from **240**; f: LiOH, MeOH, H_2O, rt; g: 2,2-dimethoxypropane, PPTS, toluene, reflux; h: Li, NH_3(l), THF, –78°C; i: I_2, Ph_3P, imidazole, THF, 0°C, 86% from **241**; j: conc. aq. HCl, MeOH, 0°C; k: (i) TBDPSCl, DMAP, Et_3N, CH_2Cl_2, –15°C, (ii) TESOTf, –15°C; l: Ph_3P, K_2CO_3, MeCN, reflux, 83% from **242**; m: Swern oxidation; n: CH_2=CH–MgBr, Et_2O, –78°C; o: LiOH, MeOH, H_2O, rt, 53% **245** and 17% **244** from **241**; p: $PhCO_2H$, Ph_3P, DEAD, THF, –20°C; q: LiOH, MeOH, H_2O, rt, 65% from **244**; r: 2,2-dimethoxypropane, PPTS, acetone, rt; s: (i) O_3, $NaHCO_3$, CH_2Cl_2, MeOH, –78°C, (ii) Me_2S, rt; t: (i) **243**, *t*-BuLi, THF, –78 to –5°C, (ii) **246**, 10°C, 85% from **245**; u: I_2, Ag_2CO_3, Et_2O, rt, 94%

overall yield of this process could be further enhanced by Mitsunobu inversion of the minor diastereoisomer **244** to afford additional diol **245**. Joining of the two subunits **243** and **246** by Wittig olefination provided the pure (*Z*)-alkene **247**, treatment of which with iodine in the presence of silver carbonate induced a completely diastereoselective cyclization to give tetrahydrofuran **248**.

Elaboration of **248** to **257** is depicted in Scheme 32. Acidic deketalization, basic epoxide formation, and silylation of the remaining hydroxyl group led to **249**, which was subjected to ring opening with lithium dimethylcuprate in a regioselective (78:15) fashion to provide the desired alcohol **250** as the major product. TES blocking of the secondary alcohol proved to be pivotal to the success of this alkylation. Simultaneous reductive deblocking of the TES and benzyl ether, acetonide protection, iodination, and substitution with triphenylphosphane achieved the conversion of **250** to **251**. Wittig coupling of **251** with aldehyde **90**, this time derived from diol **252** via periodate cleavage, subsequent deketalization, and silylation yielded γ-triethylsilyloxy alkenes **253** as an (*E*):(*Z*)=10:1 mixture. The second key cyclization of **253** using iodine/silver carbonate followed by reductive deiodination gave rise to bis-tetrahydrofuran **254** with complete stereocontrol. After reductive release of the two secondary alcohols, a chemoselective Mitsunobu reaction of the less sterically hindered

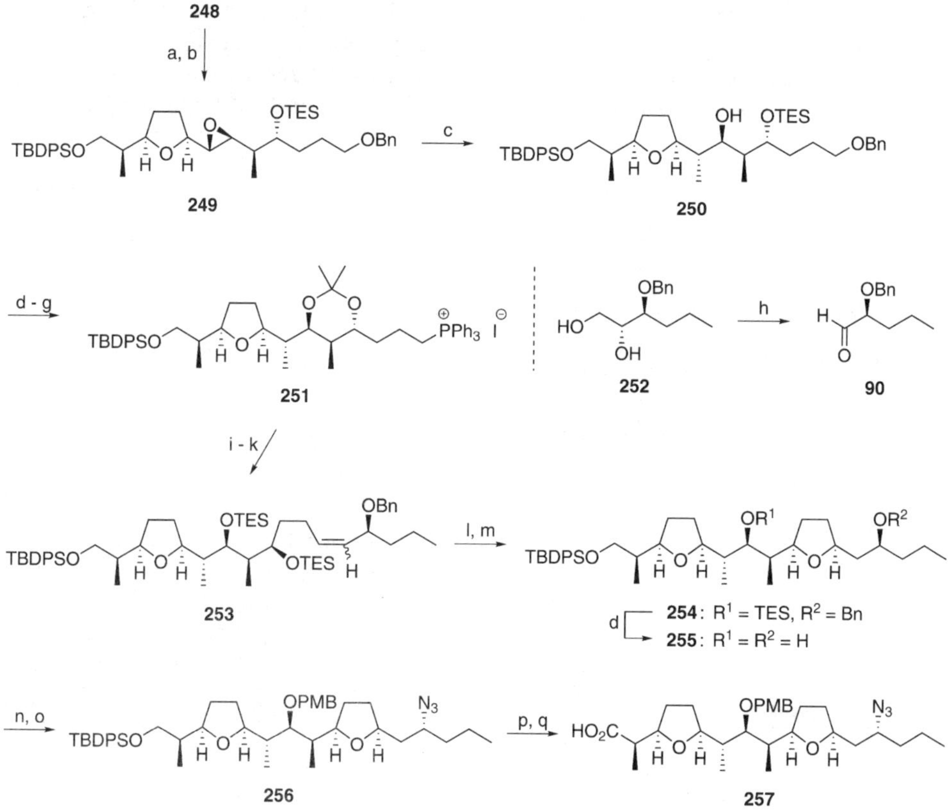

Scheme 32 Kang's alternative synthesis of the larger fragment of **1b** (part 2). *Reagents and conditions*: a: (i) PPTS, MeOH, THF, rt, (ii) K_2CO_3, rt; b: TESOTf, 2,6-lutidine, CH_2Cl_2, –78°C, 87% from **248**; c: Me_2CuLi, Et_2O, 10°C, 78%; d: H_2, $Pd(OH)_2/C$, EtOH, rt, 96% from **250**, 95% **255**; e: PPTS, acetone, rt; f: I_2, Ph_3P, imidazole, THF, 0°C; g: Ph_3P, K_2CO_3, MeCN, reflux, 90% over the last three steps; h: $NaIO_4$, acetone, H_2O, 0°C; i: (i) BuLi, THF, –78 to –5°C, (ii) **90**, 10°C, 88% from **252**; j: PPTS, MeOH, THF, rt; k: TESOTf, 2,6-lutidine, CH_2Cl_2, rt, 89% over the last two steps; l: I_2, Ag_2CO_3, Et_2O, rt; m: Ph_3SnH, Et_3B, THF, 0°C, 77% from **253**; n: HN_3, Ph_3P, DEAD, benzene, 0°C; o: PMBCl, Bu_4NI, DMF, 0°C, then KHMDS, 0°C, 86% from **255**; p: Bu_4NF, THF; q: Jones oxidation, 0°C, 85% from **256**

alcohol with hydrazoic acid, PMB protection, desilylation, and oxidation completed the synthesis of the larger fragment surrogate **257**. In total, ca. 37 steps were required to access **257** from alcohols **238** and **252** and boronate **239**.

Nagumo and Kawahara used the stereoselective ring contraction of a tetrahydropyran and a δ-lactone, respectively, via an intermediate phenonium ion to synthesize the racemic smaller fragment [54] and an enantiomerically pure C(2)-C(8) segment [55] of the larger fragment of **1b** (Schemes 33 and 34). The smaller fragment methyl ester *rac*-**144** was prepared from the δ-lactone *rac*-**258** by the reaction sequence shown in Scheme 33. Axial allylation of the lactol derived from *rac*-**258** by diisobutylaluminum hydride reduction with allylsilane **139** provided tetrahydropyran *rac*-**259**, which was transformed to a mixture of *rac*-**260** and *rac*-**261** via oxidative olefin scission and allyl Grignard addition. After separation of the epimers, they were hydrogenated to give the crucial rearrangement substrates *rac*-**262** and *rac*-**263**, respectively. Treatment of either compound with trifluoroacetic acid at 70°C and subsequent alkaline

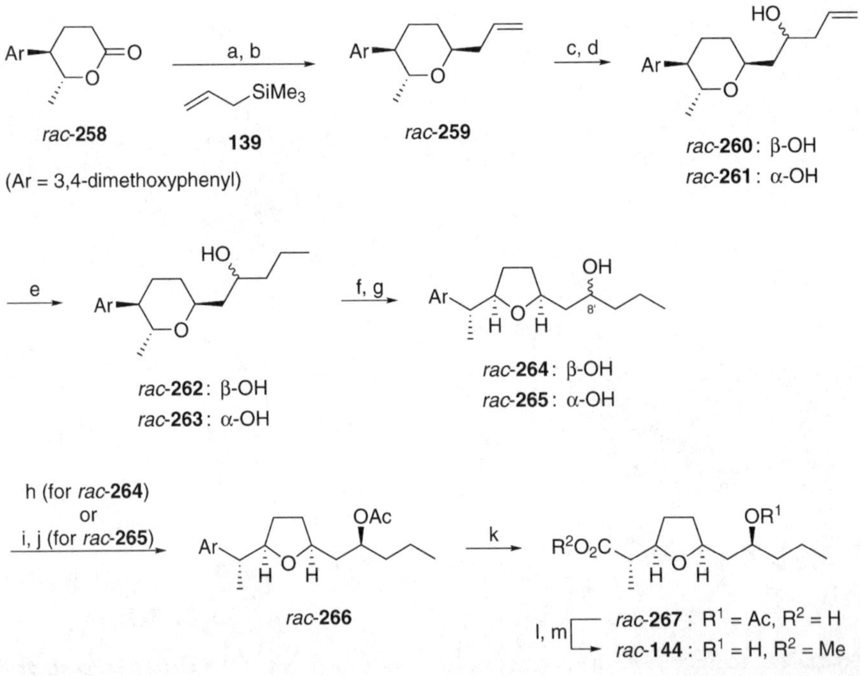

Scheme 33 Nagumo's and Kawahara's synthesis of the racemic smaller fragment of **1b**. *Reagents and conditions*: a: DIBALH, toluene, –78°C, 100%; b: **139**, $BF_3 \cdot Et_2O$, CH_2Cl_2, 93%; c: OsO_4, $NaIO_4$, *t*-BuOH, H_2O, Et_2O; d: CH_2=CH–CH_2–MgBr, $ZnCl_2$, THF, –78°C, 56% *rac*-**260** and 34% *rac*-**261** from *rac*-**259**; e: H_2, 5% Pd/C, MeOH, 91% *rac*-**262** from *rac*-**260**, 84% *rac*-**263** from *rac*-**261**; f: TFA, 70°C; g: K_2CO_3, MeOH, 54% *rac*-**264** from *rac*-**262**, 53% *rac*-**265** from *rac*-**263**; h: Ac_2O, DMAP, pyridine; i: MsCl, DMAP, Et_3N; j: CsOAc, DMF; k: $RuCl_3 \cdot nH_2O$, $NaIO_4$, CCl_4, MeCN, H_2O, 31% from *rac*-**264** and *rac*-**265**; l: CH_2N_2, Et_2O; m: K_2CO_3, MeOH, 51% from *rac*-**267**

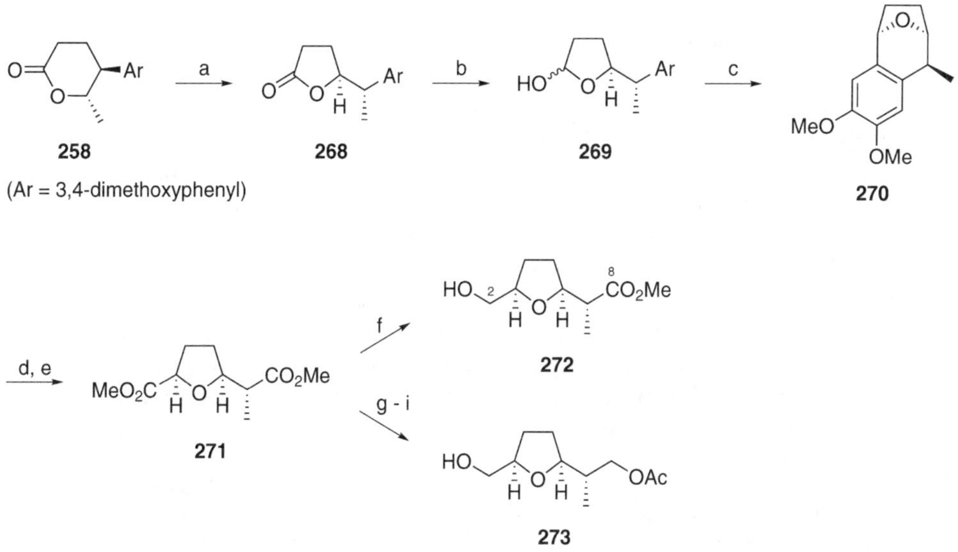

Scheme 34 Nagumo's and Kawahara's synthesis of a C(2)-C(8) segment of the larger fragment of **1b**. *Reagents and conditions*: a: $TsOH \cdot H_2O$, $MeNO_2$, 70°C, 97%; b: DIBALH, –78°C, 100%; c: $BF_3 \cdot Et_2O$, MeCN, 0°C, 98%; d: $RuCl_3 \cdot nH_2O$, $NaIO_4$, MeCN, CCl_4, H_2O, 0°C; e: CH_2N_2, Et_2O, 0°C, 40% from **270**; f: $NaBH_4$, MeOH, 0°C, 90%; g: $LiAlH_4$, THF, 0°C, 90%; h: Ac_2O, DMAP, pyridine, 0°C, 82%; i: Amano lipase PS, phosphate buffer, 33°C, 67%

hydrolysis yielded the corresponding tetrahydrofurans *rac*-**264** and *rac*-**265**, respectively, with complete diastereoselectivity along with recovered substrates. Acetylation of *rac*-**264** and inversion at C(8′) in *rac*-**265** via mesylation/acetate substitution afforded acetate *rac*-**266** from both precursors. Oxidative degradation of the electron-rich aromatic ring to carboxylic acid *rac*-**267** followed by esterification with diazomethane and chemoselective acetate cleavage finally delivered methyl ester *rac*-**144**.

Conversion of the (–)-enantiomer **258** to hydroxy esters **272** and **273**, both of which can serve as a C(2)-C(8) segment of the larger fragment of **1b** (and also as a C(2′)-C(8′) segment of the smaller fragment of **1c** and other homologs) was initiated by the key ring contraction, this time using *p*-toluenesulfonic acid in nitromethane, to furnish *γ*-lactone **268** (Scheme 34). After reduction to lactol **269**, an intramolecular Friedel–Crafts reaction yielded the tricyclic compound **270**. Oxidative degradation of the aromatic ring and esterification of the resulting dicarboxylic acid with diazomethane provided diester **271**. Differentiation between the two ester groups of **271** was accomplished either by chemoselective sodium borohydride reduction to afford **272** or by chemoselective enzymatic hydrolysis of the diacetate derived from **271** via global reduction and bis-acetylation to give **273**.

4 Concluding Remarks

The structural complexity and the biological activity of the pamamycin macrodiolides has provided a significant challenge to the synthetic community. Only recently, the total synthesis of pamamycin-607 (**1b**) was accomplished by several groups. Much interesting, important, and novel chemistry has arisen from attempts to synthesize and couple the hydroxy acid fragments and, after about a decade of effort, continues to do so. Given the high activity of the homolog **1b** against multidrug-resistant strains of *Mycobacterium tuberculosis*, more detailed structure–activity relationship studies with respect to the antimycobacterial action of the various pamamycins and unnatural analogs are clearly warranted. Surely, the strategies and methods compiled in this review will be of help in accomplishing this goal.

Acknowledgements I thank my talented coworkers that have contributed to this research program; their work is cited in the references. Financial support of our work by the Deutsche Forschungsgemeinschaft and the Fonds der Chemischen Industrie is gratefully acknowledged.

References

1. For reviews on the isolation, structural elucidation, and biological activities of the pamamycins, see: (a) Natsume M (1999) Recent Res Dev Agric Biol Chem 3:11; (b) Natsume M (1999) Actinomycetol 13:11; (c) Pogell BM (1998) Cell Mol Biol 44:461
2. For isolation of de-*N*-methylpamamycins **1o–q**, see: Kozone I, Abe H, Natsume M (2001) J Pesticide Sci 26:149
3. (a) Lefèvre P, Peirs P, Braibant M, Fauville-Dufaux M, Vanhoof R, Huygen K, Wang X-M, Pogell B, Wang Y, Fischer P, Metz P, Content J (2004) J Antimicrob Chemother 54:824; (b) Tokalov SV, Gutzeit HO, Wang Y, Fischer P, Metz P (unpublished)
4. Germay O, Kumar N, Thomas EJ (2001) Tetrahedron Lett 42:4969
5. (a) Lee E, Jeong EJ, Kang EJ, Sung LT, Hong SK (2001) J Am Chem Soc 123:10131; (b) Jeong EJ, Kang EJ, Sung LT, Hong SK, Lee E (2002) J Am Chem Soc 124:14655
6. Wang Y, Bernsmann H, Gruner M, Metz P (2001) Tetrahedron Lett 42:7801
7. Kang SH, Jeong JW, Hwang YS, Lee SB (2002) Angew Chem 114:1450; Angew Chem Int Ed 41:1392
8. (a) Thomas EJ (1997) Chem Commun 411; (b) Carey JS, Thomas EJ (1992) Synlett 585
9. (a) Raddatz P, Minck KO, Rippmann F, Schmitges C-J (1994) J Med Chem 37:486; (b) Mori K, Takaishi H (1989) Tetrahedron 45:1639
10. Lee E (2001) In: Renaud P, Sibi MP (eds) Radicals in organic synthesis, vol 2 Applications. Wiley-VCH, Weinheim, p 303
11. Evans DA, Bartroli J, Shih TL (1981) J Am Chem Soc 103:2127
12. Keck GE, Abbott DE (1984) Tetrahedron Lett 25:1883
13. For a review, see: Metz P (1998) J Prakt Chem 340:1
14. (a) Karsch S, Schwab P, Metz P (2002) Synlett 2019; (b) Plietker B, Seng D, Fröhlich R, Metz P (2001) Eur J Org Chem 3669; (c) Plietker B, Seng D, Fröhlich R, Metz P (2000) Tetrahedron 56:873; (d) Plietker B, Metz P (1998) Tetrahedron Lett 39:7827; (e) Meiners U, Cramer E, Fröhlich R, Wibbeling B, Metz P (1998) Eur J Org Chem 2073

15. Tokunaga M, Larrow JF, Kakiuchi F, Jacobsen EN (1997) Science 277:936
16. Metz P, Bernsmann H (1999) Phosphorus Sulfur Silicon 153–154:383
17. Bernsmann H, Hungerhoff B, Fechner R, Fröhlich R, Metz P (2000) Tetrahedron Lett 41:1721
18. Bernsmann H, Fröhlich R, Metz P (2000) Tetrahedron Lett 41:4347
19. Hoffmann RW (1989) Chem Rev 89:1841
20. Bernsmann H, Gruner M, Metz P (2000) Tetrahedron Lett 41:7629
21. Bernsmann H, Gruner M, Fröhlich R, Metz P (2001) Tetrahedron Lett 42:5377
22. Fleming I, Ghosh SK (1998) J Chem Soc Perkin Trans I 2733
23. Wang Y, Metz P (unpublished)
24. Seebach D, Hungerbühler E, Naef R, Schnurrenberger P, Weidmann B, Züger M (1982) Synthesis 138
25. Kiegiel K, Prokopowicz P, Jurczak J (1999) Synth Commun 29:3999
26. Blakemore PR, Cole WJ, Kociénski PJ, Morley A (1998) Synlett 26
27. Paterson I, Goodman JM, Isaka M (1989) Tetrahedron Lett 30:7121
28. Forsyth CJ, Lee CS (1996) Tetrahedron Lett 37:6449
29. Evans DA, Chapman KT, Carreira EM (1988) J Am Chem Soc 110:3560
30. Roush WR, Hoong LK, Palmer MAJ, Staub JA, Palkowitz AD (1990) J Org Chem 55:4117
31. Walkup RD, Park G (1988) Tetrahedron Lett 29:5505
32. Walkup RD, Kim SW, Wagy SD (1993) J Org Chem 58:6486
33. Walkup RD, Kim SW (1994) J Org Chem 59:3433
34. Walkup RD, Kim YS (1995) Tetrahedron Lett 36:3091
35. Heathcock CH, Pirrung MC, Montgomery SH, Lampe J (1981) Tetrahedron 37:4087
36. Mavropoulos I, Perlmutter P (1996) Tetrahedron Lett 37:3751
37. (a) Brown HC, Pai GG, Jadhav PK (1984) J Am Chem Soc 106:1531; (b) Brown HC, Pai GG (1985) J Org Chem 50:1384
38. Arista L, Gruttadauria M, Thomas EJ (1997) Synlett 627
39. Mandville G, Girard C, Bloch R (1997) Tetrahedron Asymmetry 8:3665
40. Mandville G, Bloch R (1999) Eur J Org Chem 2303
41. Cinquin C, Shaper I, Mandville G, Bloch R (1995) Synlett 339
42. Haddad M, Dorbais J, Larchevêque M (1997) Tetrahedron Lett 38:5981
43. Knight DW, Rustidge DC (1981) J Chem Soc Perkin Trans I 679
44. Solladié G, Salom-Roig XJ, Hanquet G (2000) Tetrahedron Lett 41:2737
45. Solladié G, Salom-Roig XJ, Hanquet G (2000) Tetrahedron Lett 41:551
46. Calter MA, Bi FC (2000) Org Lett 2:1529
47. Furuya Y, Kiyota H, Oritani T (2000) Heterocycl Commun 6:427
48. Kiyota H, Furuya Y, Kuwahara S, Oritani T (2001) Biosci Biotechnol Biochem 65:2630
49. Mori Y, Asai M, Okumura A, Furukawa H (1995) Tetrahedron 51:5299
50. Rama Rao AV, Reddy ER, Joshi BV, Yadav JS (1987) Tetrahedron Lett 28:6497
51. Kang SH, Jeong JW (2002) Tetrahedron Lett 43:3613
52. Morimoto Y, Yokoe C (1997) Tetrahedron Lett 38:8981
53. Roush WR, Ando K, Powers DB, Palkowitz AD, Halterman RL (1990) J Am Chem Soc 112:6339
54. Nagumo S, Ishii Y, Kakimoto Y, Kawahara N (2002) Tetrahedron Lett 43:5333
55. Nagumo S, Ishii Y, Kanno S, Kawahara N (2003) Heterocycles 59:101

Author Index Volumes 201–244

Author Index Vols. 26–50 see Vol. 50
Author Index Vols. 51–100 see Vol. 100
Author Index Vols. 101–150 see Vol. 150
Author Index Vols. 151–200 see Vol. 200

The volume numbers are printed in italics

Achilefu S, Dorshow RB (2002) Dynamic and Continuous Monitoring of Renal and Hepatic Functions with Exogenous Markers. *222*: 31–72
Albert M, see Dax K (2001) *215*: 193–275
Albrecht M (2005) Supramolecular Templating in the Formation of Helicates. *248*: 105–139
Ando T, Inomata S-I, Yamamoto M (2004) Lepidopteran Sex Pheromones. *239*: 51–96
Angyal SJ (2001) The Lobry de Bruyn-Alberda van Ekenstein Transformation and Related Reactions. *215*: 1–14
Antzutkin ON, see Ivanov AV (2005) *246*: 271–337
Anupõld T, see Samoson A (2005) *246*: 15–31
Armentrout PB (2003) Threshold Collision-Induced Dissociations for the Determination of Accurate Gas-Phase Binding Energies and Reaction Barriers. *225*: 227–256
Astruc D, Blais J-C, Cloutet E, Djakovitch L, Rigaut S, Ruiz J, Sartor V, Valério C (2000) The First Organometallic Dendrimers: Design and Redox Functions. *210*: 229–259
Augé J, see Lubineau A (1999) *206*: 1–39
Baars MWPL, Meijer EW (2000) Host-Guest Chemistry of Dendritic Molecules. *210*: 131–182
Balazs G, Johnson BP, Scheer M (2003) Complexes with a Metal-Phosphorus Triple Bond. *232*: 1–23
Balbo Block MA, Kaiser C, Khan A, Hecht S (2005) Discrete Organic Nanotubes Based on a Combination of Covalent and Non-Covalent Approaches. *245*: 89–150
Balczewski P, see Mikoloajczyk M (2003) *223*: 161–214
Ballauff M (2001) Structure of Dendrimers in Dilute Solution. *212*: 177–194
Ballauff M, see Likos CN (2005) *245*: 239–252
Baltzer L (1999) Functionalization and Properties of Designed Folded Polypeptides. *202*: 39–76
Balzani V, Ceroni P, Maestri M, Saudan C, Vicinelli V (2003) Luminescent Dendrimers. Recent Advances. *228*: 159–191
Bannwarth W, see Horn J (2004) *242*: 43–75
Barré L, see Lasne M-C (2002) *222*: 201–258
Bartlett RJ, see Sun J-Q (1999) *203*: 121–145
Bauer RE, Grimsdale AC, Müllen K (2005) Functionalised Polyphenylene Dendrimers and Their Applications. *245*: 253–286
Beifuss U, Tietze M (2005) Methanophenazine and Other Natural Biologically Active Phenazines. *244*: 77–113
Bergbreiter DE, Li J (2004) Applications of Catalysts on Soluble Supports. *242*: 113–176
Bertrand G, Bourissou D (2002) Diphosphorus-Containing Unsaturated Three-Menbered Rings: Comparison of Carbon, Nitrogen, and Phosphorus Chemistry. *220*: 1–25

Betzemeier B, Knochel P (1999) Perfluorinated Solvents – a Novel Reaction Medium in Organic Chemistry. *206*: 61–78
Bibette J, see Schmitt V (2003) *227*: 195–215
Blais J-C, see Astruc D (2000) *210*: 229–259
Bogár F, see Pipek J (1999) *203*: 43–61
Bohme DK, see Petrie S (2003) *225*: 35–73
Boillot M-L, Zarembowitch J, Sour A (2004) Ligand-Driven Light-Induced Spin Change (LD-LISC): A Promising Photomagnetic Effect. *234*: 261–276
Boukheddaden K, see Bousseksou A (2004) *235*: 65–84
Boukheddaden K, see Varret F (2004) *234*: 199–229
Bourissou D, see Bertrand G (2002) *220*: 1–25
Bousseksou A, Varret F, Goiran M, Boukheddaden K, Tuchagues J-P (2004) The Spin Crossover Phenomenon Under High Magnetic Field. *235*: 65–84
Bousseksou A, see Tuchagues J-P (2004) *235*: 85–103
Bowers MT, see Wyttenbach T (2003) *225*: 201–226
Brady C, McGarvey JJ, McCusker JK, Toftlund H, Hendrickson DN (2004) Time-Resolved Relaxation Studies of Spin Crossover Systems in Solution. *235*: 1–22
Brand SC, see Haley MM (1999) *201*: 81–129
Bravic G, see Guionneau P (2004) *234*: 97–128
Bray KL (2001) High Pressure Probes of Electronic Structure and Luminescence Properties of Transition Metal and Lanthanide Systems. *213*: 1–94
Bronstein LM (2003) Nanoparticles Made in Mesoporous Solids. *226*: 55–89
Brönstrup M (2003) High Throughput Mass Spectrometry for Compound Characterization in Drug Discovery. *225*: 275–294
Brücher E (2002) Kinetic Stabilities of Gadolinium(III) Chelates Used as MRI Contrast Agents. *221*: 103–122
Brüggemann J, see Schalley CA (2005) *248*: 141–200
Brunel JM, Buono G (2002) New Chiral Organophosphorus atalysts in Asymmetric Synthesis. *220*: 79–106
Buchwald SL, see Muci AR (2002) *219*: 131–209
Bunz UHF (1999) Carbon-Rich Molecular Objects from Multiply Ethynylated *p*-Complexes. *201*: 131–161
Buono G, see Brunel JM (2002) *220*: 79–106
Cadierno V, see Majoral J-P (2002) *220*: 53–77
Caminade A-M, see Majoral J-P (2003) *223*: 111–159
Carmichael D, Mathey F (2002) New Trends in Phosphametallocene Chemistry. *220*: 27–51
Caruso F (2003) Hollow Inorganic Capsules via Colloid-Templated Layer-by-Layer Electrostatic Assembly. 227: 145–168
Caruso RA (2003) Nanocasting and Nanocoating. *226*: 91–118
Ceroni P, see Balzani V (2003) *228*: 159–191
Chamberlin AR, see Gilmore MA (1999) *202*: 77–99
Chasseau D, see Guionneau P (2004) *234*: 97–128
Chivers T (2003) Imido Analogues of Phosphorus Oxo and Chalcogenido Anions. *229*: 143–159
Chow H-F, Leung C-F, Wang G-X, Zhang J (2001) Dendritic Oligoethers. *217*: 1–50
Chumakov AI, see Winkler H (2004) *235*: 105–136
Clarkson RB (2002) Blood-Pool MRI Contrast Agents: Properties and Characterization. *221*: 201–235
Cloutet E, see Astruc D (2000) *210*: 229–259
Co CC, see Hentze H-P (2003) *226*: 197–223
Codjovi E, see Varret F (2004) *234*: 199–229

Schöning K-U, see End N (2004) *242*: 241–271
Schöning K-U, see End N (2004) *242*: 273–317
Schröder D, Schwarz H (2003) Diastereoselective Effects in Gas-Phase Ion Chemistry. *225*: 129–148
Schwarz H, see Schröder D (2003) *225*: 129–148
Schwert DD, Davies JA, Richardson N (2002) Non-Gadolinium-Based MRI Contrast Agents. *221*: 165–200
Seikh MM, see Rao CNR (2004) *234*: 1–21
Sergeyev S, see Thilgen C (2005) *248*: 1–61
Sheiko SS, Möller M (2001) Hyperbranched Macromolecules: Soft Particles with Adjustable Shape and Capability to Persistent Motion. *212*: 137–175
Shen B (2000) The Biosynthesis of Aromatic Polyketides. *209*: 1–51
Shinkai S, see James TD (2002) *218*: 159–200
Shinkai S, see Jung JH (2005) *248*: 223–260
Shirakawa E, see Hiyama T (2002) *219*: 61–85
Shogren-Knaak M, see Imperiali B (1999) *202*: 1–38
Sinou D (1999) Metal Catalysis in Water. *206*: 41–59
Sinz CJ, Rychnovsky SD (2001) 4-Acetoxy- and 4-Cyano-1,3-dioxanes in Synthesis. *216*: 51–92
Siuzdak G, see Trauger SA (2003) *225*: 257–274
Skrifvars M, see Nummelin S (2000) *210*: 1–67
Slessor KN, see Keeling CI (2004) *239*: 133–177
Smith DK, Diederich F (2000) Supramolecular Dendrimer Chemistry – A Journey Through the Branched Architecture. *210*: 183–227
Sorai M (2004) Heat Capacity Studies of Spin Crossover Systems. *235*: 153–170
Sour A, see Boillot M-L (2004) *234*: 261–276
Spiering H (2004) Elastic Interaction in Spin-Crossover Compounds. *235*: 171–195
Spiering H, see Real JA (2004) *233*: 167–193
Spiering H, see Kusz J (2004) *234*: 129–153
Stec WJ, see Guga P (2002) *220*: 169–200
Steudel R (2003) Aqueous Sulfur Sols. *230*: 153–166
Steudel R (2003) Liquid Sulfur. *230*: 80–116
Steudel R (2003) Inorganic Polysulfanes H_2S_n with n>1. *231*: 99–125
Steudel R (2003) Inorganic Polysulfides S_n^{2-} and Radical Anions $S_n^{\cdot -}$. *231*: 127–152
Steudel R (2003) Sulfur-Rich Oxides S_nO and S_nO_2. *231*: 203–230
Steudel R, Eckert B (2003) Solid Sulfur Allotropes. *230*: 1–79
Steudel R, see Eckert B (2003) *231*: 31–97
Steudel R, Steudel Y, Wong MW (2003) Speciation and Thermodynamics of Sulfur Vapor. *230*: 117–134
Steudel Y, see Steudel R (2003) *230*: 117–134
Steward LE, see Gilmore MA (1999) *202*: 77–99
Stocking EM, see Williams RM (2000) *209*: 97–173
Streubel R (2003) Transient Nitrilium Phosphanylid Complexes: New Versatile Building Blocks in Phosphorus Chemistry. *223*: 91–109
Strojek W, see Eckert H (2005) *246*: 195–233
Stütz AE, see Häusler H (2001) *215*: 77–114
Sugihara Y, see Nakayama J (1999) *205*: 131–195
Sugiura K (2003) An Adventure in Macromolecular Chemistry Based on the Achievements of Dendrimer Science: Molecular Design, Synthesis, and Some Basic Properties of Cyclic Porphyrin Oligomers to Create a Functional Nano-Sized Space. *228*: 65–85
Sun J-Q, Bartlett RJ (1999) Modern Correlation Theories for Extended, Periodic Systems. *203*: 121–145

Sun L, see Crooks RM (2001) *212*: 81–135
Surjná PR (1999) An Introduction to the Theory of Geminals. *203*: 63–88
Sylvain I, see Drain CM (2005) *245*: 55–88
Taillefer M, Cristau H-J (2003) New Trends in Ylide Chemistry. *229*: 41–73
Taira K, see Takagi Y (2003) *232*: 213–251
Takagi Y, Ikeda Y, Taira K (2003) Ribozyme Mechanisms. *232*: 213–251
Takahashi S, see Onitsuka K (2003) *228*: 39–63
Takeda N, Tokitoh N, Okazaki R (2003) Polysulfido Complexes of Main Group and Transition Metals. *231*: 153-202
Tamao K, Miyaura N (2002) Introduction to Cross-Coupling Reactions. *219*: 1–9
Tanaka M (2003) Homogeneous Catalysis for H-P Bond Addition Reactions. *232*: 25–54
ten Holte P, see Zwanenburg B (2001) *216*: 93–124
Thiem J, see Werschkun B (2001) *215*: 293–325
Thilgen C, Sergeyev S, Diederich F (2005) Spacer-Controlled Multiple Functionalization of Fullerenes. *248*: 1–61
Thutewohl M, see Waldmann H (2000) *211*: 117–130
Tichkowsky I, see Idee J-M (2002) *222*: 151–171
Tiecco M (2000) Electrophilic Selenium, Selenocyclizations. *208*: 7–54
Tietze M, see Beifuss U (2005) *244*: 77–113
Toftlund H, McGarvey JJ (2004) Iron(II) Spin Crossover Systems with Multidentate Ligands. *233*: 151–166
Toftlund H, see Brady C (2004) *235*: 1–22
Tohma H, Kita Y (2003) Synthetic Applications (Total Synthesis and Natural Product Synthesis). *224*: 209–248
Tokitoh N, see Takeda N (2003) *231*: 153–202
Tomoda S, see Iwaoka M (2000) *208*: 55–80
Tóth E, Helm L, Merbach AE (2002) Relaxivity of MRI Contrast Agents. *221*: 61–101
Tovar GEM, Kruter I, Gruber C (2003) Molecularly Imprinted Polymer Nanospheres as Fully Affinity Receptors. *227*: 125–144
Trauger SA, Junker T, Siuzdak G (2003) Investigating Viral Proteins and Intact Viruses with Mass Spectrometry. *225*: 257–274
Trautwein AX, see Paulsen H (2004) *235*: 197–219
Trautwein AX, see Winkler H (2004) *235*: 105–136
Tromas C, García R (2002) Interaction Forces with Carbohydrates Measured by Atomic Force Microscopy. *218*: 115–132
Tsiourvas D, see Paleos CM (2003) *227*: 1–29
Tuchagues J-P, Bousseksou A, Molnàr G, McGarvey JJ, Varret F (2004) The Role of Molecular Vibrations in the Spin Crossover Phenomenon. *235*: 85–103
Tuchagues J-P, see Bousseksou A (2004) *235*: 65–84
Tuherm T, see Samoson A (2005) *246*: 15–31
Turecek F (2003) Transient Intermediates of Chemical Reactions by Neutralization-Reionization Mass Spectrometry. *225*: 75–127
Tzschucke CC, see Horn J (2004) *242*: 43–75
Ublacker GA, see Maul JJ (1999) *206*: 79–105
Uemura S, see Nishibayashi Y (2000) *208*: 201–233
Uemura S, see Nishibayashi Y (2000) *208*: 235–255
Uggerud E (2003) Physical Organic Chemistry of the Gas Phase. Reactivity Trends for Organic Cations. *225*: 1–34
Uozumi Y (2004) Recent Progress in Polymeric Palladium Catalysts for Organic Synthesis. *242*: 77–112
Valdemoro C (1999) Electron Correlation and Reduced Density Matrices. *203*: 187–200

Valrio C, see Astruc D (2000) *210*: 229–259
van Benthem RATM, see Muscat D (2001) *212*: 41–80
van Koningsbruggen PJ (2004) Special Classes of Iron(II) Azole Spin Crossover Compounds. *233*: 123–149
van Koningsbruggen PJ, Maeda Y, Oshio H (2004) Iron(III) Spin Crossover Compounds. *233*: 259–324
van Koten G, see Kreiter R (2001) *217*: 163–199
van Manen H-J, van Veggel FCJM, Reinhoudt DN (2001) Non-Covalent Synthesis of Metallodendrimers. *217*: 121–162
van Veggel FCJM, see van Manen H-J (2001) *217*: 121–162
Varret F, Boukheddaden K, Codjovi E, Enachescu C, Linarès J (2004) On the Competition Between Relaxation and Photoexcitations in Spin Crossover Solids under Continuous Irradiation. *234*: 199–229
Varret F, see Bousseksou A (2004) *235*: 65–84
Varret F, see Tuchagues J-P (2004) *235*: 85–103
Varshney DB, see MacGillivray LR (2005) *248*: 201–221
Varvoglis A (2003) Preparation of Hypervalent Iodine Compounds. *224*: 69–98
Vega S, see Vinogradov E (2005) *246*: 33–90
Verkade JG (2003) $P(RNCH_2CH_2)_3N$: Very Strong Non-ionic Bases Useful in Organic Synthesis. *223*: 1–44
Vicinelli V, see Balzani V (2003) *228*: 159–191
Vinogradov E, Madhu PK, Vega S (2005) Strategies for High-Resolution Proton Spectroscopy in Solid-State NMR. *246*: 33–90
Vioux A, Le Bideau J, Mutin PH, Leclercq D (2003): Hybrid Organic-Inorganic Materials Based on Organophosphorus Derivatives. *232*: 145–174
Vliegenthart JFG, see Haseley SR (2002) *218*: 93–114
Vogler A, Kunkely H (2001) Luminescent Metal Complexes: Diversity of Excited States. *213*: 143–182
Vogtner S, see Klopper W (1999) *203*: 21–42
Vögtle F, see Schalley CA (2005) *248*: 141–200
Voigt U, see Eckert H (2005) *246*: 195–233
Vostrowsky O, see Hirsch A (2001) *217*: 51–93
Vostrowsky O, see Hirsch A (2005) *245*: 193–237
Waldmann H, Thutewohl M (2000) Ras-Farnesyltransferase-Inhibitors as Promising Anti-Tumor Drugs. *211*: 117–130
Wang G-X, see Chow H-F (2001) *217*: 1–50
Weil T, see Wiesler U-M (2001) *212*: 1–40
Weilandt T, see Schalley CA (2005) *248*: 141–200
Wenzel B, see Dötz KH (2005) *248*: 63–103
Werschkun B, Thiem J (2001) Claisen Rearrangements in Carbohydrate Chemistry. *215*: 293–325
Wiesler U-M, Weil T, Müllen K (2001) Nanosized Polyphenylene Dendrimers. *212*: 1–40
Williams RM, Stocking EM, Sanz-Cervera JF (2000) Biosynthesis of Prenylated Alkaloids Derived from Tryptophan. *209*: 97–173
Winkler H, Chumakov AI, Trautwein AX (2004) Nuclear Resonant Forward and Nuclear Inelastic Scattering Using Synchrotron Radiation for Spin Crossover Systems. *235*: 105–136
Wirth T (2000) Introduction and General Aspects. *208*: 1–5
Wirth T (2003) Introduction and General Aspects. *224*: 1–4
Wirth T (2003) Oxidations and Rearrangements. *224*: 185–208
Wong MW, see Steudel R (2003) *230*: 117–134

Wong MW (2003) Quantum-Chemical Calculations of Sulfur-Rich Compounds. *231*: 1–29
Wooley KL, Hawker CJ (2005) Nanoscale Objects: Perspectives Regarding Methodologies for their Assembly, Covalent Stabilization and Utilization. *245*: 287–305
Wrodnigg TM, Eder B (2001) The Amadori and Heyns Rearrangements: Landmarks in the History of Carbohydrate Chemistry or Unrecognized Synthetic Opportunities? *215*: 115–175
Wyttenbach T, Bowers MT (2003) Gas-Phase Confirmations: The Ion Mobility/Ion Chromatography Method. *225*: 201–226
Yamaguchi H, Harada A (2003) Antibody Dendrimers. *228*: 237–258
Yamamoto M, see Ando T (2004) *239*: 51–96
Yersin H, Donges D (2001) Low-Lying Electronic States and Photophysical Properties of Organometallic Pd(II) and Pt(II) Compounds. Modern Research Trends Presented in Detailed Case Studies. *214*: 81–186
Yeung LK, see Crooks RM (2001) *212*: 81–135
Yokoyama S, Otomo A, Nakahama T, Okuno Y, Mashiko S (2003) Dendrimers for Optoelectronic Applications. *228*: 205–226
Yoshifuji M, Ito S (2003) Chemistry of Phosphanylidene Carbenoids. *223*: 67–89
Zablocka M, see Majoral J-P (2002) *220*: 53–77
Zarembowitch J, see Boillot M-L (2004) *234*: 261–276
Zhang J, see Chow H-F (2001) *217*: 1–50
Zhdankin VV (2003) C-C Bond Forming Reactions. *224*: 99–136
Zhao M, see Crooks RM (2001) *212*: 81–135
Zimmermann SC, Lawless LJ (2001) Supramolecular Chemistry of Dendrimers. *217*: 95–120
Zwanenburg B, ten Holte P (2001) The Synthetic Potential of Three-Membered Ring Aza-Heterocycles. *216*: 93–124

Subject Index

Printing: Krips bv, Meppel
Binding: Litges & Dopf, Heppenheim